mr £30.00

X-ray crystallography and drug action

X-ray crystallography and drug action

The ninth course of the
International School of Crystallography
Erice, Sicily, Italy
March 1983

Edited by

A.S. Horn

and

C.J. De Ranter

CLARENDON PRESS · OXFORD
1984

Oxford University Press, Walton Street, Oxford OX2 6DP

London Glasgow New York Toronto
Delhi Bombay Calcutta Madras Karachi
Kuala Lumpur Singapore Hong Kong Tokyo
Nairobi Dar es Salaam Cape Town
Melbourne Auckland

and associate companies in
Beirut Berlin Ibadan Mexico City Nicosia

British Library Cataloguing in Publication Data

X-ray crystallography and drug design.
1. Pharmacology   2. Structure-activity
  relationship (Pharmacology)   3. X-ray
  crystallography
I. Horn, Alan S.   II. Ranter, C J de
615.7   RAM301
ISBN 0 19 855185 1

Published *in the United States*
by *Oxford University Press, New York*

Printed by The Thetford Press Limited, Thetford, Norfolk

Preface

The results of X-ray crystallography over the last twenty-five years have revolutionized our insights about molecular structure, especially in the biological sciences: a full understanding of some biological processes has only become possible after the three-dimensional structure of the intervening macromolecule had been solved. Indeed, X-ray crystallography is the only method that permits the accurate determination of the molecular architecture of these compounds. A careful crystal structure analysis can provide a wealth of information concerning bond lengths and angles, intermolecular interactions, details of conformation, molecular packing, and hydrogen bonding. In addition it provides with great accuracy the relative atomic positions needed for advanced theoretical work. Careful experimental work also permits the assignment of the absolute configuration of a molecule based on the anomalous scattering from atoms no heavier than oxygen.

In comparison, molecular pharmacology—the study of drug action at the receptor level—is at the moment a much less exact science. This is, of course, understandable when one takes into account the complexity of biological systems. However, it too has made enormous advances in the past fifteen years, due mainly to our increasing understanding of the modes of interaction of drugs with hormonal and neurotransmitter systems via their receptors. The actual isolation of various receptors such as those for certain steroid hormones and acetylcholine can be considered as milestones for this discipline.

X-ray crystallography has also played an important role in increasing our understanding of drug action. It has provided structural information about drugs, hormones, and neurotransmitters; about their receptors, i.e. proteins, enzymes, and membrane-bound macromolecules; and finally in certain cases even accurate details of the nature of drug–receptor interactions derived from studies of crystalline complexes of the two entities.

In the spring of 1983 a course on X-ray crystallography and drug action was arranged under the auspices of the International School of Crystallography at the Ettore Majorana Centre in Erice, Sicily. It was felt by the organizers that we should bring together, perhaps for the first time, various people working in this multidisciplinary field in order to try and achieve a better understanding of each other's achievements, aims, and problems, and also with the hope of increasing collaboration between the various disciplines. In fact because of the multidisciplinary nature of the topic it was thought that certain other subjects such as drug design and computer graphics should also be included bearing in mind their relevance to other material in the programme.

The editors would like to thank the executive secretary of the Ettore Majorana Centre, Professor Lodovico Riva di Sanseverino, for making this course such a memorable and successful one for all of those concerned.

Groningen and   A.S.H.
Leuven   C.J. De R.
June 1983

Contents

Contributors

Federico Arcamone

Farmitalia Carlo Erba SpA,
Ricerca & Sviluppo Chimico,
Via dei Gracchi 35,
20146 Milan,
Italy

V. Austel

Department of Chemistry,
Dr Karl Thomae GmbH,
Postfach 17 55,
D–7950 Biberach 1,
West Germany

C.R. Beddell

Wellcome Research Laboratories,
Beckenham,
Kent

Joel Bernstein

Department of Chemistry,
Ben-Gurion University of the Negev,
Beer-Sheva,
Israel

Colin C.F. Blake

Laboratory of Molecular Biophysics,
Department of Zoology,
South Parks Road,
Oxford

T.L. Blundell

Laboratory of Molecular Biology,
Department of Crystallography,
Birkbeck College,
University of London,
Malet Street,
London WC1E 7HX

Mario Brufani

Istituto di Strutturistica Chimica
  'Giordano Giacomello' del Consiglio
  Nazionale delle Ricerche,
Area della Ricerca di Roma,
Montelibretti,
Rome,
Italy

Jane M. Burridge

Department of Chemistry,
University of California,
San Diego,
California 92037,
U.S.A.

Arthur Camerman

Department of Medicine (Neurology) &
  Pharmacology,
University of Washington,
Seattle,
Washington 98102,
U.S.A.

Norman Camerman

Department of Biochemistry,
University of Toronto,
Toronto,
Canada M5S 1A8

Simon F. Campbell

Pfizer Research,
Sandwich,
Kent

Luciana Cellai

Istituto di Strutturistica Chimica
  'Giordano Giacomello', Consiglio
  Nazionale delle Ricerche,
C.P. 10,
00016 Monterotondo Stazione,
Rome,
Italy

C.J. De Ranter

Laboratorium voor Analytische Chemie
   en Medicinale Fysicochemie,
K.U. Leuven,
Instituut voor Farmaceutlsche
   Wetenschappen,
Van Evenstraat 4,
B–3000 Leuven,
Belgium

W.L. Duax

Medical Foundation of Buffalo Inc.,
73 High Street,
Buffalo,
New York 14203,
U.S.A.

P.J. Goodford

Laboratory of Molecular Biophysics,
Department of Zoology,
South Parks Road,
Oxford

D.R.H. Gourley

Department of Pharmacology,
Eastern Virginia Medical School,
Norfolk,
Virginia 23501,
U.S.A.

J.F. Griffin

Medical Foundation of Buffalo Inc.,
73 High Street,
Buffalo,
New York 14203,
U.S.A

P. Gund

Merck, Sharp, and Dohme,
West Point,
Pennsylvania 19486,
U.S.A.

Thomas A. Hamor

Department of Chemistry,
University of Birmingham,
Birmingham B15 2TT

A.M. Hemmings

Laboratory of Molecular Biology,
Department of Crystallography,
Birbeck College,
University of London,
Malet Street,
London WC1E 7HX

Wim G.J. Hol

Biomolecular Study Centre (BIOS),
University of Groningen,
Nijenborgh 16,
9747 AG Groningen,
The Netherlands

Alan S. Horn

Department of Pharmacy,
University of Groningen,
Groningen,
The Netherlands

Peter Kollman

Department of Pharmaceutical
   Chemistry,
School of Pharmacy,
University of California,
San Francisco,
California 94143,
U.S.A.

G.R. Marshall

Department of Physiology
   and Biophysics,
School of Medicine,
Washington University,
St Louis, Missouri 63110,
U.S.A.

Ian L. Martin

Neurochemical Pharmacology Unit,
Medical Research Council Centre,
Hills Road,
Cambridge CB2 2QH

Donald Mastropaolo
Department of Medicine (Neurology) &
  Pharmacology,
University of Washington,
Seattle,
Washington 98102,
U.S.A.

Herbert Merz
Boehringer Ingelheim KG,
Department of Medicinal Chemistry,
D-6507 Ingelheim,
West Germany

H. Moereels
Department of Theoretical Medicinal
  Chemistry,
Janssen Pharmaceutica Research
  Laboratories,
Beerse,
Belgium

Stephen Neidle
Department of Biophysics,
King's College,
Strand,
London WC2R 2LS

Stuart J. Oatley
Laboratory of Molecular Biophysics,
Department of Zoology,
South Parks Road,
Oxford.

Patricia de la Paz
Laboratory of Molecular Biophysics,
Department of Zoology,
South Parks Road,
Oxford

L.H. Pearl
Laboratory of Molecular Biology,
Department of Crystallography,
Birkbeck College,
University of London,
Malet Street,
London WC1E 7HX

L.A. Raymaekers
Department of Theoretical Medicinal
  Chemistry,
Janssen Pharmaceutica Research
  Laboratories,
Beerse,
Belgium

W. Graham Richards
Physical Chemistry Laboratory,
South Parks Road,
Oxford

D.C. Rohrer
Medical Foundation of Buffalo Inc.,
73 High Street,
Buffalo,
New York 14203,
U.S.A.

D.S. Savage
Organon Laboratories Ltd.,
Newhouse,
Lanarkshire,
Scotland

B.L. Sibanda
Laboratory of Molecular Biology,
Department of Crystallography,
Birkbeck College,
University of London,
Malet Street,
London WC1E 7HX

John J. Skehel
National Institute for Medical Research,
Mill Hill,
London NW7 1AA

I.J. Tickle
Laboratory of Molecular Biology,
Department of Crystallography,
Birkbeck College,
University of London,
Malet Street,
London WC1E 7HX

J.P. Tollenaere
Department of Theoretical Medicinal
  Chemistry,
Janssen Pharmaceutica Research
  Laboratories,
Beerse,
Belgium

Don C. Wiley
Department of Biochemistry,
Harvard University,
7 Divinity Avenue,
Cambridge,
Massachusetts 02138,
U.S.A.

Kenneth N. Trueblood
Department of Chemistry &
  Biochemistry,
University of California,
Los Angeles,
California 90024,
U.S.A.

Ian A. Wilson
Department of Molecular Biology,
Research Institute of Scripps Clinic,
10666 North Torrey Pines Road,
La Jolla,
California 92037,
U.S.A.

Rik K. Wierenga
Biomolecular Study Centre (BIOS),
University of Groningen,
Nijenborgh 16,
9747 AG Groningen,
The Netherlands

1 Crystals, X-ray crystallography, and drugs
C.J. De Ranter

1. STEREOSTRUCTURAL PROPERTIES AND DRUG ACTION

During the last decade much evidence has been accumulated that many
drugs, such as the opiate analgesics, sympathomimetic amines, neurotransmitters and hormones, act on specific target cell receptors in
a *stereoselective* fashion to elicit pharmacological effects. By
stereoselectivity is meant the extent to which an enzyme or another
macromolecule or macromolecular structure (antibody or receptor)
exhibits affinity towards one molecule of a pair of stereoisomers in
comparison and in contrast to the other isomer.

The differences in affinity and intrinsic activity based on differences in sterical structure are not always so striking as in the
quinidine-quinine case. The molecule has four chiral carbon atoms,
but the asymmetry of only two is important for biological activity.
The most important conformers are the *erythro* epimers, quinine (8S,
9R, an antimalarial) and quinidine (8R, 9S, an antiarrhythmic); the
threo epimers are inactive.

The recognition of the fact that minor changes in the structure of
the drug can alter very substantially its pharmacodynamic uptake,
shows that the stereoselectivity displayed by pharmacological systems
need not be limited to those aspects which have to do with the
presence or reactivity of particular atoms or groups, but can include
all those aspects of molecular recognition which have to do with the
shape of the molecule (i.e. configurational as well as conformational
isomerism). The implication of the conformational aspects becomes
clear if one assumes, and this is highly probable, that one single
conformation of a flexible drug molecule is bound in the drug-
receptor complex. These facts appear to be of extreme importance in
explaining some of the intriguing SAR's encountered in the field of

the seemingly structurally unrelated morphinomimetics. It was
generally assumed that for opiate analgesics to be active a rigid
T-shaped molecular architecture with two large hydrophobic fragments
approximately at right angles to each other and/or an optical center
were essential. An exception, and difficult to explain, was the
achiral pethidine molecule, whose solid-state conformation showed a
phenyl ring in equatorial position, in contrast to the axial phenyl
position in the compounds thus far known (van Koningsveld, 1970).
The appearance of fentanyl and sufentanil, two molecules with great
conformational flexibility but with potencies of 300 and 4,500 times
that of morphine, completely disrupted the current ideas on the
opiate receptor model.

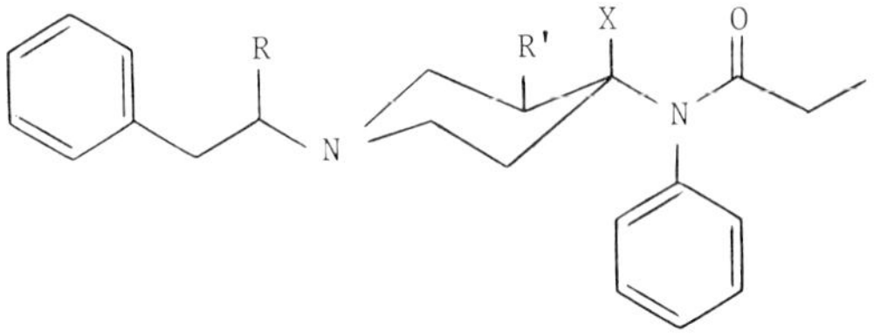

FIG. 1   Structural formula of fentanyl derivatives

Introduction of R'=CH$_3$ (R=X=H) renders the edges of the piperidine
ring enantiotopic, and the d-*cis* isomer is approximately 20 times
more potent than fentanyl. A potency as high as 10,000 times that of
morphine (tail withdrawal, rats, i.v.), was found in compound R31833
with X = COOCH$_3$, R = R' = H. Its R' = methyl-substituted derivative
(R34995) led to diastereoisomers with an extremely large enantiomeric potency ratio [*cis*(+)/*cis*(-)] of over 3,000 (Janssen and
Tollenaere, 1979). This enormous ratio suggests that the introduction
of the CH$_3$ group on the appropriate enantiotopic edge induces a
pharmacophoric orientation of the substituents at C-4 unattainable
by the *cis*(+)-diastereoisomer.
Although it is evident that the stereochemistry of the drug will play
an important role in its pharmacological profile (rate of uptake,
the rate and route of biotransformation and excretion), many drugs
having a centre of asymmetry or exhibiting geometrical isomerism are
still used in clinical practice as racemates or mixtures of *cis*- and
trans-isomers. The question then arises as to whether this is to be

considered as a blessing or a curse. The following examples will show
that this question cannot be answered very easily, but that more
attention should be given to the pharmacological, metabolic, kinetic
and toxicological properties of both isomers separately and in
combination.

1) Although in many test situations one isomer may be consistently
   less active than the corresponding isomer, the activity values
   may be reversed in a different test situation. When tested for
   heart rate, blood pressure and tracheal relaxation, (-)-isoprotere-
   nol is consistently more potent than the (+)-form, but for lowering
   the intraocular pressure of the rabbit eye, (+)-isoproterenol is
   more potent than the (-)-form (Seidehamel, Dungan and Hickey, 1975).
   In a recent study on the α- and β-adrenoceptor blocking potencies
   of labetalol and its individual stereoisomers in anaesthetized dogs
   and in isolated tissues, Brittain, Drew and Levy (1982) found that,
   although most of the α_1-adrenoceptor blocking activity of labetalol
   is attributable to the SR stereoisomer and nearly all of its β-
   adrenoceptor blocking activity resides in the RR diastereoisomer,
   each of the four isomers contributes to the overall pharmacological
   profile of labetalol.

2) All isomers may be nearly equipotent in their action on a given
   receptor, but one isomer can show unpleasant or even toxic side
   effects. To assess the intraoperative and postoperative effects of
   the optical isomers of ketamine compared with the racemic mixture
   as sole anesthetics, equianesthetic doses of racemic ketamine (RK),
   (+)-ketamine (PK) and (-)-ketamine (NK), were administered intravenously in a randomized, double-blind fashion to 60 healthy
   patients. PK was judged to produce more effective anesthesia than
   RK or NK (95 *vs* 75 *vs* 68 percent). NK produced more unwanted side
   effects (agitated behaviour, postoperative pain) than did RK or PK
   (White, Ham, Way and Trevor, 1980).
   The thalidomide tragedy (Contergan[R], Softenon[R]) would probably
   never have occurred if, instead of using the racemate, the R-enan-
   tiomer has been brought onto the market. In studies by Ockenfels,
   Köhler and Meise (1977) on the optical isomers of N-phthaloyl-
   aspartic acid, structurally related to thalidomide, and by Blaschke,

Kraft, Fickentscher and Köhler (1979) on the enantiomers of
thalidomide itself, it was shown that after i.p. administration
only the S(-)-enantiomers exert an embryotoxic and teratogenic effect.
The R(+) isomer is devoid of any of those effects under the same
experimental conditions.

3) Although from the foregoing examples it could be concluded that it
   seems sensible to use the sole, appropriate isomer in order to
   decrease the load of xenobiotics on the organism and to escape from
   the risks of unwanted, toxic side effects, considerations about the
   extra costs of the sometimes cumbersome separation of the isomers
   can justify the use of a mixture of equal proportions of the isomers
   (e.g. racemates). This is evident when the findings are such that the
   racemate is more active, less toxic or has longer (or shorter)
   duration of action than the *eutomer* (i.e. the more tightly bound
   isomer). The case is less self-evident if a given isomer, not producing any apparent side-effect, differently affects the different
   target structures of a given drug. In studying the renal actions of
   the optically active isomers of ozolinone,the main metabolite of the
   diuretic drug etozolin, Greven, Defrain, Glaser, Meywald and
   Heidenreich (1980) found that only the levorotatory isomer increased
   urine and electrolyte excretion whereas the dextrorotatory isomer
   exerted no diuretic effect, but may even be expected to antagonize,
   at high doses, low doses of the eutomer. In contrast to the diuretic
   action, both isomers are equally effective in depressing tubular
   para-aminohippurate (PAH) and in increasing renal blood flow. This
   experiment could prove that an increase in renal blood flow does not
   necessarily involve a diuretic effect.

4) That the metabolic process of drugs in living systems will always
   be subjected to a number of uncontrollable variables is demonstrated
   by the study of Kaiser, Vangiessen, Reischer and Wechter (1976).
   Enantiomeric compositions of the major urinary metabolites of ibu-
   profen, an anti-inflammatory agent in animals and humans, were
   characterized after oral administration of the racemic mixture and
   of the individual enantiomers to normal human volunteers. It was
   of interest that the *in vivo* biological activities of the individual
   enantiomers of the intact drug were equivalent, but that the R(-)-
   enantiomer was inverted to its optical antipode S(+).

Endosulfan, a non systemic contact and stomach insecticide, is a
mixture of two stereoisomers, and should, in practical use, be
harmless to wildlife and bees. The α-isomer, however, is highly
toxic for fish, outstripping the β-endosulfan by a factor of 10^3.

The conclusion is that drugs may interfere with very complicated regulatory systems which are in dynamic equilibrium, so that a dose
may initially cause a change in the system. The intensity of effects
may depend on the route of administration and the kinetics of the
drug, but the interaction of the pharmacon and its molecular sites
of action implies, in any case, conformational changes based on
intramolecular forces, resulting from a certain degree of chemical
complementarity between the interacting molecules. During the last
years great emphasis has therefore been laid on the elucidation and
analysis of the stereostructural properties of drugs. These studies
involve determination of the configurational and conformational
characteristics in the solid state (X-ray crystallography), in the
liquid state [nuclear magnetic resonance (NMR), optical rotatory
dispersion (ORD)], and in the isolated state (quantum chemical calculations).

2.   X-RAY STRUCTURE DETERMINATION

Structure analysis by X-ray diffraction techniques is the most powerful method of determining relative atomic positions in a molecular
structure. The remark that the crystal state gives only one picture
of the conformationally possible ones has been met by quantum chemical
calculations that have shown that the conformation observed in the
crystalline state is always close to one of those preferred
energetically.

2.1.   The diffraction phenomenon

Very early in the history of the natural sciences (for the first time
in mineralogy) it was recognized that crystals must be built up by
elementary particles aligned in a very regular way in what is now
called a *crystal lattice.*
Fairly soon after Röntgen's discovery in 1896, the first interference
pictures of X-rays on crystalline material were obtained in 1912 by
coworkers of Max von Laue. The most important feature in von Laue's
work was the idea that the periodical arrangement of atoms in a

crystal plane could be compared with a diffraction grating, where
the equidistant slits are replaced by arrays of atoms at exactly the
same distance being of the order of angstroms and thus of the same
order as the wavelength of X-rays. The diffraction pattern of a real
crystal, however, is more complicated than that of an optical grating,
as the crystal shows a three-dimensional periodicity and causes
diffraction maxima in all directions of space, the positions of these
maxima now being indicated by the three symbols h, k and l.

In 1913 W.L. Bragg showed that von Laue's rather complicated formulation of the diffraction phenomenon could be in essence reduced to a
reflection phenomenon of X-rays on the with atoms occupied planes of
the crystal lattice. This has led to the well-known reflection
condition of Bragg

$$2d_{hkl} \sin \theta = n \lambda, \qquad\qquad 2.1$$

where d_{hkl} is the interplanar spacing or distance between the regularly
oriented layers of a set of (hkl)-planes, θ the angle between the
striking X-rays beam of wavelength λ and the crystal plane, and n the
order of reflection (normally limited to n = 1).
The intensity of the X-rays diffracted in the direction of reflection
hkl by the N atoms in the unit cell is proportional to $|F_{hkl}|^2$, where
F_{hkl} represents the *structure factor* and is expressed by

$$F_{hkl} = \sum_{n=1}^{N} f_n \exp 2\pi i(hx_n + ky_n + lz_n) \qquad\qquad 2.2$$

with f_n being the scattering factor (or form factor) of the n-th atom.
The value of f_n depends on the number of electrons in the atom concerned, and also (for reasons that cannot be discussed here, see e.g.
Buerger, 1966) on the glancing angle θ. It can be shown that the
general expression of F_{hkl} can be rewritten in the form

$$F_{hkl} = \sum_{n=1}^{N} f_n \{\cos 2\pi(hx_n + ky_n + lz_n) +$$
$$i \sin 2\pi(hx_n + ky_n + lz_n)\} \qquad\qquad 2.3$$

If the structure is centrosymmetric the imaginary part in the formula
drops out and a much simpler formula with only cosine terms remains.
In fact the structure factor represents the hypothetical number of
electrons that, if they should scatter all in phase from the origin
of the unit cell, should scatter with the same intensity as the actual

group of electrons spread out throughout the whole cell. As a consequence the structure factor is characterized by an amplitude, given
by $\sum_{n=1}^{N} f_n$, and a relative phase ϕ_{hkl}, given by the rest of the term
in eqn. 2.3, in which x_n, etc..., are the fractional coordinates of
the n-th atom.

Measuring the reflection intensities gives us the absolute value of
the structure factors. If in any way the phases ϕ_{hkl} could also be
determined, then the electron densities throughout the unit cell
could be calculated. That, however, requires a tremendous amount of
work, because in order to describe $\rho(x,y,z)$ rather precisely the
electron density must be calculated on a grid with a size of 0.25 A
by 0.25 A. This means that for a unit cell of 10 x 10 x 10 A and for
example 2000 measured $|F_{hkl}|$-values a Fourier summation over 2000
terms has to be calculated in 40 x 40 x 40 points. Even for a modern
computer this is a formidable challenge. Fortunately the electron
density formula

$$\rho(x,y,z) = \frac{1}{V} \sum_h \sum_k \sum_l F_{hkl} \exp\{-2\pi i (hx + ky + lz)\} \qquad 2.4$$

can be simplified by rewriting the trigonometric part in terms of the
following type

$$\frac{\sin}{\cos} 2\pi hx \qquad \frac{\sin}{\cos} 2\pi ky \qquad \frac{\sin}{\cos} 2\pi lz \qquad 2.5$$

For example, by searching first for all terms with a particular h and
k a summation over l can be made giving the intermediate result
T (h,k,z). These intermediate results are then summed over k for any
particular h and z leading to new intermediate results S (h,y,z),
that in a last one-dimensional summation at given y and z results in
the expected electron density ρ (x,y,z) (Schenk, 1981). In this way
the three-dimensional summation is replaced by a large number of
one-dimensional summations giving an enormous simplification in computing work.

2.2. Collecting the reflection intensities

Although in recent years some advancement has been made in the
measuring and processing procedures for polycrystalline materials, the
main point of future structural work will still lie in single crystals.
One chooses a well-formed crystal, not too small to obtain sufficient
accuracy in measuring the diffracted X-ray beam, and not too big to

prevent high absorption in the crystal, as precise corrections for
absorption are difficult to do. As soon as all the crystal data
(space group, unit cell parameters (a, b, c, α, β, γ), volume, number
of molecules in the unit cell) are known data collection can begin.
In order for the analysis of a structure with $\pm$ 25 independent atoms
to be successful the measurement of a few thousand intensities will
be required. Two decades ago photographic registration (rotation,
Weissenberg or precession photographs; multiple film techniques) was
the only one possible. The blackness of the diffraction spots was
measured by a densitometer, a process that could take months. Today's
data collection is executed by an automated diffractometer : the
so-called *four-circle diffractometer*. The equipment consists of four
parts :

1) the X-ray source (high voltage generator and X-ray tube);
2) the goniometer, usually constructed, but not necessarily, as an
   "Eulerian cradle";
3) the detector (scintillation counter) and its counting chain;
4) the computer, for control of the diffractometer and storage of
   results from the detector.

The measuring procedure is such that for each 20 to 30 reflections a
reference reflection is remeasured in order to allow a final rescaling
of all measured intensities. During data collection time some crystals
may deteriorate. A few thousand independent reflections can thus be
measured in one week.
The fundamental difficulty in structure analysis by X-ray diffraction
is the experimental loss of the relative phases of the diffracted
X-ray beams (the "*phase problem*"). A great deal of the crystallographer's
time will therefore be taken up by finding the phases for as much as
possible of the obtained structure amplitudes. To tackle this phase
problem a variety of methods were developed during the last decennia.

2.3. Resolution of the phase problem

The present crystallographer can no longer imagine that in earlier days
crystal structures could be determined without intense use of a computer.
Nevertheless, it was only 25 years ago that the computer period begun.
The interested reader is referred to a recent interview with Nobel
laureate Dorothy C. Hodgkin (Julian, 1982), where she comments on the
resolution of the vitamin B-12 structure, the largest organic molecule

whose structure was determined before 1956. Sometimes very clever
tricks (e.g. Beevers-Lipson strips to reduce the computational work
to a one-dimensional summation) or instrumental gadgets (e.g. Bragg's
optical diffractometer, von Eller's *photosommateur harmonique*) were
used to solve the phase problem. Nothwithstanding many difficulties,
remarkable results have been obtained [e.g. penicillin (Crowfoot,
Bunn, Rogers-Low and Turner-Jones, 1949)].

Patterson Method.   An attempt to evade the phase problem was made by
Patterson (1935). Instead of using the structure factors, he used the
squares of the moduli as Fourier coefficients. These quantities are
directly related to the observed intensities and so they can always
be calculated.
Patterson defined a function $P(u,v,w)$ such that

$$P(u,v,w) = \frac{1}{V} \sum_h \sum_k \sum_l |F_{hkl}|^2 \exp -2\pi i(hu + kv + lw) \qquad 2.6$$

The function is always centrosymmetric and defines a map, not of
atomic positions, but of interatomic distances plotted from one point :
the origin. A peak in the Patterson map at u,v,w, in other words,
implies that there are two atoms in the crystal structure at $x_1 y_1 z_1$
and $x_2 y_2 z_2$ such that $x_2 - x_1 = u$, $y_2 - y_1 = v$ and $z_2 - z_1 = w$.
The height of the peak is proportional to the product of the number
of electrons in each of the two atoms involved. The corresponding
Patterson map for a simple four-atom structure is shown in Fig. 2.

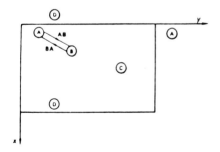

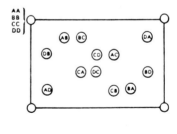

FIG. 2 Four-atom structure   Patterson map

Note that the vector map has a large origin peak, corresponding to the vector from every atom to itself. For a structure with N atoms in the unit cell there will be N^2 *Patterson peaks* in the map: N of them are piled up at the origin and 1/2 N(N-1) are related to the other 1/2(N-1) peaks by a centre of symmetry.

As in practice the atoms are not points but show a certain spread in electron density, it is not difficult to imagine that even for a rather simple structure regions of positive electron densities will overlap. This implies that the Patterson function P(u,v,w) does not necessarily reach its maximum at the point u,v,w, complicating considerably the interpretation of the map. This is especially true in structures that contain only light atoms or in structures that contain a heavy atom, but whose contribution to the structure factor is relatively too small (e.g. in proteins). The Patterson syntheses will consist of collections of unresolved peaks and the above approach will fail.

In the cases the structure contains a heavy atom (e.g. an organic molecule with a Cl, Br, S atom) the Patterson map will show up well-resolved peaks due to the dominating scattering power of the heavy atom. These peaks will permit a rather precise location of the heavy atom and the phases thus obtained are a good approximation of the phases of the structure factors. In subsequent Fourier electron density maps one can search for the missing atoms. An other approach is to use the known part of the structure for calculating difference structure factors (representing the lighter part of the structure) and to progress in applying direct methods (DIRDIF, Beurskens, Bosman, Doesburg, Gould, van den Hark, Prick, Noordik, Beurskens and Parthasarathi, 1981).

A completely different approach is to search between light atom-light atom vectors (superposition method, minimum function), but the theory of these methods, exhaustively discussed by Buerger (1951), falls beyond the scope of this text.

Although,before the general break-through of the computer a graphical Patterson analysis resulted in unwieldy amounts of paper, the Patterson method has been extensively used and important structures have been solved as a result of its use : cysteinylglycine sodiumiodide (Dyer, 1951); morphine hydroiodide (Mackay and Hodgkin, 1955); 5-bromogriseo-fulvine (Brown and Sim, 1963).

Isomorphous replacement. If the structure is so complex that even though the heavy atom can be located, the rest of the molecule can neither be unravelled from the Patterson map nor located by Fourier synthesis, the structure determination can be tackled by the method of isomorphous replacement. The method works for crystals that can assimilate different heavy atoms while leaving the light atom part of the structure unchanged (e.g. a K-salt and the isomorphous Rb-salt; a Cl- and a Br-compound). The two Patterson maps will be very analogous, but the maxima corresponding to the heavy atom-heavy atom vectors will differ, and so the positions of the heavy atoms can be located. In some cases (non-centrosymmetric structures for instance) information from a third isomorphous compound is required. The labour involved in such calculations is enormous, yet this is how the structures of proteins are unravelled.

A serious limitation in the application of the isomorphous replacement is that the derived heavy atom crystals must be precisely isomorphous. That even then unexpected results are not excluded has been illustrated by the structure determination of (+)-tubocurarine dichloride (Codding and James, 1973) and (+)-tubocurarine dibromide (Reynolds and Palmer, 1976). In the tubocurarine moiety of both salts significant differences in conformation have been observed (Reynolds, Palmer, Gorinsky and Gorinsky, 1975).

Anomalous scattering and absolute configuration. So far we have assumed that the incident X-ray beam and the diffracting electrons do not interact with each other. This assumption holds as long as the frequency of the X-rays is sufficiently distinct from the absorption edge of the scattering atoms. Under these conditions the X-ray diffraction pattern of the crystal is centrosymmetric, whether the crystal structure itself is centrosymmetric or not. In other words, the intensities of a reflection (hkl) and of its antireflection ($\overline{hkl}$) are equal (inversion of a structure will change the sign of all phase angles but the magnitudes of F_{hkl} and $F_{\overline{hkl}}$ remain the same : *Friedel's law*).

As stated before and depending on the energy of the incident X-rays, some of the atoms may scatter slightly out of phase with the others [e.g. in an organic Co-compound such as vitamin B-12 all electrons in the molecule scatter Cu K_α-radiation (λ = 1.5418 Å) as free electrons

except for the Co-K electrons, their absorption edge being at 1.6081 $\overset{\circ}{A}$].
For such atoms the simple scattering factor f_n, which was calculated
for free electrons, has to be modified to take account of the interaction of the incident X-rays with the bound electrons. The *anomalous
scattering factor* of the atom now becomes

$$f_{an} = f_n + \Delta f' + i \Delta f'', \qquad\qquad 2.7$$

with f_n, the simple atomic scattering factor, $\Delta f'$, correction for
change in magnitude, $\Delta f''$, correction for phase change. The imaginary
term $\Delta f''$ is always positive, this means that the phase of the anomalous
scattered wave runs in front of that of the normal.
The important consequence is that the presence of an anomalous scatterer
in a non-centrosymmetric structure (e.g. $P2_12_12_1$, $P2_1$, ...) will lead
to a slight difference in intensities between the anomalously scattering
reflection (hkl) and its antireflection ($\overline{hkl}$). This can be illustrated
in the following way

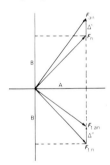

$$F_{an}^2 = \left|F_{hkl}\right|_{an}^2 = A^2 + (B + \Delta'')^2$$

$$F_{t,an}^2 = \left|F_{\overline{hkl}}\right|_{an}^2 = A^2 + (B - \Delta'')^2$$

FIG. 3   The effect of anomalous scattering on $\left|F_{hkl}\right|$ and $\left|F_{\overline{hkl}}\right|$

From Fig. 3 we see that the difference in intensities will be proportional to

$$\Delta I = I_{an} - I_{t,an} = A^2 + (B + \Delta'')^2 - \{A^2 + (B - \Delta'')^2\} \qquad\qquad 2.8$$

$$= 4\, B\Delta''$$

where Δ'' can be found from the ratio

$$\frac{\Delta''}{F_n} = \frac{(\Delta f'')_z}{(f_n)_z} \qquad\qquad 2.9$$

$(\Delta f'')_z$ and $(f_n)_z$ refer to the heavy atom present in the structure. As
can be seen F_{an} and $F_{t,an}$ as well as F_n have to be known. This can be

done by measuring the crystal at two different wavelengths (e.g. Cu Kα-
and Mo Kα-radiation for a I-salt or Cr Kα- and Cu Kα-radiation for a
Br-salt). Mostly, however, one radiation is used and the Friedel
opposites, hkl and $\overline{hkl}$, are measured. If the effect of the anomalous
dispersion is large enough, the absolute configuration can be established
by comparing the observed Bijvoet differences, $\Delta F_o = |F_{hkl}|_o - |F_{\overline{hkl}}|_o$,
with the calculated differences, $\Delta F_c = |F_{hkl}|_c - |F_{\overline{hkl}}|_c$. The correct-
ness of the configuration can be checked by inverting the coordinates
(Ibers and Hamilton, 1964) and by testing the differences between both
sets of results on a certain significance level (Hamilton, 1965;
Beurskens, Noordik and Beurskens, 1980).

The first examples of absolute configuration determination were those
of the dextrorotatory NaRb-tartrate (Peerdeman, van Bommel and Bijvoet,
1951) and of the D-isoleucine hydrobromide (Trommel and Bijvoet, 1954).
The absolute configuration of the morphinomimetics was fixed by the
determination of the codeine molecule (Kartha, Ahmed and Barnes, 1962).

Direct Methods. The interpretation of the subsequent Fourier syntheses
phased by the heavy-atom contribution is not always quite straightforward, and the method completely fails for an equal-atoms structure.
Therefore, it would be convenient to have methods of deriving phases
for the structure factors that do not depend on the presence of heavy
atoms and do not involve any structural assumption, but that reduce
the phase problem to an objective procedure to which any decisions are
of a purely mathematical nature. These methods are usually called
"direct methods".

The physical basis of the whole of present-day direct methods are the
two statements, that in a crystal

   1) the electron density is positive everywhere,
   2) the electron density consists of discrete spherically symmetric
      atoms.

The first steps in the development of direct methods were the publica-
tions by Harker and Kasper (1948) and by Karle and Hauptman (1950) on
inequality relationships between the structure factors.

Harker and Kasper pointed out that for centrosymmetric structures the
following relationship holds

$$|U_H|^2 \leqslant \frac{1}{2} (1 + U_{2H}) \qquad\qquad 2.10$$

where U_H is the unitary structure factor defined by

$$U_H = \frac{F_H}{\sum\limits_{n=1}^{N} f_n} \qquad\qquad 2.11$$

Equation 2.10 is the formal mathematical presentation of the idea that, if $|F_H|$ and $|F_{2H}|$ are both strong then the sign of F_{2H} is likely to be positive. The number of inequality relationships which may be found for the various space groups is extremely large and many have been published [e.g. the important matrix inequalities developed by Karle and Hauptman (1950), who expressed the condition that the sum of a Fourier series should be non-negative everywhere].

The next step in the development of direct methods was the derivation of simple forms of equality relations between structure factors by Goedkoop (1950) and Sayre (1952). The latter showed that the following equation is valid for a structure containing equal and resolved atoms

$$F_H = \frac{1}{V} \frac{f_n}{g_n} \sum\limits_{K} F_K F_{H-K} \qquad\qquad 2.12$$

where f_n is the atomic scattering factor of the n-th atom and g_n the scattering factor of the squared-atom. Although the equations themselves were difficult to handle (e.g. the phases associated with all the products on the right side must be known), they led Cochran and Woolfson (1955) to the formulation of a simple, but quite practical probability relationship between structure factors. For a centro-symmetric structure containing N equal atoms in the unit cell the probability equation is

$$P_+ = \frac{1}{2} + \frac{1}{2} \tanh \; \{N^{-1/2} \, |E_H E_K E_{H+K}| \} \qquad\qquad 2.13$$

which is the probability of the product of signs

$$s(H) \cdot s(K) \cdot s(H+K) \simeq + 1 \qquad\qquad 2.14$$

being positive. The larger the product of the three moduli the higher the probability of the sign relationship being true. It should be noted that, because of the presence of $N^{-1/2}$ in the argument of tanh, the more atoms the structure contains the lower will be the probability of the sign relationship. This is the main reason for the current failure of direct methods to solve very large structures. Further progress in the use of probability methods was brought about by the derivation of the corresponding formulae for non-centrosymmetric structures (Cochran, 1955; Karle and Karle, 1966) and the publication of what is now known as the *tangent formula* (Karle and Hauptman, 1956)

$$\tan \phi_H = \frac{\sum_K |E_K E_{H-K}| \sin(\phi_K + \phi_{H-K})}{\sum_K |E_K E_{H-K}| \cos(\phi_K + \phi_{H-K})} \qquad 2.15$$

where the sums are taken over all available terms.

2.4. Structure solving strategy

The first step in crystal structure solving procedures is to convert
the observed intensities into *normalized structure factors*

$$|E_H| = \left(\frac{|F_H|^2}{\varepsilon <I>}\right)^{1/2} \qquad 2.16$$

where $<I>$ is the average intensity in a certain $\sin \theta/\lambda$ interval and
ε is a factor that takes into account the effect of space group
symmetry on the value of $|F_H|^2$. With the 200 or 300 strongest
reflections all possible triples H, K, H+K are generated (the so-called
$\Sigma 2$ list). Often there are as many as a few thousand, a lot of them being
very weakly or even dubiously interrelated. Therefore extreme care is
needed in evaluating these relationships.

The most important element governing the success or failure of direct
methods is the choice of the starting reflections from which new phases
will be developed. An invalid sign relationship in the first few steps
of sign-determination will lead to a complete breakdown of the sign-
determinating process. A method of allowing for such a contingency has
been discussed by Germain, Main and Woolfson (1970, 1971). The basis of
their procedure is that for centrosymmetric structures the origin is
fixed by assigning positive signs to some reflections and symbols are
allocated to up to six others. The arbitrary signs and symbols are allocated to those reflections which occur most frequently in the triple-
product sign relationship. For non-centrosymmetric structures some
reflections, in general three, are allocated specific phases to fix the
origin of the cell; one reflection, depending on the space group, is
allocated a specific phase ($+\pi/4$, $+3\pi/4$ or its antipodes $-\pi/4$, $-3\pi/4$) to
fix the enantiomorph and a number of others are tried in all possible
combinations of ($\pm \pi/4$, $\pm 3\pi/4$), $(0, \pi)$ or ($\pm \pi/2$) depending on the
reflection type. New phase indications are found by the use of a weighted
tangent formula. These principles have been worked out in the computer
program packages MULTAN (Main, Woolfson, Lessinger, Germain and Declercq,
1974) and SHELX (Sheldrick, 1976). In other structure solving programs
(e.g. SIMPEL, Overbeek and Schenk, 1978) letter symbols are used in a

systematic way and the symbols are only exchanged for true values at an
advanced stage of phase extension. Selecting the appropriate phase
relationships can be done more properly by using Harker - Kasper in-
equalities and quartet relations as selection criteria (Schenk and
de Jong, 1973; Schenk, 1973).

A quartet is expressed by the following equation

$$\phi_4 = \phi_H + \phi_K + \phi_L + \phi_{-H-K-L} \simeq 0 \qquad\qquad 2.17$$

where the four corresponding normalized structure factors, $|E_H|$, $|E_K|$,
$|E_L|$ and $|E_{-H-K-L}|$, are all large. The relation, however, seems to be
much less reliable than the triplet phase relationship. Much more
useful information can be gained by introducing the so-called "cross-
term" magnitudes $|E_{H+K}|$, $|E_{H+L}|$ and $|E_{K+L}|$. In the case where the seven
magnitudes (main and cross-terms) are all large, then the probability
is high that $|\phi_4| \simeq 0$ (modulo 2π); if, on the other hand, the three
cross-term magnitudes are small, $|\phi_4|$ can be estimated to be near π :
the so-called "negative quartet" (Hauptman, 1974, Schenk, 1974). Use
of the negative quartets as selection criterium for reliable phase
relationships has enhanced the chance of structure determination in the
more difficult space groups (e.g. P1) to the normal 90 % success level
obtained for the other routinely solvable space groups.

Overlooking the crystallographic literature on organic and organo-
metallic compounds published in 1982, about 28 % of all structures
were solved by the heavy-atom method and 72 % by direct methods. Between
the direct methods MULTAN (in its subsequent versions) was used in 70 %
of the cases (or about 50 % of all structures published in that domain).

2.5. Refinement and interpretation

Once a model of a structure has been found by one of the previous methods,
it is necessary to improve the preliminary coordinates by some process
of refinement. The two principal numerical techniques used to refine
crystal structures are the Fourier transform method (Fast Fourier Transform algorithm) and the method of linearized least-squares. In the least-
squares approach there are two different schools of thought concerning
what is the best method to use : the unconstrained-model school and the
constrained-model school (e.g. rigid-body model, segmented-body model).
The advantage of the unconstrained-model is its direct application to a
wide variety of problems. The disadvantage is often the large number of
variable parameters that must be handled. For example, a full-matrix

refinement with anisotropic thermal parameters for a 45-atom structure will involve at least 406 variables and will require about $406^2/2$ words of core storage for the least-squares matrix alone. The common procedure is to compare the calculated structure factors, $|F_c|$, for a proposed model with the actually observed ones. Then a necessary condition for the proposed model to be correct is that the calculated values duplicate those observed. The best agreement is obtained for

$$D = \sum_{i=1}^{m} w_i (|F_o| - |F_c|)_i^2 \qquad\qquad 2.18$$

being minimal; w_i is a weighting factor for an observation $|F_o|$. It is difficult, however, to know how discrepancies in different $|F|$'s are to be weighted, or combined. No really satisfactory weighting has been proposed. Nevertheless it has become common practice to utilize a *residual* of the following form for this purpose

$$R = \frac{\sum_i (|F_o| - |F_c|)}{\sum_i |F_o|} \qquad\qquad 2.19$$

Correct structures usually have $R < 0.25$, and very well refined structures may have R in the neighbourhood of 0.05.

The structure can now be interpreted and here again a number of computing programs may be very useful for calculating bond lengths and bond angles, torsion angles, Newman projections, stereo views and conformation energies.

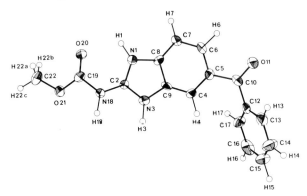

FIG. 4   ORTEP (Johnson, 1971) drawing of mebendazole, a broad-spectrum
         anthelmintic (Blaton, Peeters and De Ranter, 1980)

With due care in the experimental conditions the precision reached in bond lengths can be 0.005 to 0.003 Å and in bond angles 0.4° to 0.1°.

The output from a least-squares refinement includes the estimated
standard deviation (e.s.d.) for each atomic parameter; this is based
on the number of parameters being refined, the quality of the fit
between the observed and calculated data and the contribution made to
the latter by the atom in question (Glasser, 1977). The last part of
that statement means that in a structure containing a mixture of light
and heavy atoms, the positional parameters of the heavy atoms will be
determined more accurately than those of the light ones. As hydrogen
atoms are particularly sensitive to this effect, the positions determined for these by X-ray diffraction and usual refinement techniques
will be generally displaced towards the atom to which the hydrogen is
bonded.

In interpreting the final results one should never be tempted to accept
the thesis "the bigger and more expensive the machinery, the more
reliable the results". An innocent example will illustrate this statement. In a review article on the conformation of neuroleptic drugs and
their conformational resemblance to dopamine Tollenaere, Moereels and
Koch (1977) pointed out that the dopamine molecule was nicely super-
imposable on the spiperone conformation as found in the solid state by
Koch (1973). It is noteworthy that, although the authors mentioned the
presence of five equivalent energy minima in the PCILO energy contour
map, in the literature up till then no reference could be found to any
possible polymorphism of the butyrophenone type neuroleptics. Very
recently, however, Azibi (1982) has shown that a second polymorph of
spiperone exists and that the polymorph, that was used for the fitting
with dopamine, is the least stable one. The commercialized polymorph
was always form II (the most stable one) and it was just by accident
that polymorph I was obtained in the form of single crystals.

3. POLYMORPHISM, CRYSTAL HABIT AND DRUGS

With the spiperone example we touched the last aspect of the crystallography and drug interfaces we want to mention here. Many pharmacologists,
not being chemists, take the label as evidence for the content of the
bottle. But it is known that the production process is affected with
many variables, some of them being uncontrollable in practice : the
occurrence of impurities of about 1 % can be very hardly avoided, batch
to batch variations exist. Polymorphism is also one of these common
variables. The existence of different polymorphic forms of a rather in-

soluble drug may play an enormous role in the outcome of the experiments.
A spectacular example is chloramphenicol palmitate where the efficacy
decreases in proportion to the transformation of the pharmacologically
active metastable form into the inactive stable one (Aguiar and Zelmer,
1969). Thus it would now be wise to attempt to obtain the most active
and relatively most stable form at the moment a new compound shows
interesting pharmacological activity. In addition, in the different
Pharmacopeiae more attention should be given to the polymorphism problem.
Two examples can be mentioned. A study on different commercial samples
of erythromycin (Hoogmartens and Vanderhaeghe, 1983) showed that for
some of them the solubility data did not conform either with the in-
structions of the British Pharmacopoeia 80 nor with those of the
European Pharmacopoeia (Belgian Ed.). X-ray powder diffractograms showed
crystallographic differences indicating the presence of polymorphic
forms or hydration. The second example is from the study by Himmelreich,
Rawson and Watson (1977) on the polymorphic forms of mebendazole, a
broad-spectrum anthelmintic of the substituted benzimidazole class.
Dissolution rates and determination of the saturation solubilities proved
that the sequence for the three polymorphs was $\gamma > \beta > \alpha$, indicating the
best bioavailability potency for the γ-form. Differential thermal analysis,
however, and seeding experiments with crystals of each of polymorphs α and
β on saturated solutions of polymorphic form γ showed that the order of
thermodynamic stability is $\alpha > \beta > \gamma$, what makes α more suitable for
manufacture.

A knowledge of the different polymorphic forms is not only important to
start pharmacological and clinical investigations, but it is perhaps of
even greater importance as soon as the production process starts. An
anhydrous compound can be transformed during the preparation of a suspen-
sion or an ointment, into a solvate. The compressional behaviour, crystal
habit modifications and the final formulation (e.g. adsorption phenomena
to other compounds) are all variables that may affect the bioavailability
of the drug. For example the release of drugs from hard gelatine capsules
will not only depend on the capsule wall properties (composition, manufacturing), but also on the contact between the drug and the wall, the
particle size, the wettability of the powder bed. The more the powder is
hydrophilic, the better the availability of the drug [a modification of
the surface properties of phenacetin obtained by a crystallization in the
presence of tween 80 or by nebulisation of a hydrophilic polymer, causes
an improvement of the availability of the drug (Ludwig, 1979)] .

4. CONCLUDING REMARKS

A common assumption in the analysis of structure-activity relationships
of new drugs was that chemical similarity often implies similarity in
action. That this is not necessarily the case has been demonstrated for
different classes of drugs. A striking example is the drug Etaphylline.
The active component of the drug is acefylline, a N-7-substituted theofylline derivative. The chemical similarity with theofylline and caffeine
led to the assumption that their pharmacological and pharmacokinetic
properties would be identical. This explains the therapeutic administration of Etaphylline as a cardiotonic and smooth muscle relaxant (for
bronchial asthma patients)for the last 30 years. Studies by Zuidema and
Merkus (1979), however, have shown that the resorption of the drug after
peroral administration was insignificant and that even after intravenous
injection acefylline was rapidly and completely excreted so that its
therapeutic use is rather doubtful. Interdisciplinary fundamental research
and open communication between industrial pharmacists, medicinal chemists,
pharmacologists, clinicians and physical chemists is a must for developing
and manufacturing more selective and better drugs and this meeting will
hopefully contribute to the achievement of this goal.

REFERENCES

Aguiar, A.J. and Zelmer, J.E. (1969). *J. Pharm. Sci.*, 58, 983-987.
Azibi, M. (1982). Thèse de Doctorat, U.C. Louvain, Belgium.
Beurskens, G., Noordik, J.H. and Beurskens, P.T. (1980). *Cryst. Struct.
    Comm.*, 9, 23-28.
Beurskens, P.T., Bosman, W.P., Doesburg, H.M., Gould, R.O., van den Hark,
    Th. E.M., Prick, P.A.J., Noordik, J.H., Beurskens, G. and
    Parthasarathi, V (1982). DIRDIF. Technical Report, Crystallography
    Laboratory, Nijmegen, The Netherlands.
Blaschke,G., Kraft, H.P., Fickentscher, K. and Köhler, F. (1979).
    Arzneim.-Forsch., 29, 1640-1642.
Blaton, N.M., Peeters, O.M. and De Ranter, C.J. (1980). *Cryst. Struct.
    Comm.*, 9, 181-186.
Brittain, R.T., Drew, G.M. and Levy, G.P. (1982). *Br. J. Pharmacol.*,
    77, 105-114.
Brown, W.A.C. and Sim, G.A. (1963). *J. Chem. Soc.*, 1050-1059.

Buerger, M.J. (1951). *Acta Cryst.*, 4, 531-544.

Buerger, M.J. (1966). *X-Ray Crystallography* (7th ed.). John Wiley and
    Sons, Inc., New York, pp. 51-53.

Cochran, W. (1955). *Acta Cryst.*, 8, 473-478.

Cochran, W. and Woolfson, M.M. (1955). *Acta Cryst.*, 8, 1-12.

Codding, P.W. and James, M.N.G. (1973). *Acta Cryst.*, B29, 935-942.

Crowfoot, D., Bunn, C.W., Rogers-Low, B.W. and Turner-Jones, A. (1949).
    *The X-Ray Crystallographic Investigation of the Structure of
    Penicillin*. Oxford, Univ. Press.

Dyer, H.B. (1951). *Acta Cryst.*, 4, 42-50.

Germain, G., Main, P. and Woolfson, M.M. (1970). *Acta Cryst.*, B26,
    274-285.

Germain, G., Main, P. and Woolfson, M.M. (1971). *Acta Cryst.*, A27,
    368-376.

Glasser, L.S.D. (1977). *Crystallography and its Applications*.
    Van Nostrand Reinhold. New York, pp. 198-200.

Goedkoop, J.A. (1950). *Acta Cryst.*, 3, 374-378.

Greven, J., Defrain, W., Glaser, K., Meywald, K. and Heidenreich, O.
    (1980). *Pflügers Arch.*, 384(1), 57-60.

Hamilton, W.C. (1965). *Acta Cryst.*, 18, 502-510.

Harker, D. and Kasper, J.S. (1948). *Acta Cryst.*, 1, 70-75.

Hauptman, H. (1974). *Acta Cryst.*, A30, 472-476.

Himmelreich, M., Rawson, B.J. and Watson, T.R. (1977). *Austr. J.
    Pharmac. Sc.*, 6, 123-125.

Hoogmartens, J. and Vanderhaeghe, H. (1983). *Private Communication*.

Ibers, J.A. and Hamilton, W.C. (1964). *Acta Cryst.*, 17, 781-782.

Janssen, P.A.J. and Tollenaere, P.J. (1979). *Neurochemical Mechanisms
    of Opiates and Endorphins*. Adv. Biochem. Psychopharmacol. (ed.
    H.H. Loh and D.H. Ross) Vol. 20, pp. 103-129. Raven Press, New
    York.

Johnson, C.K. (1971). ORTEP. Report ORNL-3794, revised. Oak Ridge
    National Laboratory, Tenessee.

Julian, M.M. (1982). *J. Chem. Educ.*, 59, 124-125.

Kaiser, D.G., Vangiessen, G.J., Reischer, R.J. and Wechter, W.J. (1976).
    J. Pharm. Sci., 65, 269-273.

Karle, J. and Hauptman, H. (1950). *Acta Cryst.*, 3, 181-187.

Karle, J. and Hauptman, H. (1956). *Acta Cryst.*, 9, 635-651.

Karle, J. and Karle, I.L. (1966). *Acta Cryst.*, 21, 849-859.

Kartha, G., Ahmed, F.R. and Barnes, W.H. (1962). *Acta Cryst.*, 15, 326-333.

Koch, M.H.J. (1973). *Acta Cryst.*, B29, 379-382.

Ludwig, A. (1979). Doctoral thesis, U.I.Antwerp, Belgium.

Mackay, M. and Hodgkin, D.C. (1955). *J. Chem. Soc.*, 3261-3267.

Main, P., Woolfson, M.M., Lessinger, L., Germain, G. and Declercq, J.P. (1974). MULTAN 74. *A System of Computer Programs for the Automatic Solution of Crystal Structures from X-Ray Diffraction Data*. Univ. of York, England, and Louvain-la-Neuve, Belgium.

Ockenfels, H., Köhler, F. and Meise, W. (1977). *Arzneim.-Forsch.*, 27, 126-128.

Overbeek, A.R. and Schenk, H. (1978). SIMPEL. *Computing in Crystallography*. (ed. H. Schenk, R. Olthof, H. van Koningsveld and G.C. Bassi). Delft University Press, Delft, The Netherlands.

Patterson, A.L. (1935). *Z. Kristallogr.*, 90, 517-542.

Peerdeman, A.F., van Bommel, A.J. and Bijvoet, J.M. (1951). *Proc. Koninkl. Nederland. Wetenschap. Acad.*, B54, 16-19.

Reynolds, C.D., Palmer, R.A., Gorinsky, B.A. and Gorinsky, C. (1975). *Biochim. Biophys. Acta*, 404, 341-344.

Reynolds, C.D.and Palmer, R.A. (1976). *Acta Cryst.*, B32, 1431-1439.

Sayre, D. (1952). *Acta Cryst.*, 5, 60-65.

Schenk, H. (1973). *Acta Cryst.*, A29, 77-82.

Schenk, H. and de Jong, J.G.H. (1973). *Acta Cryst.*, A29, 31-34.

Schenk, H. (1974). *Acta Cryst.*, A30, 477-481.

Schenk, H. (1981). *Chemisch Magazine*, jul/aug, 407-408.

Seidehamel, R.J., Dungan, K.W. and Hickey, Z.E. (1975). *Amer. J. Opthalmol.*, 79, 1018-1025.

Sheldrick, G.M. (1976). *Program for Crystal Structure Determination*. University of Cambridge, England.

Tollenaere, J.P., Moereels, H. and Koch, M.J.H. (1977). *Eur. J. Med. Chem.*, 12, 199-211.

Trommel, J. and Bijvoet, J.M. (1954). *Acta Cryst.*, 7, 703-709.

van Koningsveld, H. (1970). *Recl. Trav. Chim. Pays-Bas*, 89, 375-378.

White, P.F., Ham, J., Way, W.L. and Trevor, A.J. (1980). *Anesthesiology*, 52, 231-239.

Zuidema, J. and Merkus, F.W.H.M. (1979). *Ned. T. Geneesk.*, 123, 1527-1528.

2 Crystal forces and molecular conformation
Joel Bernstein

"...All work of the crystallographers serves only to demonstrate
that there is only variety everywhere where they suppose
uniformity...that in nature there is nothing absolute, nothing
perfectly regular."

Comte Georges Louis de Clerq de Buffon (1701-1789)
Histoire Naturelle Des Mineraux, Paris 1783-88, III.

1. INTRODUCTION

The title of this paper comprises two subjects which in and of themselves
have been the basis of a great body of work. They cannot, of course, be
completely covered even in a review of this type. What can be done, however, is to examine the *connection* between crystal forces and molecular
conformation in the light of the interests of those who are using crystal-
ographic information on the one hand to understand the mode of action of
drug systems on the molecular level, and on the other hand to design new
drugs. The role of crystallographers in this field is best defined by
posing some very general questions

1)   What kind of information is provided to the drug design people?
2)   For what purpose is this information?
3)   Where does the information come from?

The data which crystallographers supply to the chemists, biologists
and pharmacologists are unit cell constants and a set of atomic coordinates. From these atomic chemical connectivity and a variety of geometric
features may be obtained for use in the interpretation of pharmacological
and biological processes. In fact, over the past few decades the major
portion of our molecular geometric information, and certainly the most
precise for larger molecules, has come from crystal structure analyses and
extensive use has been made of this information in many fields of biology,
medicine and chemistry.

In the early days of crystallography, when a crystal structure
analysis could earn one a Ph.D., just having this kind of precise molecular
geometric information made it very valuable, and lacking additional or
alternative information of such accuracy, a great deal of confidence was
placed on it. Indeed, in those days the maximum amount of information on a
molecule was extracted from a single crystal structure. As the number of

structures multiplied and consistent trends in bonding and chemical connectivity became apparent, confidence in the constancy of characteristic geometric features increased and these features were employed for model building and became the basis for a great deal of new understanding of molecules and molecular properties.

With the rapid proliferation of structures of high precision, we are witnessing with increasing frequency the appearance of exceptions to these generally accepted rules, and the time has come to take these into account when these crystallographically obtained parameters are utilized for other purposes. In this regard we should long ago have heeded the words of Comte Georges Louis de Clerq quoted at the outset.

The main premise under which we operate (and for the most part rather successfully) is that the solid state derived geometrical information can be transferred to some other medium (gas, solution, receptor, etc.). The rapidly increasing body of information indicates that this premise is basically sound, but that we must be aware of its pitfalls. My purpose in this paper is to point out a number of them in the hope that they will serve as a *caveat* to the crystallographers who provide the geometrical information culled from crystal structure analyses and to the increasingly broad scientific public who are using it. The environment of a molecule plays some role in determining its geometry, and it is the particular role of the crystal environment to which I with to address myself. The space limitations here preclude the inclusion of many of the examples now extant in the literature which illustrate the points made below; appropriate reference citing some of these will be given.

The fact that molecular geometry may be influenced by crystal forces has certainly been recognized before, and has been discussed at some length by Kitaigorodsky (1973). The most often cited and now classic example is that of biphenyl I, which is twisted 42° about the exocyclic bond in the gas phase, but apparently planar is in the solid state.

(Almenningen and Bastiansen, 1958; Robertson, 1961; Hargreaves and Rizvi, 1962). Indeed, the discovery of unexpected or unusual geometric features in a crystal structure analysis has prompted many a practicing crystallographer to include an often perfunctory remark on the possible role of 'crystal forces' in determining that geometry. To get some feeling for

the role of crystal environment in determining molecular conformation it
is helpful to examine the magnitudes of the energies and forces involved
on both the intramolecular and intermolecular levels.

2. INTRAMOLECULAR FORCES AND ENERGIES

Molecular shape is defined by three different types of geometric parameters: bond lengths, bond angles and torsion angles. Distortions of a
molecule may then be simply categorized as bond stretching or compression,
bond bending or deformation, and bond twisting or torsion. Typical force
constants for stretching different types of carbon-carbon bonds are
(Brand and Speakman, 1961):

$$C-C \text{ (ethane)} \quad 4.5 \times 10^5 \text{ dyne-cm}^{-1}$$
$$C=C \text{ (ethylene)} \quad 9.6 \times 10^5$$
$$C\equiv C \text{ (acetylene)} \quad 15.7 \times 10^5$$
$$C\because C \text{ (benzene)} \quad 7.6 \times 10^5$$

The deformation of a benzene ring such that alternate angles will become
greater and less than the trigonal angle of 120° involves a force constant
of 0.7×10^5 dyne-cm^{-1}. Similarly, for benzene, a torsional out-of-plane
libration has a value of 0.06×10^5 dyne-cm^{-1}, whereas torsional motion
about an ethylenic bond involves a force constant of 0.6×10^5 dyne-cm^{-1}.
In terms of energies the barrier to rotation about the C-C bond in ethane
is about 2.8 kcal-mole^{-1} while the barrier to free rotations of methyl
groups in dimethyl acetylene is on the order of 0.5 kcal-mole^{-1} (Brand
and Speakman, 1961). These values vary over two orders of magnitude and
it is clear that the most easily accomplished geometric perturbation
involve torsional rotation, followed by energetically more expensive
changes in bond angles and bond lengths.

3. INTERMOLECULAR FORCES AND ENERGIES IN ORGANIC CRYSTALS

Any crystal structure corresponds to a free energy minimum which can
usually be identified with a potential energy minimum. An additional
entropy term must be included for disordered structures. The potential
energy minimum represents a balance between attractive and repulsive
forces for which there is a wide variety of nomenclature in the
literature:

van der Waals interactions	dipole-dipole interactions
steric repulsions	hydrogen bonds
charge-transfer interactions	electrostatic interactions
donor-acceptor interactions	π-π interactions

polarizability   non-bonded interactions
London forces   σ-π interactions
...etc.   dispersion forces

Some of these are redundant and used interchangeably depending on the
chemical nature of the system: others are not very well defined at all.
However, for a number of them there is general agreement on their
definition and it is possible to estimate the energies involved since
these will be required in the discussion which follows.

van der Waals interactions or London forces are weak attractive forces
between uncharged atoms or molecules and may be roughly estimated from the
sublimation energy of organic crystals in which other interactions are
essentially absent. The sublimation energy, usually in the range 10-25
kcal-mole^{-1} may be considered to be that required to remove a single
molecule from a crystal. Since the structures of covalently bonded
organic molecules are characterized by 'coordination numbers' in the range
10-14, the contribution of each neighbor molecule to the total sublimation
energy is 2-3 kcal-mole^{-1}. Hydrogen bond strengths are generally estimated to be 5-10 kcal-mole^{-1} per hydrogen bond and charge-transfer interactions (e.g. in π-π complexes are in the range 0.5-5.0 kcal-mole^{-1}.
Electrostatic interactions vary over a wide range, depending on the amount
of charge involved as well as the distance.

Where do these intermolecular interactions fall on the scale of
energies involved in intramolecular distortions? For the most part, they
occupy the low end. Hence it is reasonable to suggest that the molecular
geometric parameters most likely to be affected by crystal forces will be
the torsion angles around single bonds, and these are often the most
important in our consideration of the geometry in relationship to the
chemical properties of the molecule.

4. MOLECULAR CONFORMATIONS IN CRYSTALS

To this point use of the term conformation has been avoided in this
discussion. How do we define *the conformation* of a molecule? This is
often a semantic problem, although most of us have a notion of what we
mean. The problem was best stated by Prof. J.D. Dunitz (1979) in his
recent book:

> Of all possible spatial arrangements of atoms of given constitution
> and configuration, those that correspond to potential energy
> minima are known as conformations. The definition, like the
> concept itself, is imprecise - the term *constitution* implies that
> the molecule is describable in terms of a unique classical
> structural formula, the term configuration overlaps with

conformation and how are we to know whether a proposed spatial arrangement corresponds to a potential energy minimum? However, we do know that the physical and chemical properties of many classes of compounds can be correlated with spatial factors that are not defined by specification of constitution and configuration alone, and we need a concept that encompasses these factors. Generally, conformations are arrangements that arise by rotations about bonds, and they may be described by specification of relevant torsion angles. Moreover, we can be reasonably certain that any particular arrangement of atoms observed in a molecular crystal cannot be far from an equilibrium structure of the isolated molecule.

X-ray analyses provide information about the preferred conformations [note plural] of molecules although it has nothing to say about the energy differences between conformations and the energy barriers that separate them. This information has to be obtained by other methods. Energy differences can be derived [experimentally] in principle, from measurements of equilibrium concentrations of the relevant conformational isomers, and energy barriers can be obtained from interconversion rates".

Thus, many conformations of a molecule may be energetically equivalent, or nearly so, and we should be aware of the possibility of a number of them appearing in different crystal structures, or even in the same crystal structure. Certain packing motifs are more favorable than others (e.g. planar molecules pack more efficiently than non-planar ones) and this may lead to a predominance of one confromation in a crystal structure, while in solution a different conformation may be present. The point is that both the crystal structure and the molecular conformation represent potential energy minima which are not necessarily unique and there exist in principle many possibilities of very nearly the same energy for both molecular conformation and crystal structures. One of the best pieces of information we should like to obtain is quantitative comparison of fluid and solid state conformations.

5. COMPARISON OF CONFORMATIONS IN FLUID AND SOLID STATES - TECHNIQUES
How may conformations in the solid be compared with those in other states? A variety of techniques have been employed, but the most common are NMR, IR, Raman, CD-ORD and UV-VIS spectroscopy in solution, and for the gas phase electron diffraction. The system under investigation often dictates the choice of method. Recent and continuing technological advancements in all of the traditionally solution-based techniques (NMR, CD, UV-VIS) have facilitated these measurements in the solid state with detail and precision approaching that in solution and provide us with a means for direct comparison. Traditionally, IR was one of the few methods which provided

equally detailed information both in solution and the solid state (see, for instance Byrn, Graber and Midland, 1976; Caillet, Claverie and Pullman, 1976), and a recent example applied to polypeptides has been given by Benedetti *et al*. (1981).

The recent developments in the NMR field are particularly worthy of note. In solution, by employing a combination of nuclear Overhauser enhancement (nOe) difference and decoupling techniques it has been possible to assign all the protons in a number of steroids (Hall and Sanders, 1980; Hall and Sanders, 1981; Barret *et al*., 1982) and these techniques have been further refined to obtain reasonable estimates (± 0.1 Å) of intramolecular H···H distances for the determination of the solution conformation of a series of six dihydroergopeptines (Weber, Loosti and Petcher, 1981). The development of the "magic angle" technique (Griffin, 1977; Andrew, 1976; Mehring, 1976) has permitted obtaining solid state spectra of detail equivalent to that in solution for direct comparison of solid and solution conformations. CD methods have tradition- ally been employed to determine solution conformations of natural products and other molecules (Djerassi, 1960).Light scattering poses severe problems in measuring the solid state spectra, especially at short wavelengths, but recent results (Barret *et al*., 1982) suggest that this technique also may be utilized in comparing solid state and solution conformations. In the area of UV-VIS spectroscopy, the development of sophisticated techniques for measuring solid state (Eckhardt and Pennelly, 1971; Eckhardt and Merski, 1981) spectra at high resolution has led to the direct comparison of solid and solution conformations.

Finally, we note an additional important development especially over the last decade, which provides a new tool in the armory of weapons employed to study molecular conformation especially relevant to this discussion, namely the application of theoretical methods, a situation which was for- seen by Orville-Thomas (1974), just as these methods were coming of age. Since the techniques employed in these theoretical approaches vary from *ab initio* (at various levels) through a whole range of semiempirical methods the literature is vast and increasing rapidly. However, a great deal of experience has been acquired and we are beginning to learn the advantages and limitations of the various levels of sophistication, potential functions, parameters and computer programs. There is no doubt that as these methods are refined they will prove of increasing utility for the study of conformational problems.

6.   INTRAMOLECULAR AND INTERMOLECULAR ENERGETICS - COMPUTATIONAL TECHNIQUES

Following Orville-Thomas' prophesy there has been a true evolution in the development and utilization of computational methods for investigating both molecular and crystal energetics. The field has been reviewed extensively (Ermer, 1976; Altona and Faber, 1974; Dunitz and Burgi, 1975; Allinger, 1976; Burkert and Allinger, 1982; Mislow, Dougherty and Hounshell, 1978) and only a very brief summary is warranted here. The important point for this discussion is that many of the programs developed and refined during this period have now become standard library programs which are readily transferable and easily obtained and hence may be utilized by almost any worker in the field.† Some of these programs treat conformational parameters explicitly while for others the conformational energy surface is sampled by systematically varying the parameters in question and computing the energy involved.

The latter approach applies in general to molecular orbital calculations and may include optimization procedures for producing the minimal energy conformation(s), which naturally, significantly increase the amount of computer time required. For full force field and molecular mechanics calculations the molecular energy is defined in terms of certain conformational parameters and fairly efficient procedures have been developed (Ermer, 1976) to speed these up.

Molecular orbital methods, from the simple Hückel approximation to *ab initio* methods are quite familiar and well documented elsewhere. The force field calculations which originated from vibrational spectroscopy traditionally have included specific terms for perturbations in molecular geometry which makes them advantageous for interpreting the energetics of molecular information. Recently they have been modified and generalized to include terms for the full range of torsional rotations, as well as non-bonded interpretations of the Lennard-Jones or Buckingham forms. The choice of the analytical forms for the force field, the determination of force field constants and the transferability of force fields and their constants among different chemical systems are still subjects of considerable research and interest. However, certain functions and parameters have proven useful in a variety of applications (Dunitz and Burgi, 1975;

---

†Most of these may be obtained through the Quantum Chemistry Program Exchange (QCPE), Department of Chemistry, Indiana University, Bloomington, IN. 47405.

Ermer, 1976; Warshell and Lifson, 1970; Scherago, 1971; Hagler and Lifson,
1974; Hagler, Huler and Lifson, 1974; Williams, 1981) and this now a
recognized and accepted tool in the investigation of molecular conformations.

In principle the extension of this approach to crystals is straightforward. The basic assumption is that the intermolecular interactions may
be treated as a sum of atom...atom interactions which are approximated
again by either the Lennard-Jones or Buckingham potentials (Kitaigorodsky,
1973; Ermer, 1976; Williams, 1972; Coiro, Giglio and Quagliata, 1972;
Casalone, Marione and Simonetta, 1968). Williams (1974), has demonstrated
the need for inclusion of a coulombic electrostatic term in the atom...atom
potential, even for hydrocarbon crystals, where such a contribution may
approach 30% of the total energy. The Lennard-Jones and Buckingham
potentials are then modified by adding a term which includes $q_i q_j / r_{ij}$.
In practice, net partial atomic point charges are assigned to each atom.
These may be estimated from bond dipole moments (Hagler, Huler and Lifson,
1974), quadrapole moments (Hirshfeld and Mirsky, 1979), or molecular
orbital calculations at various levels of approximations (Cox and
Williams, 1981).

7. STRATEGIES FOR INVESTIGATING THE INFLUENCE OF CRYSTAL FORCES ON MOLECULAR CONFORMATION

Kitaigorodsky (1973) has suggested four possible strategies for investigating the influence of crystal forces on molecular conformation. Each
method has advantages and drawbacks, and we proceed by discussing each in
turn with appropriate examples from the literature.

7.1. Comparison of compounds in gaseous and crystalline states

This method in its strictest definition has the rather severe
limitation that the principle techniques for determining gas phase
molecular structures and conformations are electron diffraction and micro-
wave spectroscopy. Hence, for the most part suitable substances are those
relatively small molecules with high vapor pressure and a high degree of
intrinsic symmetry. An additional complication is the fact that the
vapor usually contains a greater variety of rotational isomers than the
crystal. For instance, gaseous ethane-1,2-dithiol II at about 70°C

$$HS-CH_2-CH_2-SH$$

II

$$H_3C-C\overset{\displaystyle O}{\diagup}\underset{\displaystyle NH_2}{\diagdown}$$

III

consists of a conformational equilibrium of *anti* and *gauche* conformers
with an approximate ration of 2:1 (Schultz and Hargittai, 1973). In the
crystal spectroscopic methods indicate that only the centrosymmetric *anti*
isomer is present (Hayashi *et al.*, 1965).

However, if we were to broaden Kitaigorodsky's categorization and
include conformations determined theoretically (i.e. for the free molecule)
and experimentally in solution, then this would certainly be a very
general approach to *comparing* conformations in the solid and fluid phases.
As we will see below, it may not, however, be the most direct way for
studying the influence of crystal forces on molecular conformation. Under
the umbrella of this broadened category, a large number of studies have
been performed, and a few examples are cited here.

Comparison of gas phase and solid state conformation. The classic problem
of biphenyl I has been studied for many years. The gas phase electron
diffraction study (Almenningen and Bastiansen, 1958) indicated a rotation
about the exocyclic C-C bond of 42°. In the crystal the molecule lies on
a crystallographic center of symmetry which requires planarity, and close
contacts of the *ortho* hydrogens. The symmetry restrictions could be met
without forcing each individual molecule to be planar by a random static
disorder with a statistical distribution of molecules with relatively
small equal but opposite rotations about the central bond. An alternate
model would be a dynamic one with an intramolecular motion about the same
bond. In either case the deviation from planarity does not appear to be
very large since the room temperature structure does not exhibit large
temperature factors, so that in any event there is a significantly large
difference between the gas phase and solid state conformations. The
conformation in solutions, melt (Mayo and Goldstein, 1966; Pasguier and
Lebas, 1967) and argon matrices (LeGal and Suguki, 1977) is estimated to
be intermediate between the gas and solid phases, although apparently a
planar conformation has also been trapped in the latter medium (Baca,
Rosetti and Brus, 1979).

Comparison of computed free molecules and solid state conformation. The
rapid increase in available computing power over the past few years has
made the development, testing and use of sophisticated computational
methods feasible, and these are being applied increasingly frequently to
a variety of problems. For instance, Jeffrey *et al.* (1980) recently
reported the *ab initio* molecular orbital study of the rhombohedral form of
acetamide III at the Hartree-Fock 3-21G level with geometry optimization

together with a neutron diffraction study at 23°K. The neutron refinement
shows one C-H bond normal to the plane of all the non-hydrogen atoms,
while the lowest computed energy for the free molecule shows m symmetry
with one C-H bond eclipsed to the carbonyl bond. The conformation in the
crystal was calculated to be 0.4 kcal-mole^{-1} higher in energy than that for
the free molecule, i.e., the crystal stabilizes the higher energy conformation. Observed C=O and C-N bond lengths exceed the calculated ones by 20σ
and 30σ, respectively, and it is suggested that the differences are due in
part to hydrogen bonding in the crystal.

Comparison of solution and solid state conformation. As noted earlier, the
proliferation of crystal structure analyses has provided fertile ground
for the study of unexpected molecular conformations in a family of similar
compounds. Moreover, the increasing power and sophistication of a number
of solution-based instrumental techniques for *quantitative* determination
of molecular conformation has prompted more detailed investigations of
such problems. A recent study (Barret *et al.*, 1982) of a presumably
anomolous steroid conformation serves as a good example.

Ring A in a group of progesterone derivatives characteristically
exhibits conformations which do not differ significantly from that of the
1β, 2α-half chair (Figure 1). However, 17α-acetoxy-6α-methylprogesterone

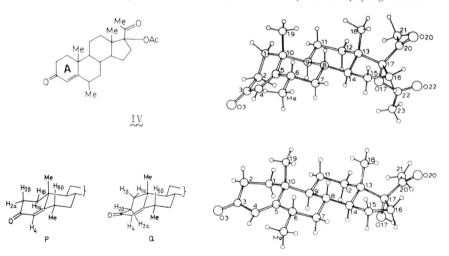

FIG. 1. Right: upper, solid state conformation of IV; lower, 17α-
hydroxy-6-α-methylprogesterone showing normal configuration of ring A.
Lower left: P; conformation of IV preferred in solution; Q, the
conformation found in the crystal.

IV has an inverted (1α,2β) half chair conformation. The CD solution
spectra strongly indicated a normal ring A conformation for IV, but some
inconsistencies in comparisons with those of "normal" compounds ruled out
a definitive determination of the solution conformation.

The use of nOe difference and two dimensional J spectroscopy NMR in
a variety of solvents led to more decisive results, showing that the
normal conformation is preferred in solution. It was shown that P in
Figure 1 is most likely the dominant conformation (> 90%) in solution
while Q in the same Figure is found in the solid. The authors suggested
that the energy difference between the two conformations is small and
that packing forces in the crystal which stabilize Q may mimic those at
the binding site for progestational activity. This is certainly a novel
idea and adds impetus to the efforts at understanding the nature of
these forces.

Another impressive combination of the use of NMR methods and crystal
structure analysis was on the earlier cited six dihydroergopeptines
(Weber, Loosti and Petcher, 1981) in which the intramolecular non-bonded
H···H contacts were determined quantitatively. For the first five derivatives the NMR results indicated a close correspondence between the solution and solid state conformations, while for the last compound the
results suggest differences between the conformations in the two media.

7.2. Comparison of the geometries of crystallographically
independent molecules in the same crystal

This approach appears at first sight to be very promising and in the
early years of crystal structure analysis was often considered a boon:
two independent determinations of the molecular geometry for one crystal
structure determination. However, the number of cases is relatively rare,
and molecules necessarily interact with one another and thus mutually
influence molecular conformation. Hence any detailed analysis of this
situation must take into account the problem of differentiating among
crystal forces between crystallographically equivalent molecules from
those which are not crystallographically equivalent. Although this is a
major disadvantage for choosing such systems for study in these early
stages of the investigation of the relationship between crystal forces and
molecular conformation, it can provide a fertile testing ground for
computational approaches which have proven successful on less complex
systems.

There are a number of cases in which the molecular conformations of

crystallographically independent molecules are different. A recently
reported case is that of 1,3,3,5,5-penta(1-aziridinyl)-1λ^6, 2,4,6,3,5-
thiatriazadiphosphine 1-oxide V (Galy $et\ al.$, 1981), one of a new class
of inorganic antitumor agents with a broad range of activity. The

V   Va   Vb

material is polymorphic and one of the forms has two independent molecules
per asymmetric unit with significantly different conformations about the
exocyclic P-N and S-N bonds which link the aziridinyl groups to the six-
membered ring N_3P_2S. At the CNDO level of approximation V_a was calculated to be the most stable conformation, being preferred over V_b by
2.5 kcal-mole^{-1} and over the conformation found in the second polymorph by
14.1 kcal-mole^{-1} (Lahana and Labarre, 1981).

In vitamin D_3 VI (Toan, DeLuce and Dahl, 1976) two different solid
state conformations of the A ring are found, corresponding to the two

VI   VII

VIII

isomers which NMR evidence (LaMar and Budd, 1974; Wing $et\ al.$, 1975)
indicates are in dynamic equilibrium in solution. The chair conformation
of the A ring is present in both molecules, but the hydroxyl group is
equatorial in one form and axial in the other. In the former the carbon
of the exocyclic double bond is below the mean cyclohexane plane while in
the latter it is above it. Differences are also evident in the other
torsion angles and in the conformation of the hydrocarbon side chain, and
the large thermal ellipsoids at the terminus of one form compared to the
second suggest more freedom of movement in the crystalline environment in
the former than in the latter.

Two recently reported cases by Birnbaum and coworkers containing three molecules in the asymmetric unit are worthy of note. The first, 9-(2-hydroxyethoxymethyl)- guanine VII, a potent inhibitor of herpes simplex viruses (Birnbaum *et al.*, 1981). The side chain in two of the molecules is partially folded, while in the third molecule all the bonds in the side chain are in a *trans* orientation. In the second, 8-bromo-9-β-D-xylofuranosyladenine VIII competition between intramolecular and intermolecular hydrogen bonding is clearly a dominant factor in determining the conformation. Again, two of the three molecules have similar conformations, while the third is significantly different (Birnbaum *et al.*, 1982).

7.3. Analysis of the structure of a molecule whose symmetry in a crystal is lower than that of a free molecule

A fundamental problem is encountered here in that the free molecule symmetry is basically a difficult quantity to measure or determine. It is at best the chemists' idealized structure and often bears little relationship to the true (even if unknown) molecular symmetry. This is especially significant for flexible molecules whose precise conformation we wish to know and which are particularly susceptible to the influence of crystal forces. Moreover, the tendency for molecules to pack as densely as possible usually overrides the retention of molecular symmetry upon crystallization. As a result, the crystallographic site symmetry of most organic molecules is significantly reduced, and in general only a center of symmetry is retained as a crystallographic symmetry element, followed in frequency by a two-fold axis and a mirror plane. The presence of these latter two symmetry elements is usually expensive in terms of packing efficiency, but if the molecular shape permits then they may be retained. A full discussion of this aspect of crystal forces has been given by Kitaigorodsky (1979).

If symmetry elements are not retained upon crystallization we may still ask if "chemical" symmetry is maintained - i.e. whether chemically equivalent parts of the molecule have identical geometric features, within experimental error. This depends on the precision of the structure determination, and even if we have precise structural information, the subsequent steps in the analysis are not straightforward, so that this proposed method seems presently to be of the least utility.

7.4. Comparison of molecules in different polymorphic modifications

Polymorphism is the ability of a compound to crystallize in more than one distinct crystal species. It was recognized quite early as widespread in organic materials (Deffet, 1942) and particularly among pharmaceuticals (Haleblian and McCrone, 1969; Haleblian, 1975; Kuhnert-Brandstatter, 1971), and it is estimated that over half of them exhibit the phenomenon (Clements, 1976). The ubiquity of polymorphism has led McCrone (1965) to state:

> "...every compound has different polymorphic forms and...the the number of forms known for a given compound is proportional to the time and energy spent in research on that compound."

The differences in lattice energy between polymorphs are usually in the range 1-2 kcal-mole^{-1} (Kitaigorodsky, 1970), which is well below the energy required to bring about changes in bond lengths and angles, but of the same order of magnitude as torsional distortions around single bonds. Since it is the torsional parameters which define the molecular conformation, it is clear that for molecules which possess torsional degrees of freedom, various polymorphs may exhibit significantly different molecular conformations, a phenomenon termed *conformational polymorphism* (Corradini, 1973; Panagiotoupoulis *et al.*, 1974; Bernstein and Hagler, 1978). A few examples of the many cases which have been observed are given by Bernstein and Hagler (1978).

In cases of conformational polymorphism changes in molecular conformation *must* be due to the influence of crystal forces since the difference in crystal environment is the only external variable which acts on the molecules. Hence utilization of this phenomenon has the distinct advantage that it provides the most direct method for investigating the influence of crystal forces on molecular conformation.

In a complete study utilizaing conformational polymorphism basically two questions are posed:

1) What are the differences in energy, if any, in the conformations observed in the various crystal forms?

2) How does the energetic environment of the crystal differ from one form to another?

To answer these questions a typical study would proceed according to the following steps:

1) Determination of the existence of polymorphism in the system under study.

2) Determination of the possibility of conformational polymorphism by appropriate physical measurements.

3) Determination of crystal structures to obtain the geometrical information - molecular geometry and packing motif - of the various polymorphs.

4) Determination of the differences in lattice energy, again by appropriate computational methods.

Polymorphism may be readily detected by calorimetric techniques (Kuhnert-Brandstatter, 1971; McNaughton and Mortimer, 1975), or microscopic hot stage methods (McCrone, 1957), in addition to X-ray powder diffraction. IR spectroscopy has been the classic spectroscopic tool for determining conformational differences in solids, although, as noted above, magic-angle NMR is proving to be a very powerful and sensitive probe of solid state molecules conformations. Even though crystal structure determinations have now become fairly routine, for polymorphic systems it is often difficult to obtain single crystals of the various forms which are suitable for structure analysis. We outlined above the computational methods employed to obtain molecular conformations and crystal lattice energies. It is important to note here that for the molecular energetics we need only investigate those points on the multi-dimensional potential energy surface which have been determined in the structure analyses. The main point of interest in these studies is the *difference* in energies of the molecular conformations observed, and not necessarily whether any one of them is the global minimum conformation. This also relieves us of the fairly serious restrictions of being concerned with absolute molecular energies. The conformational energetics of large molecules may be approximated fairly well by employing *ab initio* calculations on model compounds representing the conformational parameters in question. Alternatively, semiempirical methods often give good estimates of energy differences among conformations, even if other properties may not be approximated very well.

The total lattice energy is obtained from calculations based on atom...atom potentials. Comparison is then made with the sublimination energy which may be measured fairly readily (Daniels *et al.*, 1970) or estimated from those of analogous model compounds (Bondi, 1963). Again, *differences* in lattice energy may be determined calorimetrically for experimental verification of the computed quantities. Finally, the lattice energy may be partitioned into its component contributors to examine in detail those intermolecular interactions which lead to, say, the stabilization of an energetically less favorable molecular conformation in a particular polymorph or of the presence of a more stable conformation in a second structure.

A complete prototypical study of this type has been published (Bernstein and Hagler, 1978) on the dimorphic dichlorobenzylideneaniline system IX, which has two torsional degrees of freedom (α, β).

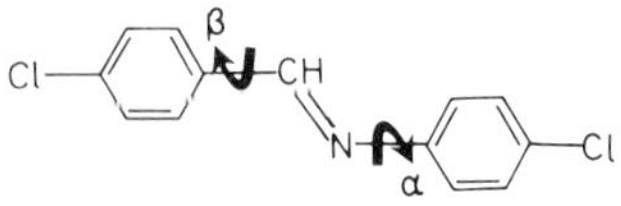

IX

The general success of this first study prompted us to attempt to refine the general approach and learn the extent of its applicability. Hence, similar techniques have been applied to investigate why a molecule does not pack in a particular crystal structure in which the lowest energy conformation would be favored (Hagler and Bernstein, 1978), and to a trimorphic system including disorder (Bar and Bernstein, 1982). These studies have all indicated that consistent results can be obtained as above and that this strategy for investigating the role of crystal forces in determining molecular conformation is indeed one of the most promising available.

8. THE CHALLENGES - AN EXAMPLE

In the light of the experimental and computational developments which now permit us to study in detail the influence of crystal forces on molecular conformation it seems appropriate to conclude this paper with the presentation of a current problem which encompasses many of the points raised earlier and typifies the challenges that remain before us in understanding the relationship between crystal forces and molecular conformation.

Two members of a class of inorganic ring systems which have exhibited significant effectiveness on all the tumors on which they were tested were hexaziridinocyclotriphosphazene X, and the earlier mentioned pentaziridinocyclophosphathiazine V (Labarre *et al.*, 1979,1980,1981). X

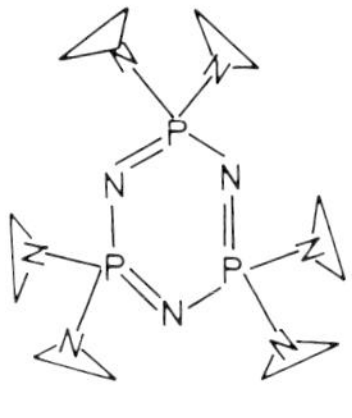

X 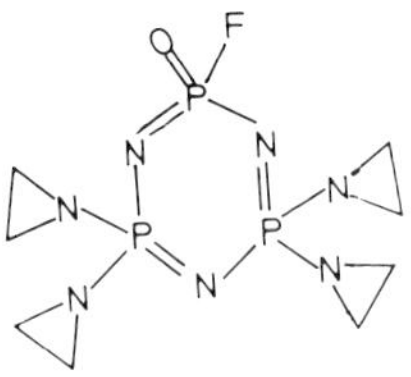   XI 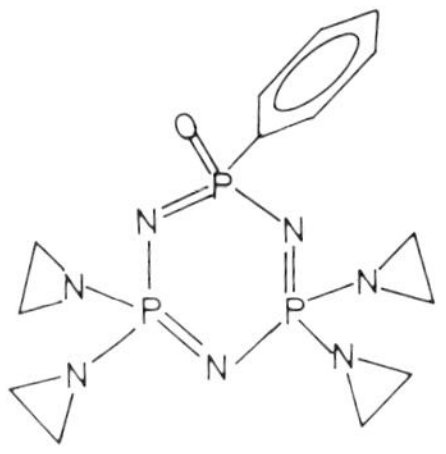   XII

crystallizes from *m*-xylene, CS_2 or water in three different polymorphs (Cameron and Labarre, 1982) with essentially identical molecular conformation (Lahana and Labarre, 1980). However, when X is crystallized from benzene a complex is obtained (Cameron, Labarre and Graffeuil, 1982) (perhaps more properly described as a solvate) while an "anti-clathrate" structure is obtained from CCl$_4$ (Galy, Enjalbert and Labarre, 1982). As noted above (V) is dimorphic (Galy *et al.*, 1981), both forms crystallizing from anhydrous ether. Form I of (V) has two independent molecules in the asymmetric unit. The anti-tumor activity of the two forms is identical. Two additional structures (XI, XII) were included in a study of the conformational features of these compounds.

An understanding of the conformation of these molecules was deemed particularly important, since the effectives of these inorganic ring systems does not require intermediate metabolism. Hence it was proposed that there might be a direct relationship between the *in vivo* antitumor activity and the *in vitro* (e.g. crystalline) geometric structure (Cameron, Labarre and Graffeuil, 1982). The solid state conformations, shown in Va, Vb and Figure 2, exhibit a variety of conformations of the aziridine

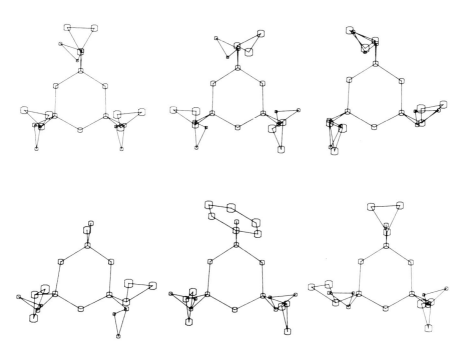

FIG. 2. Top, left to right: conformations of X from xylene, CCl$_4$, and benzene. Bottom: left to right: conformation of XI, XII and V, form I.

REFERENCES

ALLINGER, N.L. (1976). Adv. Phys. Org. Chem. 13, 2.

ALMENNINGEN, A. and BASTIANSEN, A. (1958). Kgl. Nor. Vidensk. Selk. Skr. 4, 4.

ALTONA, C. and FABER, D.H. (1974). Topics Current Chem. 45, 1.

ANDREW, E.R. (1976). MTP Int. Rev. Sci. Phys. Chem. Ser. 2 4, 173.

BACA, A., ROSETTI, R. and LE BRUS, L.E. (1979). J. Chem. Phys. 70, 5575.

BAR, I. and BERNSTEIN, J. (1982). J. Phys. Chem. 86, 3223.

BARNET, M.W., FARRANT, R.D., KIRK, D.N., MERSH, J.D., SANDERS, J.K.M. and DUAX, W.L. (1982). J. Chem. Soc. (Perkin II) 105.

BENEDETTI, E., BAVOSO, A., DI BLASIO, B., PAVONE, V., PEDONE, C., CRISMA, M., BONORA, G.M. and TONIOLO, C. (1982). J. Am. Chem. Soc. 104, 2437.

BERNSTEIN, J. and HAGLER, A.T. (1978). J. Am. Chem. Soc. 100, 673.

BIRNBAUM, G.I., CYGLER, M., EKIEL, I. and SHUGAR, D. (1982). J. Am. Chem. Soc. 104, 3957.

BIRNBAUM, G.I. CYGLER, M., KUSMIEREK, J.T. and SHUGAR, D. (1981). Biochem. and Biophys. Res. Comm. 103, 968.

BONDI, A. (1963). J. Chem. Eng. Data 8, 371.

BRAND, J.C.D. and SPEAKMAN, J.C. (1960). Molecular Structure, The Physical Approach", E. Arnold, London.

BURKERT, U.B. and ALLINGER, N.L. (1982). Molecular Mechanics, American Chemical Society Monograph No. 177, Washington, D.C.

BYRN, S.R., GRABER, C.W. and MIDLAND, S.L. (1976). J. Org. Chem. 41, 2283, and refs. 1-5 therein.

CAILLET, J., CLAVERIE, P. and PULLMAN, B. (1976). Acta Crystallogr. B32, 2740.

CAMERON, T.S. and LABARRE, J.-F. (1982). Acta Crystallogr., in press (cited by Lahana and Labarre, 1980).

CAMERON, T.S., LABARRE, J.-F. and GRAFFEUIL, M. (1982). Acta Crystallogr. B38, 168.

CASALONE, G., MARIANI, C., MAGNOLI, A. and SIMONETTA, M. (1968). Mol. Phys. 15, 339.

CLEMENTS, J.A. (1976). Proc. Analyt. Div. Chem. Soc., p. 21.

COIRO, V.M., GIGLIO, E. and QUAGLIATA, C. (1972). Acta Cryst. B28, 3601.

CORRADINI, P. (1973). Chim. Ind. (Milan) 55, 122.

COX, S.R. and WILLIAMS, D.E. (1981). J. Comput. Chem. 2, 304.

DANIELS, F., WILLIAMS, J.W., BENDER, P., ALBERTY, R.A., CORNWELL, C.D. and HARRIMAN, J.E. (1970). *Experimental Physical Chemistry*, 7th ed., McGraw-Hill, New York, p. 53.

DEFFET, L. (1942). *Repertoire des Composes Organique Polymorphs*, Editions Desoer, Liege, Belgium.

DJERASSI, E. (1960). *Optical Rotatory Dispersion; Applications to Organic Chemistry*, McGraw-Hill, New York.

DUNITZ, J.D. (1979). *X-ray Analysis and the Structure of Organic Molecules*, Corne 1 University Press, Ithaca.

DUNITZ, J.D. and BURGI, H.B. (1975). *Int. Rev. Sci. Phys. Chem. Ser. 2*, 11, 81.

ECKHARDT, C.J. and MERSKI, J. (1981). *J. Chem. Phys.* 75, 3691.

ECKHARDT, C.J. and PENNELLY, R.R. (1971). *Chem. Physt. Lett.* 9, 572.

ERMER, O. (1976). *Structure and Bonding* 27, 163.

GRIFFIN, R.G. (1977). *Anal. Chem.* 49, 951A.

HAGLER, A.T. and BERNSTEIN, J. (1978). *J. Am. Chem. Soc.* 100, 5346.

HAGLER, A.T., HULER, E. and LIFSON, S. (1974). *J. Am. Chem. Soc.* 96, 5319.

HAGLER, A.T. and LIFSON, S. (1974). *J. Am. Chem. Soc.* 96, 5327.

HALEBLIAN, J.K. (1975). *J. Pharm. Sci.* 64, 1269.

HALEBLIAN, J.K. and McCRONE, W.C. (1969). *J. Pharm. Sci.* 58, 411.

HALL, L.D. and SANDERS, J.K.M. (1980). *J. Am. Chem. Soc.* 102, 5703.

HALL, L.D. and SANDERS, J.K.M. (1981). *J. Org. Chem.* 46, 1132.

HARGREAVES, A. and RIZVI, S.H. (1962). *Acta Crystallogr.* 15, 365.

HAYASHI, M., SHIRO, Y., OSHIMA, T. and MURATA, H. (1965). *Bull. Chem. Soc. Japan* 38, 1734.

HIRSHFELD, F.L. and MIRSKY, K. (1979). *Acta Crystallogr.* A35, 366.

GALY, J., ENJALBERT, R. and LABARRE, J.-F. (1980). *Acta Crystallogr.* B36, 392.

GALY, J., ENJALBERT, R., VAN DER HUIZEN, A.A., VAN DE GRAMPEL, J.C., and LABARRE, J.-F. (1981). *Acta Crystallogr.* B37, 2205.

GILLESPIE, R.J. (1963). *J. Chem. Ed.* 40, 295.

JEFFREY, G.A., RUBLE, J.R., McMULLAN, R.K., DeFrees, D.J., BINKLEY, J.S. and POPLE, J.A. (1980). *Acta Crystallogr.* B36, 2292.

KITAIGORODSKY, A.I. (1970). *Adv. Struct. Res. Diffr. Methods* 3, 173.

KITAIGORODSKY, A.I. (1973). *Molecular Crystals and Liquid Crystals*, Academic Press, New York.

KUHNERT-BRANDSTATTER, M. (1971). *Thermomicroscopy in the Analysis of Pharmaceuticals*. Pergamon Press, Oxford, England.

degree of unique complementary relationships possible between potential
complexing partners.

Cram has adopted the terms <u>host</u> and <u>guest</u> to designate the complexing partners in synthetic organic chemical studies. We will use the
terms a little more broadly here, as has become common, to encompass analogous natural hosts, such as the cyclodextrins, as well. A highly
structured molecular complex is composed of at least one host and one
guest that possess stereoelectronically complementary binding sites.
The <u>host</u> is defined, in the present context, as an organic molecule or
ion whose <u>binding</u> <u>sites</u> <u>converge</u> <u>in</u> <u>the</u> <u>complex</u>. The <u>guest</u> is defined
as a molecule or ion whose <u>binding</u> <u>sites</u> <u>diverge</u> <u>in</u> <u>the</u> <u>complex</u>. Guests
may be organic compounds or ions, metal ions or metal-ligand assemblies.
Binding sites of monatomic ions naturally diverge.   Hosts are usually
larger than guests since positioning of convergent binding sites often
involves support structures not required for guests.   However, large
non-binding parts may be attached to either hosts or guests to manipulate their size and their properties. Simple guests are abundant, but,
with few exceptions, hosts must be designed and synthesized, and with
large and structurally intricate hosts the effort involved is usually
considerable.

Hosts may be open-chain, cyclic, bicyclic or polycyclic molecules
and frequently contain repeating units. The classic examples, stemming
from the pioneering work of Pedersen (1967), are the crown ethers,
$(CH_2CH_2O)_n$, and azacrowns, $(CH_2CH_2NH)_n$. Cryptands were introduced in
1969 by Lehn (Lehn, 1978) when polyethyleneoxy units were used to make
bridges between nitrogen atoms to create bicyclic and polycyclic hosts.
Cram and his coworkers (Cram and Trueblood, 1981) have designed and studied many hosts in which a variety of other units have been substituted
for the basic structural units of the crowns and cryptands (e.g.,
CH_2CH_2, CH_2OCH_2, $N(CH_2)_3$), and Vögtle and his collaborators have synthesized and studied many open-chain hosts and their complexes as well
(Weber and Vögtle, 1981; Vögtle, Sieger and Müller, 1981).
Structural work on many of these complexes has recently been reviewed by
Cram and Trueblood (1981) and by Hilgenfeld and Saenger (1982).

This is not a general review. Many generalized and specialized reviews already exist, including those cited in the preceding paragraph
and others found in Volumes 98 and 101 of <u>Topics</u> <u>in</u> <u>Current</u> <u>Chemistry</u>,
each volume being devoted entirely to Host-Guest Complex Chemistry.
Many earlier reviews are also cited in those volumes. Dobler (1981) has
recently published a comprehensive and bountifully illustrated review of
ionophore structures; we will not consider the structures of natural
ionophores explicitly, although some will be referred to.

The aim of the present discussion is to illuminate the kinds of interactions that are typical in host-guest complexes and the factors that
make for effective complexation and discrimination, especially with organic guests. The emphasis is almost entirely on structural results obtained by X-ray crystallography, but some relevant spectroscopic and
thermodynamic evidence that permits structural inferences is cited as
well.

Many of the structures to be illustrated contain hydrogen bonds.
The existence of N-H...O, O-H...O, O-H...N and N-H...N hydrogen bonds

has long been recognized, with much supporting spectroscopic, thermodynamic and structural evidence. Recently Taylor and Kennard (1982) have
examined many precise crystal structures determined by neutron diffraction and concluded that they contain significant numbers of C-H...O,
C-H..N and C-H..Cl interactions that can reasonably be described as
(weak) hydrogen bonds. The geometrical requirements for hydrogen bonding have been analyzed carefully, an excellent summary being that by
Olovsson and Jonsson (1976). In structural studies, a natural criterion
for the presence of a hydrogen bonding interaction in a system X-H...Y
is that the H...Y distance is (significantly) smaller than the sum of
the van der Waals radii of H and Y. There is a strong tendency for the
angle X-H..O to lie within about 15° of 180°, i.e., for the hydrogen
bond to be not far from linear, if geometrical or steric constraints do
not interfere -- e.g., because of the impossibility of simultaneously
satisfying this condition for a given donor group and several acceptor
atoms that might be in a macrocyclic ring. Occasionally, bifurcated or
even trifurcated hydrogen bonds have been observed, chiefly with
N^{+}-H..O interactions. Usually one of the bonds is distinctly shorter
and more linear than the others. Several bifurcated bonds are found in
the structures illustrated here.

2. COMPLEXES OF A SIMPLE HOST, 18-CROWN-6, WITH VARIOUS KINDS OF GUESTS

Undoubtedly, the compound 1,4,7,10,13,16-hexaoxacyclooctadecane, or
more simply 18-crown-6 ($C_{12}H_{24}O_{6}$), hereinafter $\underline{1}$, has been the most studied host. More than thirty of its complexes have been analyzed crystallographically and others have been studied in solution as well.
Furthermore, the host and some of its complexes have been the subject of
force-field calculations (Maverick, Seiler and Dunitz, 1979; Bovill,
Chadwick and Sutherland, 1980; Wipff, Weiner and Kollman, 1982) that
have permitted rationalization of the relative stabilities of different
conformations. The structure of $\underline{1}$ itself in the crystalline state (Dunitz and Seiler, 1974) is shown in Fig. 1, in two representations, that
of a "ball-and-spoke" model and that of a space-filling model.

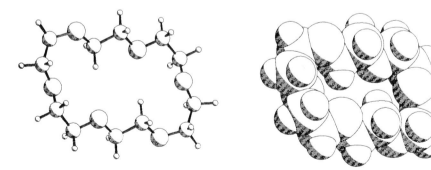

FIG. 1. The structure of 18-crown-6 ($\underline{1}$). Left: A
ball-and-spoke representation. As in all illustrations here, different atomic radii have been used to help distinguish different
kinds of atoms (O>N>C>>H). Right: A space-filling representation. The radii (in Angstrom) are: C, 1.5; O, N, 1.4; H, 1.0.

The conformation of 1 found in the crystal is that calculated to be most stable (Bovill et al., 1980; Wipff et al., 1982) in a non-polar environment, and this is consistent with infra-red and nmr evidence as well (Dale, 1974, 1980). Wipff et al. point out that the differences in energy among some of the possible conformations of 18-crown-6 are small and that the crown-like D_{3d} conformation should become more stable as solvent polarity increases, which indeed accords with some nmr evidence. This D_{3d} conformation is found in many complexes; the structure (Fig. 2a) of that with KSCN (Seiler, Dobler and Dunitz, 1974) illustrates it ideally. This conformation is not favored for the pure host because of the six >O dipoles all pointing inward; its stability is enhanced when a positive charge is situated at or not far from the center of the ring. In the pure host (Fig. 1), two methylene groups turn inward and short H...O contacts help to stabilize the structure (Dale, 1974; Maverick et al., 1979).

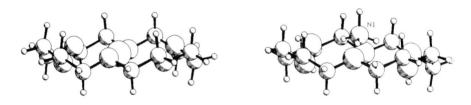

FIG. 2. Left: (a) The interaction of 1 with K^+ in the KSCN complex. The SCN^- ion is not shown; it lies above the center of the ring, about 3.2 A from the potassium ion. Right: (b) The complex of NH_4^+ with 1.

In the crown conformation, as found in many structures, the oxygen atoms are alternately above and below the median plane by a little more than 0.2 A, forming a slightly puckered hexagon of side 2.8 A, with two parallel triangles of oxygen atoms separated by nearly 0.5 A. Half the hydrogen atoms of the crown ring are oriented parallel to the axis of the ring, six pointing in each direction. The torsion angles about all C-C bonds are in the range 65-70° (gauche, g; syn-clinal, sc) while those about the C-O bonds are all near 180° (anti-periplanar, a or ap). Thus the conformation may be represented as $(aga)_6$ or alternatively as $(ap,sc,ap)_6$, with the sign of the C-C torsion angle alternating around the ring. In the pure host itself (Fig. 1), which is centrosymmetric, two C-C bonds have changed from g to a (sc to ap) and two C-O bonds in the opposite way to give the elliptical shape in which the inward-turning hydrogen atoms help to fill the cavity and stabilize the structure. Both crystallographic and spectroscopic studies of many hosts containing polyethyleneoxy units indicate that the conformational energy differences are small and consequently even in crystals alternative conformations of parts of such chains are present, leading to positional disorder. This is seldom found in complexes. In general, unless steric constraints make it impossible, the preference is for C-C bonds to be g and for C-O bonds to be a; Dale (1974) has summarized the evidence and the rationale for these preferences. Uncomplexed polyether rings (e.g., 1 in Fig. 1) contain O-C-C-O units with typical conformation g g a, i.e., one C-O bond has a gauche conformation of sign opposite to that of the gauche C-C bond. Dale refers to this as a "pseudo corner" in order to distinguish it from the "genuine corner" found in cycloalkanes, in which a C-C-C-C unit has the conformation g g a, with two adjacent gauche bonds of the same sign. The pseudo corner, stabilized by close 1,5 C-H...O interactions (evident in Fig. 1), is seldom found in open chains and is resorted to only when solvation or coordination of ether oxygen is impossible, e.g., in the gas phase or in the inaccessible interior of macrocyclic compounds.

The distance between oxygen atoms diagonally across the ring from each other in the D_{3d} conformation of 18-crown-6 is about 5.6 A. Since the van der Waals radius of an oxygen atom is near 1.4 A, there is a hole of radius about 1.4 A in the center of this ring, large enough for a K^+ ion (r = 1.33 A), but too small for Rb^+ (r = 1.5 A) or Cs^+ (r = 1.7 A) and appreciably too large for Na^+ (r = 0.95 A). In its complexes with Rb^+ and Cs^+, the 1 conformation is the same as that in the K^+ complex, with the cation situated more than 1 A above the center of the ring. In the Na^+ complex, on the other hand, the ring adopts a new, partially collapsed, conformation to adapt to the smaller cation (Dunitz, Seiler, Dobler and Phizackerley, 1974). Since the ammonium ion usually behaves as though it has a diameter similar to that of Rb^+, it was not surprising that Nagano, Kobayashi and Sasaki (1978) found that the geometrical relationship of the ammonium ion and the crown ring in the complex of 1 with ammonium bromide (Fig. 2b) was similar to that of Rb^+ to the ring in its complex.

However, the structures of the complexes of 1 with the isoelectronic series of ions, methylammonium, hydroxylammonium and hydrazinium (Trueblood, Knobler, Lawrence and Stevens, 1981) showed that the "size" of the cation is not the determining factor. The methylammonium ion, like t-butylammonium (discussed below) and many other substituted methylammonium ions, is in a perching position (Fig. 3a), like the ammonium ion (Fig. 2b), with the nitrogen atom about 1 A above the center of the ring, hydrogen bonded to the upper triangle of oxygen atoms. On the other hand, in the hydrazinium complex (Fig. 3b), the nitrogen atom of the $-NH_3^+$ group lies almost in the plane of the ring and is hydrogen bonded to the lower triangle of oxygen atoms; the amino nitrogen atom, whose attached hydrogen atoms are staggered with respect to those on the $-NH_3^+$, is hydrogen bonded to two of the oxygen atoms in the upper triangle and to an oxygen atom of the perchlorate counterion as well (not shown in Fig. 3b). The hydroxylammonium ion occupies an intermediate position in its complex, not illustrated here Clearly, the depth of penetration of the NH_3^+ group is a sensitive function of the relative strengths of the interactions within and external to the complex. It is noteworthy that in the hydrazinium complex, a second perchlorate group lies below the ring, with one of its oxygen atoms only about 3 A from the ammonium nitrogen atom, doubtless helping to draw that atom down toward the center of the ring. The ion-size concept, while useful, must be applied with caution.

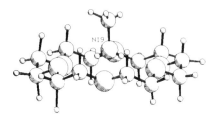

FIG. 3. <u>Left</u>: (a) The complex of methylammonium ion with <u>1</u>. <u>Right</u>: (b) The complex of hydrazininum ion with <u>1</u>.

Even the potassium ion, whose size is ideally suited for the cavity in the crown conformation of <u>1</u>, can be drawn out by a suitable counterion. Fig. 4 shows that in the complex of <u>1</u> with potassium ethylacetoacetate (Riche, Pascard–Billy, Cambillau and Bam, 1977), the strong chelating power of the counterion lifts the potassium ion about 0.9 Å above the center of the ring. A base as "soft" as SCN⁻ has no perceptible influence, an effect observed in other systems as well.

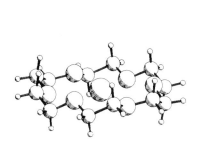
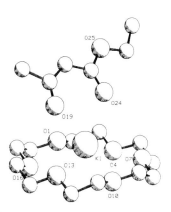

FIG. 4. Effect of counterion. <u>Left</u>: K^+ at the center of 18-crown-6 ring in SCN⁻ complex (SCN⁻ not shown). <u>Right</u>: K^+ lifted 0.9 Å above the plane of the ring by ethylacetoacetate anion.

Neutral hydrogen-bonding species also interact effectively with the 18-crown-6 ring. Complexes with amines, hydrazines, amides, phenols, water, ammonia molecules liganded to transition metals and even compounds containing CH groups of enhanced acidity have been studied crystallographically and in solution. We illustrate only a few here; many have been reviewed recently by Vögtle et al. (1981).

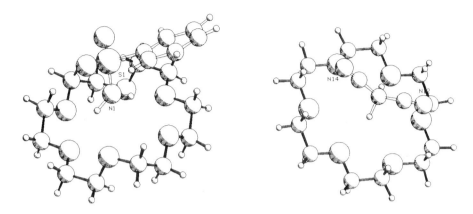

FIG. 5. <u>Left</u>: (a) Complex of <u>1</u> with benzenesulfonamide. Bonds in the guest are shown solid, those in the host unfilled. Only one of two equivalent guest molecules is shown; the other is situated below the ring. <u>Right</u>: (b) Complex of <u>1</u> with malononitrile. Another equivalent guest molecule lies below the ring. Note the C-H...O interactions.

Fig. 5a shows that in its benzenesulfonamide complex (Knöchel, Kopf Oehler and Rudolph, 1978), the <u>1</u> ring adopts once more an elliptical shape (different from that of the pure host), with no pronounced cavity. There are two equivalent guest molecules on opposite sides of the ring (only one shown here), each amide group being close to four ring-oxygen atoms. Numerous molecules containing acidic CH groups form complexes with <u>1</u>. Fig. 5b shows the structure of the complex with malononitrile (Kaufmann, Knöchel, Kopf, Oehler and Rudolph, 1977). Each equivalent guest (only one shown here) has marked C-H...O interactions with the host, which adopts the D_{3d} conformation. Even activated methyl groups interact effectively with this host. The dimethyl ester of acetylenedicarboxylic acid (Goldberg, 1975) provides a striking example. The weak C-H...O interactions of the methyl group with the ring, with doubtless a contribution from the mere space-filling capacity of the methyl group itself, suffice to change the conformation of the host to the D_{3d} crown form.

3. VARYING DEGREES OF REORGANIZATION OF HOST BY GUEST

The complexes of 18-crown-6 with different neutral and ionic guests illustrated above exhibit several distinct conformations of the host, each different from that of the pure host. Because this host is neither large nor complex, the accessible conformations, while readily distinguishable, are not grossly different in overall shape. More dramatic changes are seen with some more complex hosts, and may illustrate even better the kinds of conformational changes that can occur with many biologically significant molecules when they interact with other molecules or ions that may themselves possess conformational degrees of freedom. Fig. 6 illustrates the marked change in conformation undergone by a host

(Fig. 6a) containing an intramolecularly hydrogen-bonded carboxylic acid group when it forms a salt (Fig. 6b) with t-butylamine (Goldberg, 1975b, 1976). The carboxyl group is part of a benzoic acid moiety incorporated into a crown-ether ring, and it is seen to undergo a rotation of more than 90° when the salt is formed, so that the carboxylate ion can be most suitably hydrogen bonded to the t-butylammonium ion, while the latter is also hydrogen bonded to two oxygen atoms of the crown ring. The t-butylammonium ion is seen to "perch" in the structure.

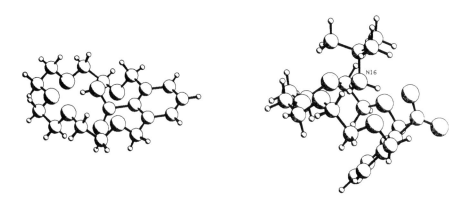

FIG. 6. Left: (a) The structure of 2,6-dimethylylbenzoic acid-18-crown-5. Right: (b) The structure of the t-butylammonium salt of the host in (a).

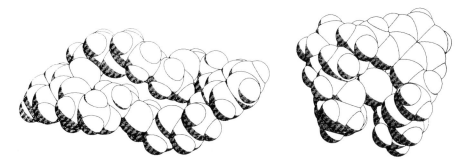

FIG. 7. Left: (a) A space-filling representation of the structure of dibenzo-30-crown-10. Right: (b) A similar representation of the structure of the Rb⁺ complex of dibenzo-30-crown-10.

In Fig. 7 an even more striking change is illustrated in the folding of the large host dibenzo-30-crown-10 when it complexes with Rb⁺. The pure host (Bush and Truter, 1972) is greatly extended (Fig. 7a), with the potential cavity of the 30-crown-10 ring collapsed like an empty balloon, the "walls" nearly meeting each other (with inward-turned methylene groups). In the Rb$^+$ complex (Hasek, Huml and Hlavata, 1979) (Fig. 7b), the 30-crown-10 ring has wrapped itself around the cation, with all oxygen atoms pointing inward and coordinated with the Rb$^+$, the whole complex now having a much more globular shape. Two K$^+$ complexes of this host have been examined, the counterions being I$^-$ (Bush and Truter, 1972) and SCN$^-$ (Hasek, Hlavata and Huml, 1980); the complexes are very similar, the K$^+$ being buried even more thoroughly than the Rb$^+$ in Fig. 7b appears to be. A similarly dramatic change occurs when a di-naphthyl-bis-16-crown-5 host reacts with KSCN; the two crown rings, which are independently extended in the crystalline host, completely en-circle the potassium ion in the complex (Knobler and Trueblood, unpub-lished; see Cram and Trueblood, 1981). In this complex, as in all of those of dibenzo-30-crown-10, the cation is completely encapsulated, having no contact with the anion or anything else external to the en-folding host.

In contrast to the hosts so far discussed, which exhibit varying degrees of conformational reorganization upon complexation, are the spherands, specifically designed for minimal reorganization upon com-plexation (Cram, Kaneda, Helgeson and Lein, 1979). The prototype spher-and, a cyclic hexamer of a 2-methoxyl-5-methylmetaphenylene unit, satis-fies this expectation remarkably (Figs. 8 and 9) (Trueblood, Knobler, Maverick, Helgeson, Brown and Cram, 1981). This host contains a cavity lined with the 12 unshared electron pairs of six octahedrally arranged oxygen atoms. The cavity diameter is about 1.6 A and can be diminished only at the expense of O...O repulsions and enlarged only by forcing aromatic rings and their attached atoms out of their natural coplanari-ty.

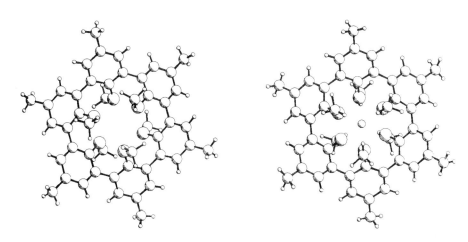

FIG. 8. Left: (a) Uncomplexed spherand $C_{48}H_{48}O_6$. Right (b) Li$^+$ complex of host shown in (a).

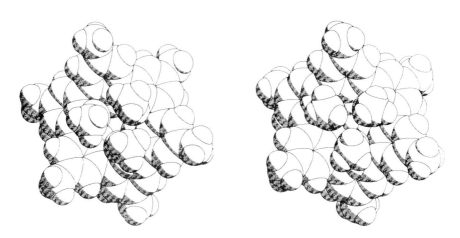

FIG. 9. Left: (a) Space-filling representation of the uncomplexed spherand $C_{48}H_{48}O_6$. Right: (b) Space-filling representation of Li^+ complex of host in (a). The lithium ion can just be
seen in the cavity.

This cavity is too small to accomodate any solvent molecule or
inward-turned methyl group. Consequently the methoxyl methyls are directed axially, three in each direction. This host complexes only Li^+
and Na^+ and the complexes are very similar in structure, with the host
conformation hardly distinguishable from that in the uncomplexed state.
The uncomplexed molecule is slightly flatter, the aryl-aryl dihedral angles being a little smaller and the methoxyl methyl groups being bent a
little away from the axial direction, as is evident in Figs. 8a and 9a.

This prototype spherand is the strongest complexer of Li^+ known, by
a factor of more than 10^4, and the second strongest complexer of Na^+,
the strongest being a related spherand (Cram, Lein, Kaneda, Helgeson,
Knobler, Maverick and Trueblood, 1981). Detailed studies of the complexing properties of various spherands have   shown   that   their
preorganization of binding sites, i.e., organization during synthesis
and prior to complexation, is a principal factor in their   strength and
selectivity.   These spherands are, in fact, too strong as complexers to
be practical as ion-carriers even if they were made water-soluble with
appropriate atached groups: their equilibrium constants are so high
that the rate at which the most powerful of them release Li^+ or Na^+ is
negligibly slow. Examination of Fig. 9 makes the reason for this clear.
Because of the rigidity of these structures, metal ions that are to be
complexed or decomplexed must pass through a lipophilic sleeve of methyl
groups whose diameter is so small that it can accomodate the ion itself
only with strain, thus making it impossible for there to be more than at
most one solvent molecule liganded to the cation. The binding sites of
the host cannot sequentially displace solvent, or be displaced by solvent, as with natural ionophores such as valinomycin or nonactin (see

Hilgenfeld and Saenger, 1982).

Because of these limitations, spherand chemistry has proceeded in the direction of providing a little additional flexibility in the design of the cavity and making the hosts more compatible with aqueous solutions, while maintaining as much preorganization of binding sites as possible. We return to consideration of molecules of this kind near the end of this presentation.

4. INTERACTIONS OF LARGER HOSTS WITH ORGANIC GUESTS

While the forces at play in the examples already discussed are no different from those involved in the interaction of larger hosts and guests, examination of some structural results for systems of the latter kind may help to bridge the transition to the complex kinds of molecules and receptors to be considered later in these sessions.

A number of research groups are now devoting considerable efforts to the synthesis of molecules containing enforced cavities, analogous to those in cyclodextrins and other natural hosts. Recently Vögtle, Puff, Friedrichs and Müller (1982) reported the structure of a 30-membered hexalactam host, $C_{72}H_{66}N_6O_6$, that selectively includes a molecule of $CHCl_3$ and Cram, Karbach and Maverick have recently created and determined the structure of a bowl-shaped host, $C_{36}H_{32}O_8$, that contains a molecule of CH_2Cl_2. Fig. 10 shows the structure of a water-soluble host-guest complex involving only van der Waals interactions (Odashima, Itai, Iitaka and Koga, 1980).

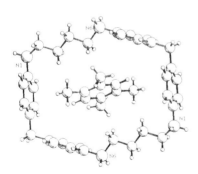 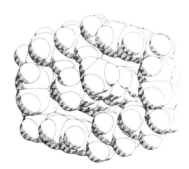

FIG. 10. A complex of durene with a tetraazaparacyclophane.$4H^+$

The host, 1,6,20,25-tetraaza[6.1.6.1]paracyclophane, solubilizes many hydrophobic benzene and naphthalene derivatives in aqueous solution. A number of crystalline complexes have been isolated; that shown in Fig. 10 is the durene complex of the tetraprotonated host. The space-filling representation shows that the guest molecule fits rather snugly in the cavity in the host; there are six C...C contacts below 3.8 A and short H...C contacts from host to guest as well. It is common to say that "hydrophobic bonding" plays a role in this kind of complex. The term "bonding" does not, however, imply a force related to the minimization of a potential energy but rather an interaction that results from the minimization of a free energy. A large entropy increase occurs when separated non-polar molecules or regions of molecules aggregate in aqueous solution, arising from the liberation of water molecules initially ordered around the non-polar moieties. Consequently, even though the accompanying enthalpy change is actually <u>positive</u>, the aggregation proceeds spontaneously because the change in free energy is negative (for a lucid discussion of this point, see Eisenberg and Crothers, 1979). Such "hydrophobic interactions", i.e., the tendency of non-polar molecules or portions of molecules to aggregate in aqueous solution, play a role in the formation of almost all complexes of organic hosts and guests in aqueous solution.

Most biologically important complexes have several loci of specific interactions. Fig. 11 illustrates, in part in stereo, a synthetic host-

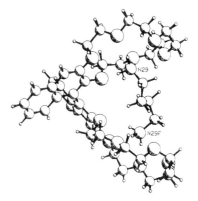

FIG. 11. <u>Right</u>: (a) The complex of tetramethylenediammonium ion with a dinaphthyl-bis-18-crown-6 host. The dinaphthyl group is at the left.

<u>Below</u>: (b) A stereoview of a space-filling representation of this complex.

guest complex (Goldberg, 1977a) in which a tetramethylenediammonium ion fits snugly into a dinaphthyl-bis-18-crown-6 host, just as it was designed to do (Tarnowski and Cram, 1976). The complex has a two-fold axis, so the interactions of the two $-NH_3^+$ groups with the host are equivalent. There are two nearly linear hydrogen bonds and one that is quite distinctly bifurcated, presumably because of the constraints on the conformation of an 18-crown-6 ring with one C-C torsion angle fixed near $0°$. Pascard, Riche, Cesario, Kotzyba-Hibert and Lehn (1982) have recently reported the structure of a cryptate complex of $H_3N(CH_2)_5NH_3^{2+}$ in which the cation is similarly preferentially complexed, in this case because of its structural complementarity with a cryptand composed of two 18-crown-N_2O_4 macrocycles linked through the N atoms by two naphthalene rings bonded at the 2,6-positions. The cavity of the free host is more compact than that of the host in the complex, the inflation of the latter coming about in order to accommodate the tubular guest. Their structural results confirm earlier nmr, binding selectivity and dynamic rigidity evidence.

Dinaphthyl units have been employed in hosts by Cram and his co-workers because of the rigidity they confer and because they are inherently chiral. Hosts based on dinaphthyl units have been resolved and used extensively in studies of chiral recognition with amino acids and their esters and in studying asymmetric induction in catalysis of organic reactions (Wilson and Cram, 1982, and references therein). Discrimination between enantiomers by factors as large as 22 has been achieved. The structure of one of the most effective of these hosts and its t-butylammonium complex have been reported by Goldberg (1980), who also determined the structure (Goldberg, 1977b) of a diastereomeric complex of S,S-bis-dinaphthyl-22-crown-6 with R-phenylglycinium methyl ester. This structure is illustrated in Fig. 12; the only hydrogen atoms shown are the three on the $-NH_3^+$ group and that on the adjacent carbon atom of the guest.

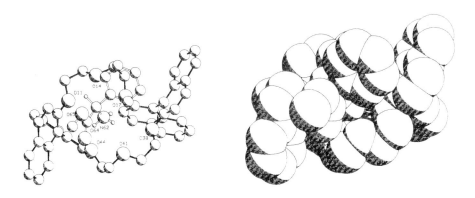

FIG. 12. Complex of a chiral host $C_{48}H_{40}O_6$ with a chiral guest, phenylglycinium methyl ester, whose bonds are shown unfilled.

The $-NH_3^+$ group approaches five of the six oxygen atoms, three of
them close enough to form hydrogen bonds (although two are distinctly
non-linear, those to O41 and O44 in Fig. 12). There is also a close
C-H...O contact (with O14), and although the angle is poor (116°), this
hydrogen atom is made somewhat acidic by each of the three groups bonded
to its attached carbon atom so that there may be some stabilizing influence contributed by this interaction. Of presumably greater importance
is the parallel and partly overlapping disposition of the ester grouping
with one naphthalene ring (at the left in Fig. 12). This presumably
adds some stabilization energy due to a weak charge-transfer interaction
between the naphthaleneoxy group acting as a pi-acid and the ester moiety acting as a pi-base. The fit of the guest into the cleft in the host
provides some other short contacts that also help to stabilize this
structure. With substituents placed in the 3 and 3′ positions, the
steric barriers that help to provide discrimination between enantiomeric
guests are enhanced, so that the fit of the less-favored enantiomer becomes even poorer.

Modification of spherand structures by incorporation of cyclic urea
units in place of some of the anisyl groups and adding two methylene
groups in the macroring to provide pivoting flexibility for one of the
anisyl units makes these hosts more compatible with aqueous solutions
while preserving many of the advantages of preorganization. In particular, the rate at which a complexed ion can be exchanged between organic
and aqueous media is greatly enhanced and it becomes possible to have
complexes not only with encapsulated ions but also with guests in a
perching position.

The formula of such a modified spherand is given as 2 (on the next
page) and Fig. 13 shows the structure of its t-butylammonium complex

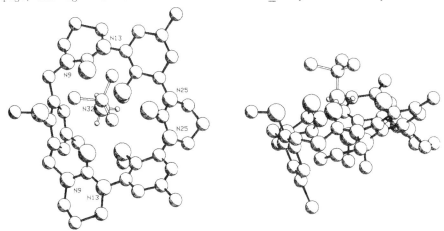

FIG. 13. Two views of the t-butylammonium complex of a modified spherand containing three cyclic urea groups. The only hydrogen atoms shown are those on the $-NH_3^+$ group. Left: View almost normal to the plane of the macroring. Right: View from the
side, showing the nearly perpendicular arrangement of the pivoting
anisyl ring and the perching position of the guest.

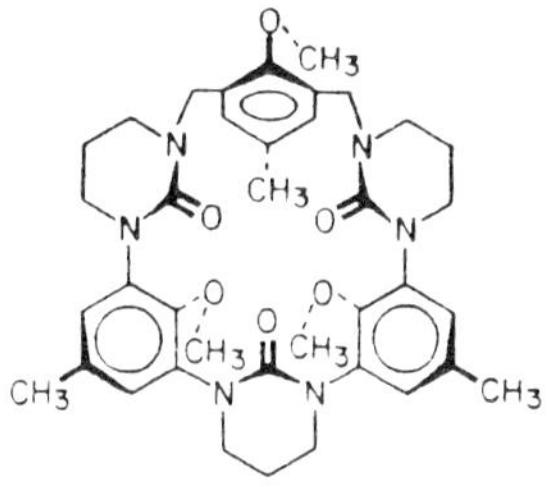

2

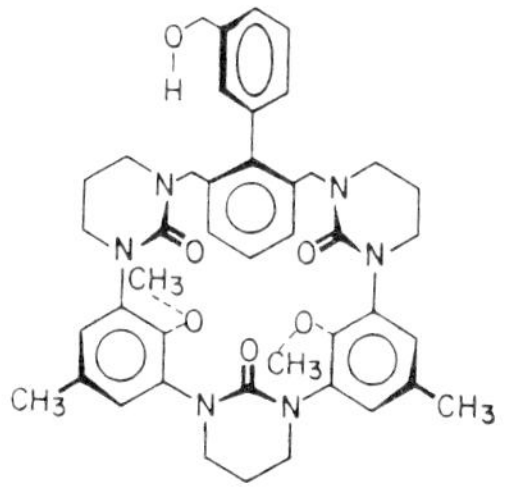

3

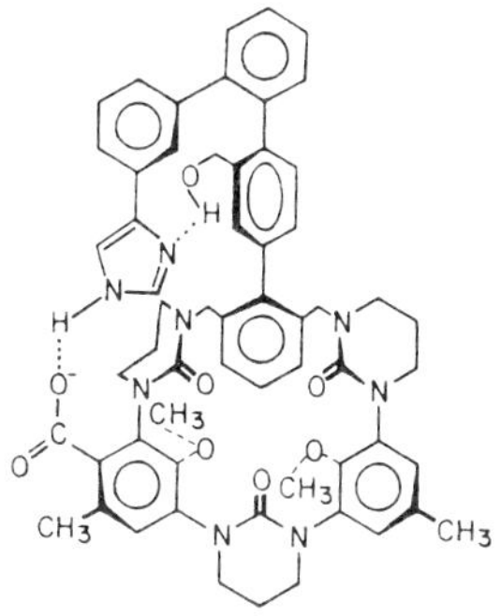

4

(Cram, Dicker, Knobler and Trueblood, 1982). The only hydrogen atoms shown are those on the $-NH_3^+$ of the guest. Because the three urea oxygen atoms of 2 can all project in the same direction and because, unlike the anisyl oxygen atoms, they are not blocked by methyl groups, this host provides an ideal conformation for a perching RNH_3^+ ion (Fig. 13, right). The N^+-H...O hydrogen bonds to the urea oxygen atoms are short and not far from linear; the methoxyl oxygen atom of the methylene-linked anisyl unit is about 0.5 A farther from the ammonium nitrogen atom than are the oxygens of the other two anisyl units. As indicated in Fig. 13, this unique anisyl unit is turned over relative to the other two, an orientation made possible by the two $-CH_2-$ links. The methyl of this methoxyl group is directed <u>away</u> from the center of the cavity, the cavity being occupied in the complex by the <u>t</u>-butylammonium nitrogen atom. Presumably in the pure host (whose structure has not yet been determined) this methyl group projects toward the center, partially filling the cavity; such an orientation is possible in this structure, although not in spherands that do not have the additional $-CH_2-$ spacers or larger ring systems. In solution, host 2 has four independent conformations, which equilibrate slowly on the nmr time scale; the complex has only a single conformation, presumably that found in the crystal.

Spherand $\underline{2}$, depicted in Fig. 13, is of particular interest because it provided confirmation of structural ideas deduced from the examination of space-filling models that were used in the design of a molecule ($\underline{3}$ on the previous page) recently synthesised and tested by Cram and Katz (1983) that enormously accelerates the rate of acyl transfer. Some have popularly termed $\underline{3}$ a "synthetic catalyst" although it is not a true catalyst since it is not automatically regenerated during the reaction. Compound $\underline{3}$ represents a stage in the development of a serine protease mimic, one conception of which (Cram and Katz, 1983) is that represented by formula $\underline{4}$ (on the previous page). The reaction that $\underline{3}$ accelerates is the acyl transfer of an L-alanyl residue from its $\underline{p}$-nitrophenyl ester to 3-phenylbenzyl alcohol, the comparison being made between the rate of transfer to 3-phenylbenzyl alcohol itself and the rate of transfer to the 3-phenylbenzyl alcohol moiety incorporated in $\underline{3}$. The acceleration is by the remarkable factor of between 10^9 and 10^{11}. Examination of a CPK model of the complex of L-alanyl $\underline{p}$-nitrophenyl ester with this host shows the ester carbonyl group to be oriented almost ideally for attack by the (potential) CH_2O^- of the host.

Many other groups are working to develop enzyme mimics, chiefly through modification of cyclodextrins and other natural hosts (see the recent reviews by Saenger (1980) and Tabushi (1982), and the references therein). To date, the largest accelerations seem to have been by factors smaller than 10^3, for acyl transfer reactions. There has been a good deal of structural work on cyclodextrins (e.g., Czugler, Eckle and Stezowski, 1981; Hamilton and Sabesan, 1982), some of it bearing on the efficacy of these hosts in complexing different guests and catalyzing reactions such as acyl transfer. Nonetheless, it seems still true today that we know as much (and sometimes more) structural detail about the intermediates presumably involved in the mechanisms of enzyme reactions (Lipscomb, 1982) as about those of most simpler reactions. This situation is now changing as various techniques, including crystallography, provide further structural detail about molecules that at least seem to resemble intermediates along inorganic, organic and biochemical reaction pathways. Much more rapid change is to be expected in the near future.

It is a pleasure to acknowledge my debt to Professor Donald Cram and his coworkers without whose imagination and prodigious synthetic efforts many of the compounds discussed here would not exist, to Drs. Israel Goldberg, Carolyn Knobler and Emily Maverick for their many essential contributions, to the National Science Foundation and the Department of Energy for support of much of the work described, and to Dr. Olga Kennard and her colleagues in Cambridge who have created and maintained the magnificent Cambridge Database and the associated computer programs, which were used in finding some and drawing all of the crystal structures shown here.

References

BOVILL, M. J., CHADWICK, D. J., SUTHERLAND, I. O. and WATKIN, D. (1980). J. Chem. Soc. Perkin II 1529
BUSH, M. A. and TRUTER, M. R. (1972). J. Chem. Soc. Perkin Trans. 2, 345
CRAM, D. J., DICKER, I. B., KNOBLER, C. B. and TRUEBLOOD, K. N. (1982). J. Amer. Chem. Soc. 104, 6828

CRAM, D. J. and KATZ, H. E. (1983). J. Amer. Chem. Soc. 105, 135
CRAM,D. J., KANEDA, T., HELGESON, R. C. and LEIN, G. M. (1979). J. Amer.
     Chem. Soc. 101, 6752
CRAM, D. J., LEIN, G. M., KANEDA, T., HELGESON, R. C., KNOBLER, C. B.,
     MAVERICK, E. and TRUEBLOOD, K. N. (1981) J. Amer. Chem. Soc. 103,
     6228
CRAM, D. J. and TRUEBLOOD, K. N. (1981) Topics in Curr. Chem. 98, 43
CZUGLER, M., ECKLE, E. and STEZOWSKI, J. J. (1981). J. Chem. Soc. Chem.
     Commun., 1291
DALE, J. (1974). Tetrahedron 30, 1683
DALE, J. (1980). Israel J. Chem. 20, 3
DOBLER, M. (1981). Ionophores and Their Structures. John Wiley, New
     York
EISENBERG, D. and CROTHERS, D. (1979). Physical Chemistry with Applica-
     tions to the Life Sciences. Benjamin/Cummings, Menlo Park,
     California
GOLDBERG, I. (1975a). Acta. Cryst. B31, 754
GOLDBERG, I. (1975b). Acta. Cryst. B31, 2592
GOLDBERG, I. (1976). Acta. Cryst. B32, 41
GOLDBERG, I. (1977a). Acta. Cryst. B33, 472
GOLDBERG, I. (1977b). J. Amer. Chem. Soc. 99, 6049
GOLDBERG, I. (1980). J. Amer. Chem. Soc. 102, 4106
HAMILTON, J. A. and SABESAN, M. N. (1982). Acta Cryst. B38, 3063
HASEK, J., HLAVATA, D. and HUML, K. (1980). Acta Cryst. B36, 1782
HASEK, J., HUML, K. and HLAVATA, D. (1979). Acta Cryst. B35, 330
HILGENFELD, R. and SAENGER, W. (1982). Topics in Curr. Chem. 101, 1
KAUFMANN, R., KNOCHEL, A., KOPF, J., OEHLER, J. and RUDOLPH, G. (1977).
     Chem. Ber. 110, 2249
KNOCHEL, A., KOPF, J., OEHLER, J. and RUDOLPH, G. (1978). J. Chem. Soc.
     Chem. Commun., 595
LEHN, J. M. (1978). Accts. Chem. Res. 11, 49
LIPSCOMB, W. N. (1982). Accts. Chem. Res. 15, 232
MAVERICK, E. , GROSSENBACHER, L. and TRUEBLOOD, K. N. (1979) Acta Cryst.
     B35, 2233
MAVERICK, E., SEILER, P., SCHWEIZER, W. B. and DUNITZ, J. D. (1980).
     Acta Cryst. B36, 615
NAGANO, O., KOBAYASHI, A. and SASAKI, Y. (1978). Bull. Chem. Soc. Japan
     51,790
ODASHIMA, K., ITAI, A., IITAKA, Y. and KOGA, K. (1980). J. Amer. Chem.
     Soc. 102, 2504
OLOVSSON, I. and JONSSON, P.-G. (1976). Ch. 8. in The Hydrogen Bond.
     II. Structure and Spectroscopy, ed. by Schuster, P., Zundel, G.
     and Sandorfy, C. North-Holland Publishing Co., Amsterdam
PASCARD, C., RICHE, C., CESARIO, M., KOTZYBA-HIBERT, F. and LEHN, J. M.
     (1982). J. Chem. Soc. Chem. Commun., 557
PEDERSEN, C. J. (1967). J. Amer. Chem. Soc. 89, 2945, 7017
RICHE, C., PASCARD-BILLY, C., CAMBILLAU, C. and BRAM, G. (1977). J.
     Chem. Soc. Chem. Commun., 183
SAENGER, W. (1980). Angew. Chem. Int. Ed. Engl. 19, 344
TABUSHI, I. (1982). Acc. Chem. Res. 15, 66
TARNOWSKI, T. L. and CRAM, D. J. (1976). Chem. Commun., 661
TAYLOR, R. and KENNARD, O. (1982). J. Amer. Chem. Soc. 104, 5063
TRUEBLOOD, K. N., KNOBLER, C. B., LAWRENCE, D. S. and STEVENS, R. S.
     (1982). J. Amer. Chem. Soc. 104, 1355
TRUEBLOOD, K. N., KNOBLER, C. B., MAVERICK, E. HELGESON, R. C., BROWN,
     S. B. and CRAM, D. J. (1981). J. Amer. Chem. Soc. 103, 5594

VOGTLE, F., PUFF, H., FRIEDRICHS, E. and MULLER, W. M. (1982). J. Chem.
    Soc. Chem. Comm. 1398
VOGTLE, F., SIEGER, H. and MULLER, W. M. (1981). Topics in Curr. Chem.
    98, 107
WEBER, E. and VOGTLE, F. (1981). Topics in Curr. Chem. 98, 1
WILSON, J. M. and CRAM, D. J. (1982). J. Amer. Chem. Soc. 104, 881
WIPFF, G., WEINER, P. and KOLLMAN, P. (1982). J. Amer. Chem. Soc. 104,
    3249

4 Drug–receptor binding forces

Peter Kollman

ABSTRACT

We analyze in detail the energy components of intermolecular (non-covalent) interactions between small molecules in the gas phase. Combining these association energies with entropy changes upon association allows one to see why even moderately strong H bonded complexes might not have a favorable free energy of association in the gas phase. We then describe the likely changes in the thermodynamic parameters for molecular association in going from the gas phase to solution. We suggest that one can approach drug design (at least in the case when one knows the structure of the receptor) by considering the three most important energy components in solution phase ligand receptor interactions: van der Waals forces (both attractive and repulsive components), electrostatic (ion–pairing and H–bonding) forces, and hydrophobic forces. We give specific examples from computer graphics/QSAR analysis and molecular mechanics simulations where each of these terms has been used to rationalize/predict relative binding affinities of ligands to receptors.

1. INTRODUCTION

Non-covalent association of molecules provides the basis for chemical and biological specificity (Bloomfield et al, 1974). Base pairing in DNA, protein–nucleic acid recognition (Anderson et al, 1981; Ptashne et al, 1980) and ligand–protein interactions (Blaney et al, 1982) are phenomena which depend on non-covalent forces for their strength and specificity. Even in enzyme catalysis, such attractive non-covalent forces are important in the formation of the Michaelis complex (Platzer et al, 1972; Fersht, 1977; Wipff et al, 1983) which preceeds changes in covalent bond structure. Thus, the importance of these non-covalent forces and their role in biological chemistry are clear. It is these

same forces which play an essential role in all drug/receptor interactions.

In this acticle we discuss the energy and entropy of non-covalent
association in the gas phase and aqueous solution. Our focus in the
first part of the manuscript will be on the association of small
molecules, where the details of molecular association are easier to sort
out and to understand. We then turn to specific examples of association
between macromolecules and small ligands, in cases where the X-ray
structure of the macromolecule is known to modest resolution. These
specific examples are suggestive of the analysis one must go through in
design of optimal ligands for a given receptor. They also illustrate
the complementary role of X-ray crystallography, computer graphics and
molecular mechanical simulations in drug design.

2.   FORCES IN MOLECULAR ASSOCIATION

2.1.   Gas Phase

2.1.1. Energy Terms

Perturbation theory and detailed ab initio quantum mechanical theories
have been used to study gas phase molecular association. Ab initio
calculations have been a very powerful tool in both predicting and
understanding the nature of gas phase molecular association (Kollman,
1977a). The structure and properties of water and hydrogen fluoride
dimer, $(H_2O)_2$ and $(HF)_2$ were predicted prior to experimental determination (Dyke and Muenter, 1974; Dyke et al, 1972) of their structure.
It is reasonable to say that for moderate sized molecules it is easier
to determine the structure and (more qualitatively) energies of gas
phase association from ab initio theory (Kollman, 1977b) than from
experiment.

Of equal importance have been the development of methods to break
down the total ab initio calculated intermolecular interaction energy
into components (Morokuma, 1971). These components (Kitaura and
Morokuma, 1976) are closely related to those suggested by perturbation
theory analysis of such interactions. For the purpose of this discussion, we consider the five most important components: electrostatic,
exchange repulsion, polarization, charge transfer and dispersion. The
first four terms can be calculated in a well defined way from Hartree
Fock SCF ab initio calculations and the fifth requires ab initio
calculations which include electron-electron correlation energy.

We now briefly describe the physical nature of such components and give examples of their magnitude for typical non-covalent interactions (Table 1).

SOME EXAMPLES OF INTERACTION ENERGIES OF NONCOVALENT COMPLEXES

TABLE 1

Interaction	Interaction Energies[a] (kcal/mole)					
	$-\Delta E$	ΔE_{ES}	ΔE_{POL}	ΔE_{CT}	ΔE_{EX}	ΔE_{DIS}
He...He	0.02^b	0	0	0	+0.008	−0.028
$H_2O...H_2O$	8.9^c	−9.2	−0.5	−2.2	4.0	(−1)
$Li^+...OH_2$	48.9^d	−51.1	−7.8	−1.7	12.7	(−1)
$NH_4^+...F^-$	164.7^d	−181.4	−8.3	−35.6	61.6	(−1)

[a]Symbols: $-\Delta E$; calculated (or experimental) total interaction in kcal/mole, ΔE_{ES} electrostatic energy; ΔE_{POL}, polarization energy; ΔE_{CT}, charge transfer energy; ΔE_{EX}, exchange repulsion energy; ΔE_{DIS}, dispersion energy (values in parentheses are estimated).

[b]See Schaefer et al (1970).

[c]See Umeyama and Morokuma (1977); this value for ΔE is somewhat too large.

[d]See Kollman (1977c)

The electrostatic interaction energy is the classical electrostatic interaction between the isolated molecule charge distributions as they associate. One can also consider each molecule's electron distribution as being an infinite expansion of all the electric multipole moments, beginning with the first non-vanishing one (e.g., for H_2O, its dipole moment). Then, the electrostatic interaction energy between water molecules (see Table 2) can be thought of as the sum of dipole-dipole, dipole-quadrupole, quadrupole-quadrupole.......etc., interactions. It is clear from our work (Kollman, 1977c) that to understand H-bond strengths and directionality, one must qualitatively consider at least the first two non-vanishing multipole moments. The distance dependence of some of the various electrostatic interactions and their classical functional form are summarized in Table 2.

SUMMARY OF IMPORTANT CONTRIBUTIONS TO NON–COVALENT ASSOCIATION

TABLE 2

Energetic (Enthalpic) Contributions	Functional Form	Magnitude of ΔH(gas phase)	Magnitude of ΔH(soln)
1. Electrostatic			
ion–ion	$\dfrac{q_A q_B}{r_{AB}}$	---	~ 0
ion–dipole	$\dfrac{\mu_A \mu_B \cos\theta}{r_{AB}}$	---	~ 0
dipole–dipole	$\dfrac{\mu_A \mu_B}{r_{AB}} \; f(\theta_A, \theta_B)$	–	~ 0
2. Exchange Repulsion	$A e^{-\alpha r_{AB}}$ or $\dfrac{A}{r_{AB}^{12}}$	+	+
3. Polarization of B by A	$\dfrac{-\frac{1}{2} \mu_A^2 \alpha_B}{r_{AB}^4}$	–	–
4. Charge Transfer	$Ce^{-Br_{AB}}$	–	–
5. Dispersion	$\dfrac{-B}{r_{AB}^6}$	–	–

Entropic Contributions		ΔS (gas phase)	ΔS (aqueous soln)
1. Trans/rotation/vib. freq. changes		--	–
2. Hydrophobic Effects		0	+

The magnitudes of the electrostatic interaction energies are worth noting for different complexes. For rare gas interactions, the electrostatic interaction energy is rigorously zero; the magnitude of the electrostatic attraction at the minimum energy geometry is 5-10 kcal/mole for neutral hydrogen bonded complexes, 15-35 kcal/mole for ion-neutral interactions such as $F^-...H_2O$ and 100-200 kcal/mole for ion-pairing interactions such as $NH_4^+...F^-$. Of course, the electrostatic energy can be attractive or repulsive, depending on the complex geometry.

The second energy component, exchange repulsion, comes from the quantum mechanical Pauli principle, and represents the repulsion due to the overlap of electron clouds accompanying complex formation. It is always repulsive and its magnitude is a very steep (exponential) function of the intermolecular separation (Margenau and Kestner, 1970), although historically it has often been represented as dying off as R^{-12}. This component is the major determinant of van der Waals radii of atoms, in that it prevents much closer approach between atoms than the sum of their radii.

The third and fourth components, polarization and charge transfer, represent the energy due to the charge redistribution which accompanies complex formation. Although interdependent, it is of conceptual value to separate them, since polarization energy represents the redistribution of electron charge within the molecules during complex formation and charge transfer the transfer of charge from one molecule to the other. The polarization energy can be thought of classically as the induction of dipoles (or higher moments) in one molecule by the electric moments of the second molecule. Although this component is quite small in magnitude for typical neutral-neutral interactions (~5% of the electrostatic for $(H_2O)_2$), it is more important in ion-neutral interactions, being typically ~15% of the electrostatic energy. The polarization component is always attractive, no matter what the molecular orientation.

The charge transfer component is an inherently quantum mechanical term and is also always attractive. For example, for H-bonded complexes such as $(H_2O)_2$ at its minimum energy geometry, the charge transfer contributes typically 2-5 kcal/mole to the intermolecular attraction.

Finally, the dispersion attraction represents the instantaneous induced dipole-induced dipole attraction between molecules, although

first term on the right side of equation (4) and of the opposite
sign. Thus, because of the much greater interactions of water with A
and B than AB, there is relatively little tendency for ion-pairing in
water, in contrast to the situation in the gas phase. Water has a
great leveling effect, in that the stronger the non—covalent
interactions between A and B in the gas phase, the stronger they
interact with water. Thus, electrostatic dominated interactions such
as ion-pairing, ion—dipole and dipole-dipole, which are the
intrinsically strongest gas phase interactions, often contribute
little to aqueous solution free energies of association, because the
ion-water and dipole-water interactons which are lost when A and B
associate are approximately equal to those gained in the association.

On the other hand, if A and B are non-polar molecules, their
association is likely to be significantly more favorable in aqueous
solution than in the gas phase, for reasons that are rather subtle.
This hydrophobically driven association (Kauzmann, 1959; Tanford,
1973) which has stimulated much research, is due to the "release" of
water molecules from A and B when the molecules associate and shows up
mainly in the change in entropy (ΔS_T) upon association. This effect
must be initially analyzed by examining thermodynamic data for the
transfer of a hydrocarbon from a non-polar solvent or the gas phase to
water. The $\Delta G(\text{soln},A)$ for such a process is positive (unfavorable),
despite the fact that ΔH_{soln} for the process is either near zero or
slightly negative (favorable). On the other hand $\Delta S(\text{soln},A)$ for such
a process is very negative. Initially, one might expect $\Delta H(\text{soln},A)$ to
be positive because, by placing a hydrocarbon in water, one is
replacing a strong attractive electrostatic water-water interaction
with a less attractive (dispersion and polarization only) hydrocarbon-
water interaction. The very negative $\Delta S(\text{soln},A)$ gives a clue.
Computer simulations of such processes (Owicki and Scheraga, 1977;
Swaminathan et al, 1976) indicate that the water molecules in the
vicinity of the hydrocarbon compensate for the weak hydrocarbon-water
interactions by forming stronger, more favorable water-water interactions and thus become more ordered because of stronger water-water
hydrogen bonding. Thus, little net H-bonding energy is lost, but the
motion of the waters are restricted and the $\Delta S(\text{soln},A)$ is very
unfavorable (negative). On the other hand, when A and B associate and
cause less of their non-polar groups to be exposed to water, this

release of water has a favorable entropy and free energy and the
association occurs. This hydrophobic effect is extremely important in
many biological processes.

Table 3 contrasts the free energy cycle for association of a proton
with NH_3 (a very strong gas phase electrostatic dominated interaction)
with that of two methane molecules (a very weak gas phase
association). As one can see, the aqueous solution free energies are
very different from those in the gas phase.

ESTIMATE OF FREE ENERGIES IN EQN (1) FOR PROTONATION OF NH_3 AND
DIMERIZATION OF METHANE AT 298K

TABLE 3

	(kcal/mole)				
Reaction	ΔG_T(aq)	ΔG_T solv(AB)	ΔG_T solv(A)	ΔG_T solv(B)	ΔG_T(gas)
$H_3N+H^+ \rightarrow NH_4^+$[a]	-4	-78	-2	-270	-198
$2CH_4 \rightarrow (CH_4)_2$	(-1.8)	(6.1)[b]	$+6.3$[c]	$+6.3$[c]	(4.7)[d]

[a]See Aue et al, 1976 and discussion therein

[b]Using the solvation free energy of ethane as a model for $(CH_4)_2$
solvation. The solvation free energy of butane is also very similar
to this value (Battino, 1977)

[c]Battino, 1977

[d]Estimated assuming a $\Delta H = -1$ kcal/mole in the gas phase and a ΔS
identical to water dimer (-19eu for a Mole/liter standard state)
(Kollman, 1979, Curtis et al, 1978)

The greater tendency for hydrocarbons to associate in water than found
for association in the gas phase comes from the very unfavorable free
energy of solution of the methane. A methane dimer, by being associated
and disturbing much less of the water structure than two monomers, has a
much less unfavorable solvation free energy than two monomers. Because
of this solvation effect, the solution free energy
of ~ -2 kcal/mole is much more favorable than that in the gas phase.
This "hydrophobic" association due to the solvation effects compensates
for the unfavorable translational/rotational entropy loss, which makes

gas phase association so unfavorable.

The situation is completely different for the strong ion/polar association of NH_3 with H^+. Here the gas phase association is very favorable, but because of the much greater solvation free energy of H^+ than NH_4^+, the solution free energy of association is only slightly favorable, and, at pH values above the pK_a, ΔG is no longer negative.

Let us now summarize the important energetic aspects of non-covalent interactions of A and B in solution, stressing the differences between gas phase and solution association. Polar and ionic molecules interact strongly with each other, but, because A-B interactions are compensated by A-H_2O and B-H_2O interactions, the net $\Delta H(soln)$ may be near zero even for strong A...B interactions. On the other hand, these strong A...B interactions are important for biological specificity, since any electrostatically unfavorable A...B interactions would extract a heavy energy penalty.

One might expect that exchange repulsion/dispersion attractions (van der Waals forces) would contribute little to AB solution association, because any A...B dispersion attraction would be compensated by loss of H_2O...A dispersion and H_2O...B dispersion upon association. However, the density of atoms in proteins is greater than that of water and so that ligand-protein dispersion attraction may be larger than the sum of ligand-H_2O and protein-H_2O and thus dispersion attraction may play an important energetic role in A...B interactions in aqueous solution. Exchange repulsions will be important in determining molecular shape complementarity of drug and receptor.

The loss of translational and rotational entropy upon AB association is likely to be less important in solution than in the gas phase, but is still a significant thermodynamic contributor to a negative ΔS for AB association. On the other hand, for the association of non-polar molecules A and B, the hydrophobic effects provide a significantly positive ΔS for AB association which is often larger than the negative ΔS due to translational and rotational entropy loss.

Finally, we should note that the ΔS change on ring closure is likely to be comparable in solution to that in the gas phase. This supports the basis of using locked analogues in drug design. Of course, if the locked analogues do not have very nearly the optimal orientation of its pharmacophores that the flexible molecule can adopt then the less favorable enthalpy of association may cancel the more favorable entropy of association. We must stress how difficult it is to find an unambiguious interpretation of aqueous solution association thermodynamics in terms of specific atomic interactions (Sturdevant, 1977 and Ross and Subramanian, 1981). However, one has some circumstantial evidence that the most important energy terms which determine the strength and specificity of non-covalent aqueous solution associations are electrostatic, dispersion and hydrophobic. We now give examples of cases where one of these terms is the key determinant of the strength and/or the specificity of the association.

3.   EXAMPLES OF BIOLOGICAL ASSOCIATION

3.1.   Electrostatic forces

Fersht (1977) has nicely summarized the relative binding energies (or relative catalytic rates kcat/Km) for different amino acids and des-NH_3^+ amino acids interacting with aminoacyl tRNA synthetases. The absence of the NH_3^+ group in the des-NH_3^+ amino acids causes a loss in binding free energy of 3.4-4.5 kcal/mole relative to the amino acids. Although this is far smaller than a "gas phase" loss of an ion pair, it is larger than one might expect, in view of our earlier discussion that ionic effects are largely damped out in solution.

An even larger loss in binding free energy (6.1 kcal/mole) occurs upon binding phenylalanine rather than tyrosine to tyrosyl-tRNA synthetase. Again, the most logical explanation of this result is that the enzyme contains hydrogen bond donors or acceptors for the tyrosine OH, which are unable to form hydrogen bonds (or fewer, more distorted ones) when phenylalanine binds because it lacks the aromatic OH group.

Andrea et al (1980) have determined the binding affinities of thyroxine analogs to the plasma protein prealbumin and have noted the order of binding affinities is des-NH_3^+ analog > L amino acid > des-COO^- analog, consistent with the preference of the protein for more anionic ligands. The electrostatic potential around prealbumin is illustrated in Fig. 1 and the binding channel is clearly mainly blue (positive). We have also shown the importance of electrostatic (H-bonding) interactions in differentiating the tetrahedral intermediates of D and L N-acetyl tryptophanamide interacting with α-chymotrypsin (Wipff et al, 1983). In the case of the L isomer, the strong His57 N-H...NH_2 substrate H-bond in the tetrahedral intermediate not present in the D isomer, nicely rationalizes the much greater rate of catalysis of α-chymotrypsin for the L rather than the D substrate (Fig. 2).

3.2.   Dispersion Forces

The unequivocal demonstration of the importance of dispersion
forces in protein-ligand association is difficult, since individual

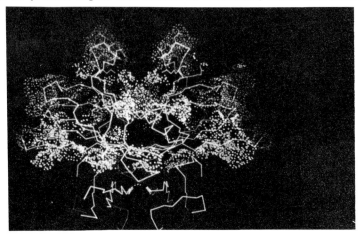

FIG. 1. The electrostatic potential of one monomer of the prealbumin
tetramer. The binding site (bottom of molecule) is made up from this
monomer and a C_2 related monomer that brings the two equivalent "bottom
of molecules" together.

atom-atom energies are rather small and it is the sum of many atom-atom
interactions which determines whether an association is driven by the
dispersion interactions or not.

Again, the comparison of the binding of isoleucine and valine with
isoleucyl-tRNA synthetase is illuminating (Fersht, 1977). The
difference in binding energy is 3 kcal/mole, whereas the difference in
octanol/water partition coefficient (thought to be a reasonable model
for the hydrophobic contribution) only amounts to 0.7 kcal/mole of free
energy difference. Placing valine in isoleucyl-tRNA synthetase
apparently leaves a hole the size of a $-CH_2-$group in the complex;
since the dispersion force dominated sublimation energy/CH_2 group of
crystalline hydrocarbons is ~2 kcal/mole, this interpretation is
reasonable.

Wolff et al (1978) have suggested, on the basis of the thermodynamics of steroid binding to glucocorticoid receptor protein, that

both hydrophobic and dispersion contributions are required to explain
the binding energies observed.

Studies with linear free energy (QSAR) methods have noted many
examples where polar group contributions to binding free energy correlates with the molar refractivity of the group (Silipo and Hansch,

a.

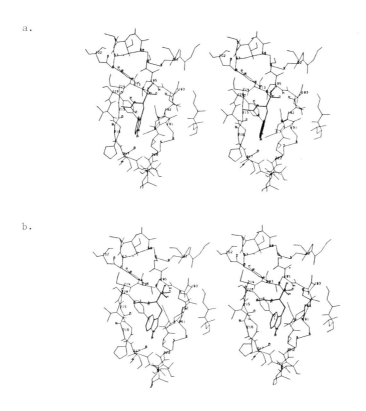

b.

FIG. 2. (a) Stereo representation of the L-N-acetyltryptophan amide-α-
chymotrypsin covalent complex from Wipff et al (1983).
(b) Stereo representation of the D-N acetyltryptophan amide-α-chymo-
trypsin covalent complex from Wipff et al (1983).

1975, Grieco et al, 1977). This has been interpreted as being due to
dispersion attraction, since both the dispersion attraction coefficient
and molar refractivity increase as the polarizability and number of

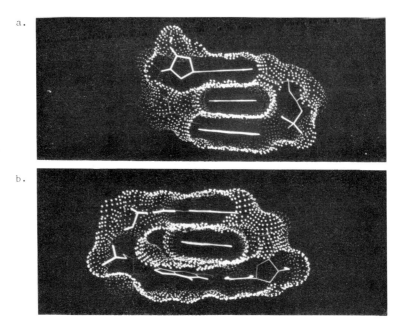

FIG. 4. (a) Molecular surface representation of the most stable nitroquinoline N-oxide-d $(C_pG)_2$) complex from Lybrand <u>et al</u> (1981). (b) Same as (a) for a less stable configuration.

$$R - \overset{\text{O}}{\underset{\|}{C}} - O - \left\langle O \right\rangle \overset{X_3}{\underset{X_5}{\diagdown}}$$

of the more hydrophobic of the two substituents, not the sum of the two. Computer graphics visualization showed the non-equivalence of the two substituents when in contact with the enzyme, with one group in contact with hydrophobic atoms on the protein and the other sticking into solution (Fig. 5). Large R substituents bind in a large hydrophobic pocket and their binding affinity does correlate with hydrophobicity.

A second very interesting example was the difference in the QSAR for identical triazine analogs binding to two different (<u>L. casei</u> and chicken liver) dihydrofolate reductases. In one case, the dependence of binding on octanol partition coefficient π for long o-alkyl chains $-(CH_2)_n$ was approx. 1, indicating an active site similar in hydropho-

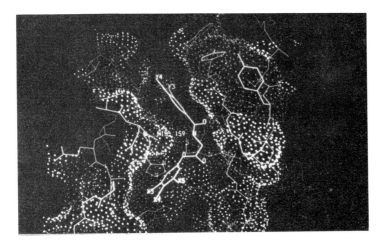

FIG. 5. Molecular surface representation of papain active site and phenyl hippurate bound to it with Y substituents in the long hydrophobic pocket, X_3 pointing toward a hydrophobic patch on the protein and X_5 pointing out towards the solvent. Red=hydrophobic; blue=polar and ionic residues.

bicity to octanol, but beyond n=6, binding affinity leveled off and there was no dependence on hydrophobicity. In the other isozyme, the dependence of binding on π was much less steep, but extended to n=11. The reasons for the differences in the QSAR became very clear upon observation of the binding geometries (one from X-ray crystallography, the other model built). In the first case, the alkyl group was in a tight hydrophobic channel, but for chains longer than n=6, the hydrocarbon stuck out into solution. In the second case, the alkyl chain sat on a long hydrophobic "floor", long enough to accomodate chains up to n=11, but with only one side of the chain in contact with the protein and the other side solvated.

The role of the hydrophobic concept in drug design is quite clear. Ariens (1971) has noted that receptors have hydrophobic and hydrophilic regions, and that one could design pharmacologically interesting molecules by adding hydrophobic groups to an existing active molecule and varying the chain length between the initial structure and the hydrophobic group. Muscarinic acetylcholine antago nists (Pauling and Datta, 1980) and histamine H_1 antagonists (Ganelin, 1980) are examples of molecules which contain the basic features of the parent molecule (prolonged amine and ester in the case of muscarinic antagonists and protonated amine in the case of histamine H_1 antagonist with large hydrophobic groups on the molecule providing a substantial binding affinity.

4.   SUMMARY

It appears that the most important attractive non-covalent forces for biological association in aqueous solution are electrostatic, dispersion/repulsion and hydrophobic. Electrostatic interactions are probably more important in providing specificity than in contributing to the overall thermodynamic driving force for association. It is likely that dispersion is important and hydrophobic terms essential in most protein-ligand interactions.

It is disappointing, but must be emphasized, that the connection between the experimental thermodynamic data and an unambiguous molecular interpretation of these in complex systems in solution is circumstantial at best. Nonetheless, thermodynamic studies on model system and QSAR/computer graphics studies of ligand protein structures provide strong evidence to support the role of electrostatic, dispersion/repulsion and hydrophobic interactions in protein-ligand association.

ACKNOWLEDGEMENTS: We would like to thank the NIH (GM-29072) for research support and Jeff Blaney and Terry Lybrand for Figures 3-5.

REFERENCES

ANDERSON, F., OHLENDORF, D., TAKEDA, Y. and MATHEWS, B. (1981). Nature (London) **290**, 754.

ANDREA, T., DIETRICH, S., MURRAY, W., KOLLMAN, P., JORGENSEN, E. and ROTHENBERG, S. (1979). J. Med. Chem. **22**, 221.

ANDREA, T., CAVALIERI, R., GOLDFINE, I. and JORGENSEN, E (1980). Biochemistry **19**, 55.

ARIENS, E. (1971). Drug Design, E. Ariens, ed., Academic Press, N.Y.

AUE, D., WEBB, H. and BOWERS, M. (1976). J. Amer. Chem. Soc. **98**, 311, 318.

BATTINO, R. (1977). Chem. Rev. **77**, 219.

BLANEY, J., WEINER, P., DEARING, A., KOLLMAN, P., JORGENSEN, E., OATLEY, S., BURRIDGE, J. and BLAKE, C. (1982a). J. Amer. Chem. Soc. **104**, 6424.

BLANEY, J., JORGENSEN, E., CONNOLLY, M., FERRIN, T., LANGRIDGE, R., OATLEY, S., BURRIDGE, J. and BLAKE, C. (1982b). J. Med. Chem. 25, 785.

BLOOMFIELD, V., CROTHERS, D. and TINOCO, I. (1974). Physical Chemistry of Nucleic Acids, Harper and Row, New York.

BLOW, D. (1976). Acc. Chem. Res. 9, 145.

CURTIS, L., FRURIP, D. and BLANDER, M. (1978). J. Chem. Phys. 71, 2703.

DYKE, T., HOWARD, B.J. and KLEMPERER, W. (1972). J. Chem. Phys. 56, 2442.

DYKE, T. and MUENTER, J. (1974). J. Chem. Phys. 60, 2929.

FARNELL, L., RICHARDS, W. and GANELLIN, C. (1974). J. Theor. Biol., 43, 389.

FERSHT, A. (1977). Enzyme Structure and Mechanism, W. Freeman, Reading and San Francisco.

GANELLIN, C. (1980). Burger's Medicinal Chemistry, vol. 2, Chapter 48, M.E. Wolff, ed., J. Wiley, New York and London.

GRIECO, C., SILIPO, C., VITTORIA, A. and HANSCH, C. (1977). J. Med. Chem. 20, 586.

HANSCH, C., LI, R., BLANEY, J. and LANGRIDGE, R. (182). J. Med. Chem. 25, 777.

KAUZMANN, W. (1959). Adv. Prot. Chem. 14, 1.

KITAURA, K. and MOROKUMA, K. (1976). Int. J. Quant. Chem. 10, 325.

KNOWLES, J. (1965). J. Theor, Biol. 9, 213.

KOLLMAN, P. (1977a). Acc. Chem. Res. 10, 365.

KOLLMAN, P. (1977b). Modern Theoretical Chemistry, vol. 4: Applications of Electronic Structure Theory, H.F. Schaefer, ed., Plenum Press, p. 109.

KOLLMAN, P. (1977c). J. Amer. Chem. Soc. 99, 4875.

KOLLMAN, P. (1979). Burger's Med. Chem., part 1: The Basis of Medicinal Chemistry, M.E. Wolff, ed., Wiley, N.Y.

KUNTZ, I. (1979). Burger's Med. Chem., vol. I, M. Wolff, ed., John Wiley, N.Y., p. 285.

LI, R., HANSCH, C., MATHEWS, D., BLANEY, J., LANGRIDGE, R., DELCAMP, T., SUSTEN, S. and FREISHEIM, J. (1982). Quant. Struc. Act. Relat. 1, 1.

LOW, B. and RICHARDS, F. (1954). J. Amer. Chem. Soc. 76, 2511.

LYBRAND, T., DEARING, A., WEINER, P. and KOLLMAN, P. (1981). Nucl. Acid. Res. 9, 6995.

MARGENAU, H. and KESTNER, N. (1970). The Theory of Intermolecular Forces, Pergamon Press, Oxford.

MOROKUMA, K. (1971). J. Chem. Phys. 55, 1236.

OWICKI, J. and SCHERAGA, H. (1977), J. Amer. Chem. Soc. 99, 7413.

PAULING, P. and DATTA, N. (1980). Proc. Nat. Acad. Sci., 77, 708.

PLATZER, K., MOMAMY, F., SCHERAGA, H. (1972). Int. J. Peptide Protein Res. 4, 187.

PTASHNE, M., JEFFREY, A., JOHNSON, A., MAURER, R., MEYER, B., PABO, C., ROBERTS, T. AND SAUER, R. (1980). Cell 19, 1.

ROSS, P. and SUBRAMANIAN, S. (1981). Biochemistry 20, 3096.

SCHAEFER, H.F., MCLAUGHLIN, D., HARRIS, F. and ALDER, B. (1970). Phys. Rev. Lett. 25, 988.

SILIPO, C. and HANSCH, C. (1975). J. Amer. Chem. Soc. 97, 6849.

SMITH, R., HANSCH, C., KIM, K., OMIYA, B., FUKUMURA, G., SELASSIE, C., JOW, P., BLANEY, J. and LANGRIDGE, R. (1982). Arch. Biochem. Biophys. 5, 319.

STURDEVANT, J. (1977). Proc. Nat. Acad. Sci. (US), 74, 2236.

SWAMINATHAN, S., HARRISON, S. and BEVERIDGE, D. (1978) J. Amer. Chem. Soc. 100, 5705.

TANFORD, C. (1973). The Hydrophobic Effect, J. Wiley, New York and London.

UMEYAMA, H. and MOROKUMA, K. (1977). J. Amer. Chem. Soc. 99, 1316.

WIPFF, G., DEARING, A., WEINER, P., BLANEY, J. and KOLLMAN, P. (1983). Molecular Mechanics Study of Enzyme—Substrate Interactions: The Interaction of L- and D-N acetyltryptophanamide with α-Chymotrypsin. J. Amer. Chem. Soc. (in press).

WOLFF, M., BAXTER, J., KOLLMAN, P., LEE, D., KUNTZ, I, BLOOM, F., MATULICH, D. and MORRIS, J. (1978). Biochem. 17, 3201.

5 Drug conformation: a comparison of X-ray and theoretical results
W. Graham Richards

Crystallography and computational chemistry are not rival techniques.
If it is possible to answer questions of conformation by X-ray structure
determination then this must be the method of choice. Theoretical
methods, whether quantum mechanical or the empirically based molecular
mechanics variety are complementary and their value lies in extensions
to experimental investigations. In terms of nuclear conformation, theory
will be the chosen approach in cases where crystallography is not
possible; for the study of non-equilibrium conformations and above all
for answering questions about the disposition of electrons rather than
of nuclei.

1. KNOWN STABLE CONFORMATIONS

   In order to assess the capabilities of theoretical methods, they
must first be judged against data provided from crystallography. For
model problems the calculations of quantum mechanics can reproduce
experimental structures to limitless accuracy although at the cost of a
great deal of computer time. These high-level calculations are of the
ab initio molecular orbital type, (that is without empirical parameters)
and employ gradient techniques. The data for a standard molecular orbital
calculation are the coordinates of the nuclei in the molecule, their
atomic numbers and the number of electrons in the system. Solving the
Hartree-Fock self-consistent field equations then yields the energy of
the molecule for the given geometry and conformation. Stable structures
are found by varying this geometry and seeking the energy minimum. Far
more effectively there are now available [1] molecular orbital programs
which take the gradient of the energy and seek out the minimum automatically rather than using a step-wise search with points on the energy
surface being plotted manually.

   For more realistic problems in the area of drug conformation the
molecules are normally too large to permit the application of the sure
but costly gradient techniques. Instead a short cut is taken. Bond
lengths and angles are derived from crystallographic data and theory is

used to calculate flexible torsion angles and possibly some other lengths
and angles. In most cases the calculated global minimum energy does
correspond to the stable conformation observed in crystals. This is
true whether the energy calculation be totally empirical molecular
mechanics, semi-empirical quantum mechanical or the non-parameterized
ab initio variety. Figure 1 summarizes the agreement between calculation
and experimental data for acetylcholine.

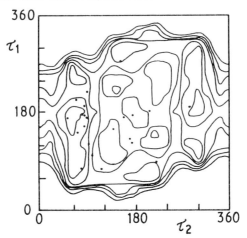

FIG. 1 Conformational potential surface for acetylcholine calculated
using the semi-empirical PCILO approximation, with the experimental
crystal structures shown as points which cluster round the energy minima.

The agreement ought not to overwhelm us. In the cases of theoretical
methods which incorporate empirical parameters, these very parameters have
been introduced so as to reproduce experiments such as the known crystal
structure of similar compounds. Nonetheless it is true that the stable
conformation computed for an isolated molecule is, in general, very close
to the one found for that molecule in the solid state.

The agreement is not, however, invariably found. Even crystal
structures vary, perhaps as a function of the choice of counter ion.
Where calculations are most often in striking disagreement with crystal
data are in instances where the conformation of the gas phase molecule
is strongly influenced by intramolecular effects which are absent in the
crystal, or where crystal structures are dominated by intermolecular
interactions which would not be present in the gas, or indeed in solution.
Far and away the most important instance of such effects is in the case
of hydrogen-bonding. A well-known example is that of histamine (Figure 2).

Apart from this type of almost pathological example, the general
level of agreement encourages the use of theory for problems not
amenable to crystallography.

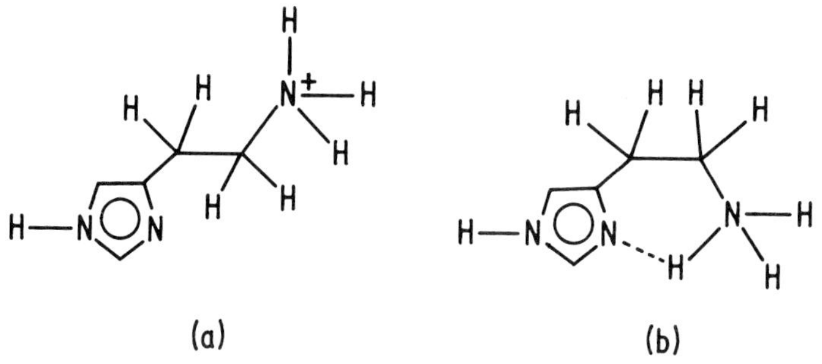

FIG. 2.   Approximate shapes of the histamine monocation
(a) experimental   (b) calculated for the isolated species.

2.   UNKNOWN STABLE STRUCTURES

Stable molecular conformations which are not susceptible to X-ray
techniques fall into two classes. Firstly, molecules which do not
crystallize and those which cannot or have not been synthesised.
Secondly, less-favoured tautomers and transition states. The former
sub-category may be treated just like stable molecules. Bond lengths
and angles are taken from crystallographic results on similar compounds
[2], or from standard values [3]. Only torsional angles are varied and
the results displayed as energy contours.

The second category demands much more care if the results are not
to be biased by the use of inappropriate experimental results. The costly
gradient techniques are essential. In the literature there have been a
number of erroneous investigations of tautomeric equilibria. It is not
sufficient merely to take the crystal geometry of one form and then to
transpose a proton. Major structure change may often result as in
Table 1 for imidazoles [4].

TABLE 1

Bond Angles in Imidazoles Derivatives, found by X-ray Crystallography

Compound	Imidazole form	$C^4C^5N^\pi$	$C^5N^\pi C^2$	$N^\pi C^2 N^\tau$	$C^2N^\tau C^4$	$N^\tau C^4 C^5$
Histamine phosphate, H_2O	Cation	106.5	108.6	108.5	109.1	107.2
sulphate I	Cation	105.8	107.9	109.9	109.0	107.4
II	Cation	105.0	111.3	105.6	109.8	108.3
dichloride	Cation	105.6	109.7	107.6	109.9	107.2
Histidine HCl, H_2O	Cation	106.2	109.5	108.7	109.6	106.9
Histamine bromide	N^τ-H	109.6	105.9	110.0	108.4	106.1
6-Histamino purine	N^τ-H	110.4	104.0	112.6	106.7	106.3
L-Histidine (monoclinic)	N^τ-H	110.0	103.1	113.1	108.0	104.5
(orthorhombic)	N^τ-H	109.6	104.9	112.1	106.9	106.4
Cyclo-(L-threonyl-L-histidyl) $2H_2O$	N^τ-H	109.4	105.8	111.2	106.9	106.4
Burimamide	N^τ-H	108.4	105.7	112.3	106.5	107.1
Thiaburimamide	N^τ-H	109.4	105.0	112.0	107.6	106.0
Metiomide	N^τ-H	109.0	105.0	111.7	108.1	105.4
Imidazole	N^τ-H	109.8	105.4	111.3	107.2	106.3
Histamine base	N^π-H	105.1	107.2	112.3	104.5	110.8
Imidazole	N^π-H	106.3	107.2	111.3	105.4	109.8

3. UNSTABLE STRUCTURES

The realm where calculations come into their own is in answering questions about the nuclear conformations of unstable species. In pharmacological problems we may be interested in the shape adopted by a flexible molecule when binding to a receptor or necessary in gaining

access thereto. To answer questions about the possible shapes one is
forced to go to computation. The results may then be displayed as energy
contour diagrams or probability maps which indicate at a glance just how
flexible a molecule may be and over what range of torsion angles (see
Figure 3).

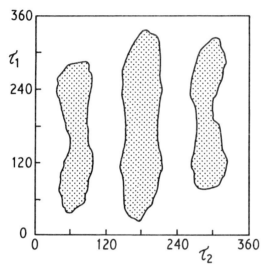

FIG. 3. Conformational percentage map for the histamine monocation at
37 ^{o}C. The shaded area which indicates the range of flexibility contains
99 per cent of the molecules at this temperature.

The probability maps are derived from the energy map by taking Boltzmann
factors over a regular grid of points.

      The calculations may be performed using molecular mechanics if the
molecule concerned falls within their range of parameterization or by
any of the quantum mechanical alternatives if necessary. The latter
on the other hand are more costly in computational terms but do have
the possible bonus of providing not only an energy but also a wave
function. The wave function has its utility in revealing electronic
as opposed to nuclear properties.

4.   ELECTRONIC PROPERTIES
      Crystallography measures the distribution of electron density in a
molecule but since this electron density is concentrated round the
positive nuclei the density maps are interpreted as indicators of nuclear
position. Quantum mechanics yields the wave function, Ψ, whose square
modulus at any point is a measure of electron density. The calculations

have an advantage in that one can ask a question about electron density
at any point without worrying about heavy density in a nearby position.
In addition the density can be broken down into orbital contributions.

Until the relatively recent advent of computer graphics devices,
however, the difficulty for the theoretician was just how to display the
data. Contour diagrams such as Figure 4, are only useful in instances
where there is an obvious plane of symmetry in the molecule; a feature
rare in drug molecules. Graphics, particularly with colour to facilitate
the interpretation, permit quasi-three dimensional displays.

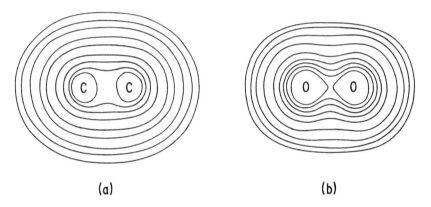

(a)   (b)

FIG. 4. Molecular charge density contours for C_2 (a) and for O_2 (b).

In the example shown in Figure 5, the net drawn on the molecule, like
the crystallographers "chicken-wire" diagrams, is drawn at a specified
electron density contour.

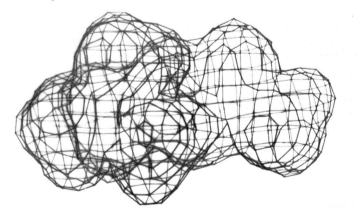

FIG. 5. A net drawn over the acetylcholine icn at an electron density
of 0.01 a.u.

This has been done to try to answer the question of how big a molecule
is in electronic terms. Exploratory calculations on the interaction of
pairs of molecules which model a real system show that molecular 'size'
is a function of the colliding partners so that the over-facile use of
models or standard covalent radii is of very dubious validity

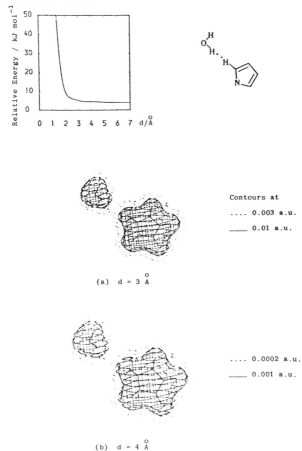

(a)   d = 3 Å

Contours at

.... 0.003 a.u.

____ 0.01 a.u.

.... 0.0002 a.u.

____ 0.001 a.u.

(b)   d = 4 Å

FIG. 6. The energy of the pyrrole-water system for a given direction
of approach with indications of electron densities at two separations.
(1 au of charge density = 67.49 e/Å³).

Nonetheless these chicken-wire plots of density can be used in comparing
the shapes of similar molecules.

Marginally more revealing than the overall electron density or the
density due to particular orbitals is the electrostatic molecular
potential. This is the energy of interaction between the molecule of
interest and an isolated proton. Again using graphics devices, these

calculations may be displayed in a manner which is comprehensible to an
experimentalist who has not been intimately involved in running the
calculation.

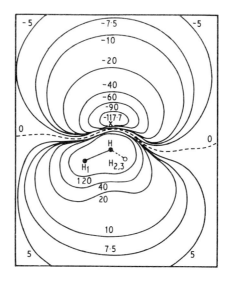

FIG. 7. A cross-section through the electrostatic potential of ammonia.

A third way of displaying similar electron density data is to
calculate the superdelocalizability of electrons at any point in the
molecular space. The reactivity of a molecule at any position is likely
to depend on just how much electron density there is at that point and
on how tightly that electron density is held. At least for nucleophilic
reactions the superdelocalizability at a point r, can be defined as

$$\sum_i n_i q_{ri}/\varepsilon_i$$

with q_{ri} being the electron density at r in orbital i and ε_i the appropriate orbital energy with occupancy n_i. Yet again such quantities can
be displayed using graphics devices.

A major criticism of the type of work is that despite the use of
graphics the pictures are still confusingly complex. To obviate this
difficulty the current trend is to display quantities such as density,
electrostatic potential or superdelocalizability on a molecular surface,
itself defined using standard spherical radii for all the atoms. With
raster graphics and colour it is becoming possible to present relatively

simple pictures for qualitative comparisons, but the optimum way of
displaying the data has not yet been achieved.

If we desire quantitative comparisons between the charge distributions of molecules, either total or orbital by orbital then the
device of integrating the charge over defined spherical regions of the
molecules may be adopted [5]. This may be done for regions surrounding
nuclei, regions in bonds, lone-pair areas or in fact in any place which
seems to be important for a quantitative comparison. As with the more
qualitative contour pictures, the sphere charges may be broken down
into orbital contributions or into superdelocalizabilities. These
quantities have been used in a number of successful correlations with
organic chemical reactivities [6-9].

5.   THE INFLUENCE OF THE RECEPTOR

Both crystallography and theoretical calculations on molecules
are generally performed in the absence of any influence of the macromolecular receptor. The effect of the protein binding site could be
overwhelming so that extrapolation of results from idealised conditions
to the biological environment is hazardous. Generally, insufficient
is known about the receptor for them to be incorporated into either the
X-ray structure determination or into computer-based calculation. The
single glaring exception is the case of enzymes whose crystal structure
is known, even in some instances with blockers bound into the active
site.

From the point of view of calculation it is not feasible to use all
the crystal data and merely to incorporate all the atoms even of the
binding site into a monster quantum mechanical calculation. Tentative
attempts are being made to use simple empirical potentials to investigate
small molecule binding to the receptor site. The difficulty is the
range of potentials required. An alternative, first used by Heyes and
Kollman [10] uses quantum mechanical formalism but achieves the reduction
in the computation required by replacing the atoms of the enzymic site
by point partial charges.

From the point of view of the quantum mechanical calculation we then
have a possibility of studying a small molecule in a binding site. The
site is represented by perhaps two or three hundred extra nuclei with
non-integral nuclear charges but no electrons. The actual values of
these partial point charges can be derived from calculations on model
systems but the values are not critical providing each amino-acid residue

has the correct overall charge. The small molecule will experience a
fairly good representation of the electrostatic nature of the binding
site; the small molecule may be polarized by the site but not the site
by the substrate or blocker.

Binding energies may be calculated as the difference in energy of
the smaller partner bound and free. The electrostatic contribution to
this binding energy will be revealed by the energy from the first cycle
of the calculation while at convergence we can also learn about the
polarization contribution, as indicated schematically in Figure 8.

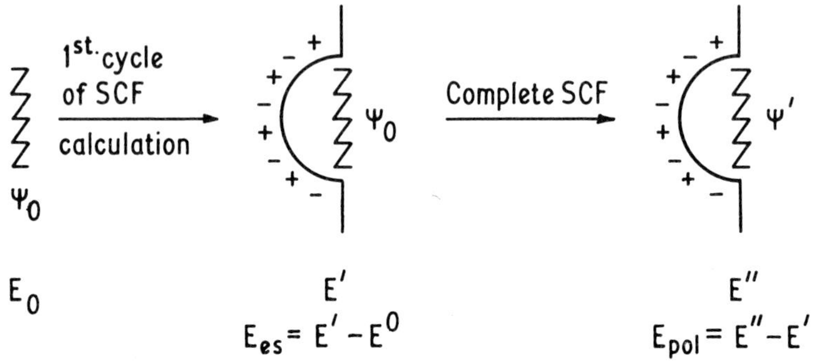

FIG. 8. Schematic description of the derivation of electrostatic and
polarization energy terms.

An illustrative application of this approach has been made by
Robins [11] who studied the relative binding energies of a series of
alkyl boronic acids to α chymotrypsin [12,13]. Each member of the series
was placed in the active site by eye using a computer graphics display
and the position adjusted using empirical molecular mechanics calculations. The final optimising was performed using the modified quantum
mechanical program with partial charges acting as dummy nuclei. Such
calculations take approximately 5% more computer time than that required
for an isolated molecule. The only imponderable remaining in the
calculation is the effect of solvation and desolvation. In the work on
alkyl boronic acids, the quantitative effect of desolvation of the
boronic acid was estimated as being 0.85 kcal per mole for each $-CH_2-$
group in the molecule [14]. Although crude such a contribution is probably a reasonable estimate of relative solvation effects. Figure 9

shows the correlation between observed and calculated binding energies
of the series $B(OH)_2(CH_2)_nH$ binding to chymotrypsin.

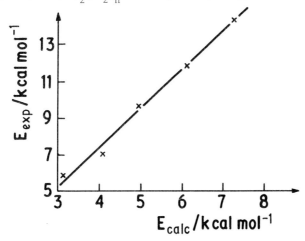

FIG. 9. Comparison of experimental and calculated binding energies of
n-alkyl boronic acids to chymotrypsin.

It may be seen that the relative binding energies do seem amenable to
investigation in this crude fashion.

    If the partial point charge idea is capable of revealing binding
energies which are reasonable, it ought similarly to be able to be
utilised in studying the influence of the receptor on the small molecule.

    Currently we are investigating the influence of protein sites known
from crystallography on molecular properties such as electron density,
electrostatic potential and superdelocalizability. Electrostatic potentials may need to mirror those of the site; superdelocalizabilities
should relate to dispersion forces. As in so many areas, progress
depends on the combination of X-ray crystal data coupled with theoretical
calculation.

ACKNOWLEDGEMENTS

    The research upon which this account is based has been supported
by the National Foundation for Cancer Research.

REFERENCES

[1] An example is the program HONDO. DUPUIS, M. (1981) QCPE 13, 403.
     Available from Quantum Chemistry Program Exchange, Department of
     Chemistry, Indiana University, Bloomington, Indiana 47401, U.S.A.

[2] KENNARD, O., et al. (1972) Molecular Structures and Dimensions
     International Union of Crystallography, Chester.

[3] SUTTON, L.E. (ed) Tables of Interatomic Distances and Configur-
     ations in Molecules and Ions. Special publication of the
     Chemical Society, London, Vol. 11 (1958); Supplement, Vol.18 (1965).

[4] RICHARDS, W.G., WALLIS, J. and GANNELLIN, C.R. (1979) Eur. J. med.
     Chem. 14, 9.

[5] WALLIS, J., (1981) QCPE 13, 411.

[6] ELLIOTT, R.J., SACKWILD, V. and RICHARDS, W.G. (1982) J. molec.
     Struct. THEOCHEM. 86, 301.

[7] ELLIOTT, R.J. and RICHARDS, W.G. (1982) J. molec.Struct.THEOCHEM.
     87, 211.

[8] ELLIOTT, R.J. and RICHARDS, W.G. (1982) J. molec. Struct.THEOCHEM.
     87, 247.

[9] ELLIOTT, R.J. and RICHARDS, W.G. (1982) J.chem.Soc.Perkin II, 943.

[10] HEYES, D.M. and KELLMAN, P.A. (1976) J.Am.chem.Soc., 98, 7811.

[11] ROBINS, S.A. (1982) D.Phil. Thesis, Oxford University.

[12] ANTONOV, V.K., IVANINA, T.V., BEREZIN, I.V. and MARTINEK, K. (1970)
     FEBS Letters, 7, 23.

[13] ANTONOV, V.K., IVANINA, T.V., BEREZIN, I.V. and MARTINEK, K. (1970)
     J. Molec. Biol., 4, 451.

[14] SMITH, R. and TANFORD, C. (1973) Proc.Nat.Acad.Sci., USA, 70, 289.

6 Receptors—a review of recent progress
D.R.H. Gourley

Receptors are those entities in or on animal cells with which biologically active molecules interact initially. It is reasonable to assume that receptors evolved to interact with and mediate the effects of molecules endogenous to the body (hormones, neurotransmitters, autacoids, and so on), but they also interact with many xenobiotics (naturally occurring in plants and synthetic compounds) which are foreign to the body. All of the endogenous molecules are agonists and cause activation when they interact with the receptor. Many of the xenobiotics are antagonists and have been extremely useful in receptor research because of their ability to block specific receptors.

Receptors perform two vital functions. The first function is to recognize and discriminate, by a process of high affinity specific binding, particular biologically active molecules. The second function is the transduction of a signal, presumably generated as a result of the specific binding interaction. This signal must be conveyed to some appropriate effector, such as an enzyme or an ion channel, and thereby alter its activity in such a way that there is initiated a cascade of biochemical reactions that leads ultimately to a characteristic physiologic or pharmacologic response.

Receptors have been divided into three major classes according to their location in the cell. Many receptors are located in the cell membrane. A partial list includes receptors for adenosine, adrenergic compounds, cholinergic compounds, dopamine, GABA, glutamate, glycine, histamine, insulin, opioid compounds, and serotonin. A second location is in the cytosol. Receptors for all classes of steroid hormones are cytosolic. Finally, there are receptors on the cell nucleus, e.g., those for the thyroid hormones. Progress in defining these and other receptors and in unravelling the mechanisms by which drug–receptor interactions alter biologic processes has been explosive in the past decade. It is obviously impossible to review all the details of this progress here. Instead, by reviewing the current status of selected

examples, an attempt will be made to illustrate the general directions
that successful receptor research has followed and the progress that
has been made. The examples are the adrenergic receptors which bind to
small endogenous and exogenous compounds, the insulin receptor which
binds to relatively large polypeptide endogenous compounds and the opioid
receptor which binds to endogenous peptides of various sizes and to
relatively smaller exogenous compounds.

As new information about receptors has been generated, it has become
necessary to designate subclasses of receptors. An extreme example is
the GABA receptor for which it has been suggested recently that as many
as 36 possible subclasses may exist. The terminology for these subclasses
is becoming cumbersome, if not confusing. In some cases, different
investigators are using different nomenclature for the same receptor
subclasses. It has been suggested that the time may be ripe for the
establishment of a Commission on Receptor Nomenclature analagous to the
one established several years ago by the International Union of
Biochemistry to systematize the naming of enzymes. An official naming
system for receptors would have a salutary effect on communication between
and among receptor pharmacologists.

The advent in recent years of the technique of direct radioligand
binding has altered the entire complexion of the experimental approach
to the study of receptors. The basis of a radioligand binding assay is
quite simple. The radioligand is incubated with a tissue, the unbound
radioactivity is removed by filtration or centrifugation, and the radioactivity remaining in the tissue is measured. Numerous control
experiments must be included to demonstrate that the tissue-bound radioactivity represents radioligand bound to receptors and not to other
sites that degrade, transport, or nonspecifically retain it. The amount
of nonspecific binding is determined by incubating the radioligand and
tissue in the presence of an appropriate nonradioactive ligand that will
bind to virtually all the receptors being studied. With all of the
receptors occupied by the unlabelled compound, the radioligand should
bind only to nonspecific sites. Specific binding to receptors is
represented by the difference in radioactivity observed in the absence
of the unlabelled ligand (total binding) and in its presence (nonspecific
binding).

Radioligand binding assays have been used primarily in kinetic
studies, binding isotherms, and competitive binding experiments. Kinetic
studies have yielded the rate constants characterizing the radioligand-

receptor interaction. Binding isotherms quantitate the concentration of receptors and the affinity of these receptors for the radioligand. Competitive binding experiments determine the ability of various compounds to compete with the radioligand in binding to receptors. The affinity of the receptors for the competitor is determined from the competitive binding curve.

The widespread use of radioligand binding assays has led to the detection of binding sites which may have no physiological significance. Binding sites may appropriately be termed receptor sites only when they are shown to exhibit the dual functions noted earlier, namely ligand recognition and cell activation.

One of the most exciting results of receptor research is the growing evidence that the receptor can be the focus of a disease process. In a number of instances, the pathophysiology appears to reside not in the receptor itself but rather in the presence of autoantibodies that can interact with the receptor either as a competitive antagonist or as an agonist. Diseases in which antireceptor antibodies play a role include: Graves' disease (receptor for thyroid stimulating hormone), myasthenia gravis (nicotinic acetylcholine receptor), the syndrome of a rare skin lesion called acanthosis nigricans with insulin resistance (insulin receptor), and a population of patients exhibiting beta-adrenergic hyporesponsiveness (beta-adrenergic receptor). In addition to disorders related to antireceptor antibodies, a number of diseases have been linked to defects in the receptor system itself. For example, there are at least thirteen diseases which appear to be related to alterations in human adrenergic receptors. As more details become available concerning the multiple steps leading from receptor occupation to cell activation, it is highly likely that receptor system components will be found to play a role in the pathophysiology of still more diseases.

Not only is receptor research helping to explain the molecular defects in certain diseases, but it may also lead to improved drug therapy for disease conditions. Although there are difficulties involved in defining the conformation of a dynamic macromolecule like a receptor, three-dimensional modelling of receptor sites for which the amino acid sequence is known provides the best systematic approach to enhancing the potency and specificity of current drugs, and designing new ones.

Before turning to a brief discussion of receptors for three distinct drug groups, it may be instructive to review the important milestones in the characterization of the nicotinic acetylcholine receptor for which research has progressed further than it has for any other receptor.

1. Physiologic and pharmacologic experiments established that acetylcholine triggers a rapid movement of small cations (sodium, potassium,
   and calcium) across the postsynaptic membrane. Agonists can substitute for acetylcholine; competitive antagonists block the ion
   movement.

2. It was postulated that the acetylcholine *regulator* is made up of at
   least two distinct structural elements: the acetylcholine *receptor*
   (binding) and the site for ion translocation, or *ionophore* (the
   active unit).

3. Snake venom alpha-toxins were found to be selective labels for the
   nicotinic acetylcholine receptor and were used to study, first
   qualitatively then quantitatively, the distribution of the acetylcholine receptors at the neuromuscular junction and other synapses.

4. A tissue which was exceptionally rich in cholinergic synapses, the
   electric organ of the *Torpedo* fish, was available as a source from
   which high yields of receptor protein were obtainable.

5. Acetylcholine receptor protein was solubilized with detergents and
   was purified to homogeneity in milligram quantities by different
   biochemical techniques including chromatography using cholinergic
   ligands covalently coupled to a solid matrix. It was found that the
   purified acetylcholine receptor is composed of four subunits, alpha,
   beta, gamma and delta, in a molar ratio of 2:1:1:1. The alpha subunits
   comprise part or all of the acetylcholine binding site, while the
   ionophore is formed by one or more of the four subunits. Purified
   acetylcholine receptor protein was morphologically characterized by
   electron microscopy.

6. An excitable membrane was reconstituted from chemically defined
   components in solution and was shown to be capable of regulating
   agonist-sensitive sodium flux.

7. By cloning genes for each of the four subunits of the acetylcholine
   receptor, it has been possible to determine the nucleotide sequence
   of the genes and hence the amino acid sequence of each subunit.
   Another tool for the study of the acetylcholine receptor has been
   the development of monoclonal antibodies against the receptor and
   each subunit. These developments have enormous potential for determining the structure of the receptor and understanding how acetylcholine binding to the receptor signals the cell to make a response.

8. Injection of acetylcholine receptor protein into animals caused an

immune response resembling myasthenia gravis. Thus was stimulated a
vital interaction between basic research into acetylcholine receptor
structure, function, and metabolism, and chemical research, which is
producing practical medical benefits.
The nicotinic acetylcholine receptor thus serves as a model for receptor
identification and isolation. In the selected examples that follow,
many of these steps have already been achieved, but much more remains to
be done.

1. ADRENERGIC RECEPTORS

1.1. Introduction

Noradrenaline and adrenaline are key endogenous regulators of many
physiologic events in animals and humans. Noradrenaline acts primarily
as a neurotransmitter released from sympathetic nerve terminals, and
adrenaline functions as a circulating hormone released from the adrenal
medulla. The actions of these two catecholamines are very similar at
some sites in the body but differ significantly at others. To explain
these differences, it was proposed in 1948 that there were two distinct
receptors for the catecholamines and they were given the names alpha and
beta. It was demonstrated that the two subclasses were differentially
sensitive to adrenergic agonists. The rank order of potency of adrenergic
agonists in target tissues was, at alpha-adrenergic receptors, adrenaline
> noradrenaline >> isoproterenol, and at beta-adrenergic receptors,
isoproterenol > adrenaline ≥ noradrenaline. Adrenergic receptors are
characterized by marked stereospecificity with the levorotatory
enantiomers of both agonists and antagonists being considerably more
potent than the dextrorotatory enantiomers. The dual receptor concept has
had a significant influence on adrenergic receptor research. It stimulated the development of more selective agonists and antagonists, led
to the development of the second messenger concept, opened new areas of
study of clinical disorders such as hypertension, and indicated the
existence of further subclasses of adrenergic receptors which has in
turn stimulated the development of clinically useful drugs.

1.2. Subclasses of adrenergic receptors

Division of the beta-adrenergic receptors into two subclasses
appeared to be appropriate on the basis of observations that the rank
order of potency of catecholamine agonists fell into two distinct categories depending upon the tissue response examined. The receptors in
heart and adipose tissue were classified as beta$_1$ whereas those in

bronchi and vascular smooth muscle were designated beta$_2$. Observations
of a similar nature supported the belief that alpha-adrenergic receptors
might also belong to two distinct subclasses. It became clear, however,
that precise analysis of receptor subclasses using classical pharmacologic
approaches in whole tissue had serious limitations. It was only with
the development of the radiolabelled ligand binding techniques noted
earlier that it became possible to characterize and quantify adrenergic
receptors. By use of these techniques, the dual beta-receptor concept
was confirmed and two classes of alpha-receptor were also delineated.
The four subclasses of adrenergic receptors, alpha$_1$, alpha$_2$, beta$_1$, and
beta$_2$, can be distinguished by the relative potencies of particular
agonists and antagonists in radioligand binding studies just as in
pharmacologic studies.

     In radioligand binding studies, alpha$_1$- and alpha$_2$-adrenergic
receptors can each be directly identified with subclass selective radiolabelled ligands such as prazosin (alpha$_1$) and yohimbine (alpha$_2$).
There are no selective ligands for beta$_1$- and beta$_2$-adrenergic receptors
as yet. All available radioligands bind equally to both beta-adrenergic
receptor subclasses. However, beta$_1$- and beta$_2$-adrenergic receptors can
be separately identified in competitive binding experiments using a non-
selective radioligand such as dihydroalprenolol and relatively selective
beta antagonists. Similar techniques have been applied also to quantitate
the alpha-adrenergic receptor subclasses. With appropriate analyses,
these techniques can indicate the relative proportion of each receptor
subclass present in a given tissue.

     Adrenergic receptors are widely distributed in the body. Beta-
adrenergic and alpha$_1$-adrenergic receptors are typically postsynaptic.
Alpha$_2$-adrenergic receptors are located presynaptically on adrenergic
terminals in the sympathetic nervous system and inhibit the release of
noradrenaline. In some instances, alpha$_2$-adrenergic receptors may also
be located postsynaptically.

1.3. Mediation of effects of a second messenger

     Both the beta$_1$- and beta$_2$-adrenergic receptors are coupled to the
membrane-bound enzyme adenylate cyclase and their occupancy by agonists
leads to stimulation of the enzyme with consequent elevation of the
intracellular concentration of a second messenger, cyclic AMP. Occupancy
of alpha$_2$-adrenergic receptors by agonists is also coupled to adenylate
cyclase but inhibits it, which in turn decreases cellular cyclic AMP
levels. By contrast, agonist activation of alpha$_1$-adrenergic receptors

increases levels of intracellular calcium and in some cases increases
phosphoinositol hydrolysis.

Radioligand binding studies have proved to be effective in determining the role of other elements in the activation of adenylate cyclase
by beta-adrenergic agonists. One approach has been to look for differences between the binding of agonists which initiates a cascade of
events leading to altered cell function, and the binding of antagonists,
which does not. This approach has indicated that an important modulator
of binding to adrenergic receptors is the guanine nucleotide, guanosine
triphosphate (GTP). Beta-adrenergic receptors are coupled to adenylate
cyclase by a GTP-binding regulatory protein. The interaction of the
receptors with this protein is thought to explain how GTP modulates
receptor affinity for agonists. Alpha$_1$-adrenergic receptors have not
been shown to be affected by guanine nucleotides but alpha$_2$-receptors
do appear to be regulated by nucleotides. How such seemingly similar
mechanisms can account for stimulation of adenylate cyclase in the case
of beta-receptors and inhibition of the other (alpha$_2$) remains an
unsolved problem.

1.4. Progress in receptor purification

A long-range goal in receptor research is to purify the receptors
and determine the details of their molecular structure. The task is
challenging because receptors are present in such minute quantities in
plasma membranes, and because they are membrane bound and need to be
solubilized before purification can begin. It would be a tremendous
advantage if a tissue could be found which is as rich in adrenergic
receptors as the electric organ of *Torpedo* fish is in nicotinic acetylcholine receptors. Some progress, however, has been made using as the
source of beta-adrenergic receptors avian or amphibian erythrocytes.
The receptors seem to be efficiently solubilized with digitonin. The
best results thus far have been obtained using affinity supports consisting of the beta-adrenergic antagonist alprenolol covalently linked
to a Sepharose inert support. By this technique, beta-receptors have
been purified 22 000- to 55 000-fold. Subjection of the purified
receptors to polyacrylamide gel electrophoresis has revealed that the
beta-receptor is composed of more than one protein subunit. The receptors
are distinct from the catalytic moiety of adenylate cyclase.

Purification of alpha-adrenergic receptors, has not yet proceeded
as far. Starting with liver membranes, solubilization was again achieved
by treatment with digitonin. The soluble extracts were passed over

alprenolol-Agarose affinity columns to selectively adsorbed the beta-
adrenergic receptors without adsorption of alpha$_1$-adrenergic receptors.
This demonstrated that alpha$_1$- and beta-adrenergic receptor binding
sites do not reside on the same macromolecule. In other experiments, the
alpha$_1$-receptors in liver membranes were prelabelled with the irreversible
ligand [^{3}H]-phenoxybenzamine and then solubilized. The alpha$_1$-receptor
had an apparent molecular weight of 96 000. The alpha$_2$-adrenergic
receptor has apparently not yet been solubilized.

Some progress has been achieved in attempts to reconstitute an
adrenergic receptor system from components which are as yet unpurified.
It has been shown that two soluble membrane fractions, one containing
the beta-adrenergic receptor and the other a GTP-binding protein, can
be recoupled in a phospholipid environment. When the resulting receptor-
GTP binding complex was implanted into plasma membranes containing
hormone-insensitive adenylate cyclase, it was found to respond to
isoproterenol stimulation. If this reconstitution experiment can be
repeated with purified components, it will mean that hormonal stimulation
of the cyclase catalytic unit requires both the receptor for binding and
the GTP-binding protein for transmission.

1.5. Receptor adaptations

One of the major insights about adrenergic receptors to come from
ligand binding studies is that they are not static entities in the
plasma membrane but are subject to very dynamic regulation by a variety
of hormonal and other influences. Not only the catecholamines themselves but a variety of hormones such as thyroid hormones, progesterone,
and cortisone can strikingly alter the number and the properties of the
adrenergic receptors, thereby influencing cellular responsiveness to
the catecholamines.

Continued exposure to a drug or hormonal agonist often leads to a
blunted response to that agonist. This phenomenon has been termed
desensitization, refractoriness, tolerance, or tachyphylaxis. Desensitization to catecholamines in animal tissues and cultured cells is
often accompanied by a decrease in the affinity of the receptors for
the hormone (uncoupling), followed by a decrease in the number of
receptors (down-regulation). Furthermore, in many studies in animals,
depletion of catecholamines or treatment with adrenergic antagonists
leads to a supersensitivity of the tissues to catecholamines and up-
regulation of the adrenergic receptor number. Little is known about
the cellular mechanisms controlling adrenergic receptors which presumably

mediate these changes.

The changes in receptors with abnormal catecholamine levels make it difficult to conclude whether in disease states such receptor changes are primary or secondary pathogenic events. Perhaps too much attention has been focussed on alterations in receptor number. In future clinical studies, the affinities of receptors for agonists, the effects and concentrations of modulators, and the potential for receptors to be dynamically regulated should also be examined. Since receptors can change during drug treatments, radioligand binding can be used as a method to monitor these changes and perhaps allow individualization of therapy. Already, radioligand binding is providing interesting suggestions of hitherto unsuspected mechanisms of action of certain drugs.

2. INSULIN RECEPTOR

2.1. Introduction

The primary sites of action of insulin are liver, muscle and fat. In these tissues, insulin influences the metabolism of carbohydrate, protein and fat, i.e., it promotes net synthesis of glycogen, protein and triglyceride, and activates cell membrane transport systems for nutrients and ions. Much is known about the activation of specific metabolic enzymes in the presence of insulin but the mechanism by which insulin triggers such molecular changes is still unknown. Studies of insulin actions in intact cells in the 1950s clearly indicated that a necessary first step was the binding of insulin to the cell surface. It seemed reasonable therefore to focus attention on the component of the cell surface membrane with which insulin binds, and in the last decade much progress has been achieved.

2.2. Characterization of the insulin receptor

The binding characteristics for the insulin-receptor interaction have been exhaustively investigated. Less than 0.01 per cent of the protein in plasma membrane preparations is associated with the insulin receptor and the number of high affinity receptors is low in all cell types. Even so, there appears to be more receptors than are required to elicit maximal physiologic responses in cells. For example, binding of ^{125}I-insulin to about 5 per cent or less of the receptors on adipocytes causes maximal activation of glucose transport.

The first successful isolation of receptors was accomplished by means of affinity-purification techniques involving insulin immobilized on Agarose or Sepharose supports. The insulin receptors were eluted from the affinity columns after the passage of nonreceptor membrane proteins. This approach, even when refined, greatly diminished the binding activity of the receptor material. New techniques of affinity-labelling in combination with affinity purification have led to a deduction of the receptor's subunit composition and stoichiometry. These techniques include the use of photo-activatable analogues of ^{125}I-insulin which affinity-label the insulin receptor after flash photolysis, and the use of a cross-linking reagent, disuccinimidyl suberate, which covalently links the bound ligand to the receptor protein. A major advance has resulted from the availability of antireceptor antibodies from patients with insulin resistance accompanying acanthosis nigricans for the specific immunoprecipitation of insulin receptors from detergent solutions. These antibodies compete with insulin for the insulin receptor.

The insulin receptor isolated by means of the above techniques is composed of at least four subunits that are linked by disulfide bridges into a large receptor complex of apparent molecular weight 350 000. Two of the subunits, referred to as alpha subunits, have an apparent molecular weight of about 135 000 and are thought to be linked by one or more disulfide bridges. The other two subunits, designated beta, have an apparent molecular weight of about 90 000 and are disulfide linked to the alpha subunits. Because the sensitivity of the disulfide bridges linking the alpha and beta subunits differs from that of the disulfide bridges linking two alpha-beta fragments, it was possible to deduce that the insulin receptor is composed of two alpha subunits and two beta subunits arranged in the sequence

(beta-S-S-alpha)-S-S-(alpha-S-S-beta).

This arrangement appears to fit observations in all tissues studied from the rat and human. The exact location of specific insulin binding sites on this structure is not known. The reason that the receptor complex exhibits a lower apparent molecular weight than would be expected from its subunit composition is thought to be partly because intrachain disulfide bridges maintain a compact structure in detergent solution.

It is now established that several hormones and growth factors share sequence homologies with insulin, and have significant insulin-like biological activity. Such peptides include insulin-like growth factors I and II, nerve growth factors, and relaxin. When the ability of these substances to bind to the insulin receptor was examined, it was found that there is some degree of mutual competition for receptor binding. Thus it appears that naturally occurring substances in addition to insulin have the ability to bind to insulin receptors (as well as to their own receptors) and thereby trigger the physiologic responses usually associated with insulin.

It is still not known how the interaction of insulin with its receptor elicits changes in cellular functions. Since insulin can regulate the phosphorylation of several proteins, one possibility now being investigated in several laboratories is that insulin exerts its effects through stimulation of a protein kinase or inhibition of a protein phosphatase. Recent evidence indicates that insulin stimulates phosphorylation of at least the beta, and perhaps also the alpha subunit of its own receptor. The receptor may in fact be a protein kinase. It is too early to decide the functional implications of the phosphorylation of the insulin receptor, but it is possible that receptor phosphorylation is an early step in insulin action.

2.3. Internalization of the receptor-insulin complex

If binding of insulin to its cell surface receptor produces a signal which initiates the various biological effects associated with insulin action, as is supposed, the signal must in some way be terminated. It has been shown that the receptor-insulin complex is freely mobile in the membrane. By this means, it may reach a specific location in the membrane where it is taken into the interior of the cell by a process of receptor-mediated endocytosis. By quantitative electron microscopic autoradiography, it has been shown that insulin then associates with different forms of lysosomes for degradation, presumably by lysosomal proteases. This mechanism provides a rapid means of terminating the hormone signal by removing it from the surface.

2.4. Intracellular mediation of insulin action

Until recently, the mechanism by which insulin's intracellular effects are triggered as a result of binding of the hormone to its receptor has eluded investigation. Now it has been found in several laboratories that when insulin binds to its receptor it generates a chemical mediator that regulates the state of phosphorylation of various key metabolic enzymes. The mediator was first obtained by extraction of muscle and adipocytes after exposure to insulin, but it has also been found in other cell types. It can be assayed by its ability to stimulate the activity of mitochondrial pyruvate dehydrogenase and the conversion of glucose 6-phosphate dependent glycogen synthase to the glucose 6-phosphate independent form of the enzyme. A number of other enzyme systems that also modulate insulin-sensitive intracellular processes, including both glucose and lipid metabolism and calcium transport, have also been found to respond appropriately to the mediator.

The insulin-sensitive chemical mediator has been partially characterized by conventional chromatographic methods including gel filtration, ion exchange, and hydroxylapatite chromatography and has been partially purified by high-pressure liquid chromatography. It is acid-stable, peptide in nature, and has a molecular weight of about 1000–1500. Evidence suggests that the mediator is derived from the plasma membrane by proteolytic cleavage. Insulin itself does not appear to be the substrate for the proteolytic reaction that generates the mediator since two other ligands, concanavalin A and antibody to the insulin receptor, appear to be able to bind to the insulin receptor and mimic the effects of insulin. However, the mechanism by which the insulin-receptor interaction is coupled to the generation of the mediator is not known.

The concept developed from these studies is that the mediator may act by altering various protein kinases and phosphoprotein phosphatases that modulate the state of phosphorylation and activity of a variety of insulin-sensitive enzymes. This will remain a hypothesis until chemical characterization of the mediator is complete and its structure is synthesized and tested.

2.5. <u>Some clinical implications</u>

Insulin deficiency in experimental animals is associated with an elevation in the number of insulin receptors. Conversely, cautious administration of increasing doses of insulin to experimental animals provides insulin excess which is associated with a diminution in receptor number and a reduction in sensitivity to administered insulin.

The most common disorders of insulin and glucose metabolism in humans are those that involve moderate insulin resistance, i.e., patients require or produce two or three times the normal amount of insulin. In humans, as in experimental animals, when the amount of circulating insulin is high, the number of insulin receptors on cell surfaces generally is low, and vice versa. Thus overweight patients with normal glucose tolerance often have hyperinsulinemia and a reduction in sensitivity to insulin associated with a decrease in the number of insulin receptors throughout the body. Likewise, many patients with Type II diabetes have decreased sensitivity to insulin, elevated levels of insulin, and a reduction in receptor concentrations. Therapy that is effective in relieving insulin resistance and ameliorating the hyperinsulinemia also produces improvement at the level of the receptor.

In some patients with extreme insulin resistance associated with acanthosis nigricans, insulin binding to its receptors is markedly reduced and circulating polyclonal antibodies specific for insulin receptors are detectable. The antibodies bind to the insulin receptor, reduce the affinity of the receptor for the hormone, and thereby impair binding of insulin. As in the patients with moderate insulin resistance, there is excellent correlation between the receptor and the clinical state for patients as a group, for individual patients, and during changes in the clinical course including the response to treatment. In some patients with other syndromes characterized by severe insulin resistance and hyperinsulinemia without autoimmunity, insulin binding to receptors is normal which suggests that the defect is beyond the receptor or in the receptor but beyond the binding step. As the molecular events subsequent to insulin-receptor binding become better understood, the molecular basis of still other disorders of insulin and glucose metabolism will be clarified.

3. OPIOID RECEPTORS

3.1. Introduction

In the cases of hormones and neurotransmitters, the endogenous ligands were discovered long before their receptors were postulated, searched for, and identified. For the opioid narcotic drugs, the procedure was reversed: an opioid receptor was identified first and because there was no logical reason for animals, including humans, to have developed receptors to interact specifically with molecules produced by a plant, endogenous ligands for opioid receptors were postulated, searched for, and identified.

Early indirect evidence for the existence of an opioid receptor was compelling. Opioid narcotics are optically active. For most opioid narcotics, the levorotatory enantiomer is pharmacologically active. The dextrorotatory enantiomer has little or no activity, indicating conformational specificity of a brain molecule with which the drug interacts. Moreover, some parts of the opioid molecule could be drastically altered with little or no change in pharmacologic activity, whereas modifications of other parts dramatically changed the activity. For example, substitution of an allyl group for the methyl group attached to the nitrogen atom of the morphine molecule changes the pharmacologic activity from agonistic to antagonistic. Such specificities were most easily explained

other neuroactive agents, encouraged a number of investigators to search
for endogenous opioids in animal brains and associated glands. In a
relatively short period of time, these studies culminated in the identification of two pentapeptides which differed by only one amino acid
residue. The pentapeptides were named methionine (met) enkephalin and
leucine(leu) enkephalin. Longer polypeptides with opioid activity also
have been isolated from pituitary glands. One of these corresponds to
the C-terminal fragment of beta-lipoprotein (LPH 61-91) and has been
named beta-endorphin. Later, an opioid peptide of seventeen amino acids
was isolated from pig neurohypophysis and gut extracts, dynorphin 1-17.
Subunits of dynorphin 1-17, dynorphin 1-13, 1-9, and 1-8, also have
opioid activity.

All of the endogenous opioid peptides have some opioid-like activity
when injected intravenously. This activity includes analgesia, respir-
atory depression, and a variety of behavioral changes. The effects of
the enkephalins are brief, presumably because they are rapidly destroyed
by peptidases. The longer chain endorphins are more stable and produce
effects of longer duration. The same is true of the dynorphin series:
dynorphin 1-17 and dynorphin 1-13 are relatively resistant to the action
of peptidase and have a longer duration of action, whereas dynorphin 1-8
and dynorphin 1-9 are readily degraded by peptidases and their duration
of action is much shorter. It seems possible, if not likely, that
membrane bound peptidases or even enkephalinases exist in the vicinity of
the opioid receptors specifically to inactivate the enkephalins and
other short-chain peptides and thereby terminate the signal presumably
activated by the peptide-receptor interaction. However, the possibility
that activity is terminated by internalization of the peptide-receptor
complex has not been ruled out.

There have been many attempts to link opioid peptides with specific
receptor subclasses. For example, it has been suggested that the
enkephalins and beta-endorphin are the main endogenous ligands for the
mu- and delta-receptors while the dynorphins are selective ligands for
the kappa-site. The opioid alkaloids bind with high affinity to the mu-
subclass. Others believe that it is unlikely that one receptor subclass
interacts specifically with one ligand. Which ligand combines with which
receptor subclass may depend upon proximity. The enkephalins are found
in smaller cells generally with shorter processes and their distribution
corresponds fairly well with that of opioid receptors. Beta-endorphin
is located mainly in the hypothalamus and appears to be closely

associated with endocrine function. Beta-endorphin and pituitary adrenocorticotropin are apparently synthesized together in the common precursor, pro-opiocortin, and are released together in response to stress. Although it contains the amino acid sequence for met-enkephalin, beta-endorphin is not the source of the brain enkephalins.

The enkephalins, unlike morphine and beta-endorphin, are not potent analgesics. The relative ease with which enkephalin analogues can be synthesized has led to a formidable number of them and to extensive studies of structure-activity relationships. These studies have produced analogues with analgesic activity and metabolic stability much greater than those of the endogenous pentapeptides. The achievements from this line of investigation have practical significance in opening up new possibilities in drug therapy. Another application, interesting from a theoretical viewpoint, is the recent use in both bioassays and binding assays of an enkephalin analogue in cyclic form and in its corresponding open-chain form to demonstrate that the various opioid receptor sub-classes have different conformational requirements. This suggests that it may be possible to design receptor-specific peptide ligands.

3.5. Some clinical implications

Since all brain regions implicated in the conduction of pain impulses have high levels of opioid receptor, and the injection of all known endogenous opioid peptides produces analgesia, it was natural to postulate that the receptors and their endogenous ligands are involved in pain modulation. Many observations using opioid antagonists in animals and humans support this concept but direct evidence still is lacking.

Upon repeated administration, all opioid peptides will produce tolerance and physical dependence similar to that produced by opioid drugs. The phenomena of tolerance and dependence have never been explained satisfactorily. The identification of opioid receptors made it possible to test the hypothesis that tolerance and dependence follow-ing chronic opioid administration was the result of changes in receptor number. Numerous investigations of this question have now been done and it is apparent that no major change in the number of opioid receptors occurs during either chronic opioid administration or withdrawal. More-over, the affinity of the receptors for either agonists or antagonists does not significantly change during the development of tolerance. However, there is evidence that the responsiveness of individual neurons to opioids changes during tolerance. It may well be that in experiments

performed so far changes in number or affinity of binding sites were
masked by studying homogenates of whole brain rather than of discrete
brain regions.

As noted earlier, beta-endorphin is released under conditions of
extreme stress or injury. This observation has led to intriguing studies
of shock. The opioid antagonist naloxone has been found to decrease or
reverse many of the symptoms of endotoxin shock, hypovolemic shock,
spinal shock and spinal injury in experimental animals. Opioid agonists
generally exacerbate the symptoms accompanying shock indicating involve-
ment of opioid receptors at some level. Potentially those observations
will lead to a new understanding and treatment of shock in humans.

*   *   *

This brief review of recent developments in research on the membrane-
bound receptors for the adrenergic compounds, insulin and the opioid
compounds illustrates how intensive research on different receptors has
proceeded more or less in parallel in many laboratories. Although
understanding of each of these receptors has reached different stages
and in no case has it reached the stage achieved with the nicotinic
acetylcholine receptor, progress has been linked generally to the
development of new biochemical techniques. When a new technique has been
successful with one receptor, it has been adapted rapidly for research
with other receptors. With the large number of investigators that this
field has attracted, it is anticipated that the next decade will see
the isolation and purification of many other receptors as well as
elucidation of their mechanisms of action.

SUGGESTED READINGS

Lamble, J. W., ed. (1981). *Towards Understanding Receptors*, Elsevier,
     Amsterdam.
Lamble, J. W., ed. (1982). *More About Receptors*. Elsevier, Amsterdam.

7 Haemoglobin as a model receptor
P.J. Goodford

It is about a hundred years since Ehrlich discovered that certain dye-stuffs could stain bacteria selectively, and his observations led to the suggestion that parts of the dye-stuff molecules might attach themselves to bacterial side-chains. At much the same time Langley (1878) proposed that biologically active chemicals such as atropine might combine with some specific substance in living cells, and by 1905 Langley wrote that there is a "receptive substance" in muscle which "is the recipient of the stimuli which it transfers to the contractile material". Thus the concept of receptors was introduced, but it remained little more than a concept for over fifty years.

The end of the 19th Century was also the time when the Law of Mass Action was being developed, first for relatively straightforward chemical reactions and then for biological systems. By the turn of the 20th Century it was being applied to the reaction of oxygen with haemoglobin which may be most simply represented by the chemical equation:-

$$Hb + O_2 = HbO_2 \qquad (1).$$

On Mass Action principles this equation would lead to a hyperbolic relationship between oxygen pressure and the proportion of haemoglobin in the oxy-form:-

$$\frac{[HbO_2]}{[HbO_2] + [Hb]} = \frac{[O_2]}{K + [O_2]} \qquad (2)$$

but it was soon discovered that the experimental dissociation curve of blood was in fact a sigmoid. A fascinating series of papers by many of the most famous scientists of the day (Ostwald, 1908a,b; Bohr, 1909; Hill, 1910; Douglas, Haldane and Haldane, 1912) were then published as attempts to explain the sigmoidal relationship. However this work was cut short by the Great War, and it was not until 1925 that Adair brought the classical studies of haemoglobin to a conclusion with his famous series of reactions:-

$$
\begin{aligned}
Hb_4 + O_2 &= Hb_4O_2 \\
Hb_4O_2 + O_2 &= Hb_4(O_2)_2 \\
Hb_4(O_2)_2 + O_2 &= Hb_4(O_2)_3 \\
Hb_4(O_2)_3 + O_2 &= Hb_4(O_2)_4
\end{aligned}
\qquad (3)
$$

which can give a mass action equation that can account for all observed oxygen dissociation curves.

Meanwhile a predilection for the products of fermentation had led to the detailed study of yeasts, and the discovery of enzymes as biologically active substances which could be separated from living cells and still retain their biological activity. Emil Fischer (1894) proposed his well-known analogy that small-molecule substrates may interact with enzymes as a key fits a lock, and the isolation and purification of enzymes started in earnest after the first world war. As each enzyme became available in a reasonably pure state its properties were studied, and it soon became clear that every enzyme had highly individual properties although they had much in common as a class of substances. Their functional organization into pathways was slowly appreciated, and the infinitely subtle interplay between enzymes, substrates, cofactors, products, inhibitors and activators in enzyme pathways was slowly unravelled.

Thus by the middle of the 20th Century it was generally recognized by workers in different branches of science that small-molecule:large-molecule interactions were widespread in biological systems. The Chemistry of dyeing; the Pharmacology of drugs; the Physiology of haemoglobin and the Biochemistry of enzymes all pointed in the same direction, and yet there was a tendency amongst scientists to emphasize the distinctive differences between their disciplines, and to express their observations in different ways. Pharmacologists plotted log-dose:response curves (Schild, 1947); Physiologists studied oxygen dissociation curves; Biochemists used Scatchard (1949) plots, and these different modes of expression corresponded to different types of observation and to different ways of thinking about the results. Relatively little emphasis was placed on common themes which might draw the various interpretations closer together.

RECEPTORS

In 1937 the pharmacologist A. J. Clark made a notable attempt to improve his own pharmacological ideas about drug action by studying a parallel scientific discipline. Clark had previously (1926a,b) introduced the Occupancy Theory according to which the pharmacological response of a tissue to a stimulating compound was not only initiated when the compound bound to tissue receptors, but the size of the response was also related to the proportion of receptors at which binding occurred. Clark now (1937) suggested that the binding of oxygen to haemoglobin might be a comparable binding system, with the advantages that haemoglobin could be purified in quantity and its properties studied much more accurately than a conventional pharmacological response. However, Clark's analogy was not followed up, and pharmacologists tended to regard receptors as convenient concepts around which they could organize and classify their observations. If they thought of receptors as real structures at all, it was as the first of a series of black boxes on the unknown pathway from stimulus to response. The understanding of "stimulus-response coupling" was an accepted research challenge, but the prospect of studying the black boxes one by one was too daunting at that time.

However, biochemistry was making systematic progress. It was shown that enzymes were proteins of defined molecular weight, amino-acid composition and three-dimensional structure. Their biological function was logically related to this structure by applying and extending traditional concepts from organic chemistry. The occurrences of α-helices, β-sheets and other features were recognized as recurrent structural motifs, and the arrangement of subunits into larger assemblies was discovered. Enzymes were classified into families and were related to structural proteins (e.g., collagen) and binding proteins (e.g., haemoglobin), and it was inevitable that comparisons should be drawn between this substantial body of biochemical information and the bare concept of the pharmacological receptor. In 1970 Cuatrecasas started to apply the novel method of affinity chromatography to the isolation of receptors, and a new era of pharmacological research began.

It is still too early to draw firm and final conclusions about the details of stimulus-response coupling in pharmacology, but some general principles are gaining widespread acceptance. The initial compound which stimulates the cell is called a "first messenger", and is thought to interact with relatively superficial receptors near the outside of the cell membrane. There are different first messengers such as acetylcholine, adrenocorticotrophic hormone, dopamine, glucagon, histamine, noradrenaline, opiate peptides, prost-

aglandins, serotonin and vasopressin, and this wide range of different compounds allows different stimuli to be transmitted in vivo with a diminished risk
of cross-talk because each first messenger seems to have its own specific
receptor on the target cell. Once that receptor has been triggered, however,
the need for specificity is substantially reduced because the message has
already reached its target. The next stage of transmission is apparently for the
triggered receptor to react with a "regulatory protein" in the cell membrane,
and this protein does not seem to discriminate significantly between receptors.
Whether the cell is stimulated by histamine at an H_2-histamine receptor,
prostaglandin at a prostaglandin receptor, serotonin at a serotinin receptor or
vasopressin at a vasopressin receptor the next event appears to be an
interaction between the receptor and the regulatory protein which results in
raised adenylate cyclase activity and an elevated intracellular concentration of
cyclic adenosine-3,5-monophosphate. This is the "second messenger", and as its
cytoplasmic concentration increases it becomes available to react with the
intracellular constituents which more directly cause the final response.

        The above sequence of events is not unique, but in broad outline a
wide range of different pharmacological first messengers trigger their receptors to interact with a much more limited number of membrane constituents,
and these finally alter the concentration of an intracellular second messenger.

        To give another specific example, when histamine interacts with
an H_1-histamine receptor there are apparently changes in the phospholipid
metabolism of the cell membrane, and the intracellular concentration of a
different second messenger (ionised calcium, Ca^{++}) increases from about
10^{-7} M to 10^{-6} M. This change of intracellular Ca^{++} concentration promotes a
shift of the intracellular calcium distibution and thereby initiates the cellular
response. The overall sequence of events may be represented as:-

$$
\begin{array}{ccccc}
\text{First} & \longrightarrow & \text{Membrane-Receptor} & \longrightarrow & \text{Second} \\
\text{Messenger in} & & \text{Complex} & & \text{Messenger out}
\end{array}
\qquad (4)
$$

and the detailed structure and function of the membrane-receptor complex is
now receiving serious study.

HAEMOGLOBIN

        At about the time when Adair was publishing his reactions (3) for
the oxygenation of haemoglobin, another worker (Greenwald, 1925) first iso-

lated 2,3-diphosphoglycerate (DPG) from red blood cells. This salt is a product of glucose metabolism and is present in millimolar concentrations, but it was not until 1963 that Sugita and Chanutin showed by electrophoresis that it forms a reversible complex with haemoglobin, and another four years passed before Chanutin and Curnish (1967) found that DPG distorts the shape of the haemoglobin dissociation curve.   This was more than forty years after Greenwald's isolation of DPG, and more than half a century since Douglas et al (1912) first predicted that "the actual dissociation curve given by the oxy-haemoglobin of blood must presumably be a rectangular hyperbola distorted in some manner which is dependent on the presence of salts".   It is almost unbelievable that such a period could elapse in a mainstream area of physio-logical and biochemical research before this prediction was explored and vindicated.

The effect of DPG on haemoglobin is to reduce its affinity for oxygen, and thereby promote the liberation of oxygen from oxyhaemoglobin. In 1972 Arnone showed exactly where DPG reacted with the haemoglobin molecule, and in 1973 a working model of haemoglobin was built by Beddell, Goodford, Norrington and Wilkinson as an analogue for drug-receptor systems. These workers regarded DPG as a "first messenger" which reacted with a receptor (the globin) causing the final release of the "second messenger" (oxygen) from its binding sites on the iron atoms of the haem groups (Fig 1). The process can be represented as:-

$$\text{First Messenger in} \longrightarrow \text{Oxyhaemoglobin} \longrightarrow \text{Second Messenger out} \tag{5}$$

which may be compared directly to the pharmacological scheme (4) above (Goodford, St-Louis and Wootton, 1980).

An important property of pharmacological receptor systems is amplification. One molecule of first messenger (histamine) does not necessarily make a single molecule of second messenger (calcium) available but many second messenger molecules may be evoked.   There are apparently different amplification mechanisms which include the activation of enzyme systems, the opening of pores in the cell membrane so that second messenger ions can enter the cell from outside, and the release of bound second messenger. For example there is evidence for a calcium-binding subunit on the acetylcholine receptor

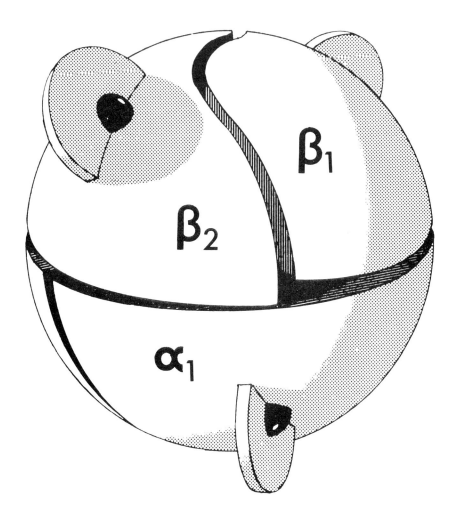

FIGURE 1   A diagrammatic representation of the structure of
oxyhaemoglobin. Two α and two β globin subunits fit together to make a
tetramer with a planar haem group embedded in each subunit. An iron atom
lies at the centre of each haem where oxygen is held. When the first messenger
DPG binds between the β subunits this structure is distorted and the second
messenger oxygen is released. See text.

itself (Rubsamen, Montgomery, Hess, Eldefrawdi and Eldefrawdi, 1976). Moreover these different mechanisms need not be mutually exclusive and could even
be synergistic. Thus enzyme activation could release calcium which might open
membrane pores and thereby allow more calcium to enter the cell.

Clearly any receptor model should make provision for amplification, and the haemoglobin model is acceptable in this respect. One DPG
molecule could release up to four oxygens from a haemoglobin tetramer, and
although this amplification factor may be modest it allows some principles of
receptor amplification to be examined. Of course, Beddell et al (1973) did not
suggest that the components of their haemoglobin model had structural features
identical to pharmacological membrane-receptor systems. Nor was their model
the same as Clark's (1937) analogy, although they shared his hopes that the
ready availability of pure haemoglobin and the vast amounts of published
information on the haemoglobin system would facilitate the use of this as a
model receptor. Furthermore the central physiological role of haemoglobin
raised the interesting possibility that its function might be amenable to
therapeutic manipulation. Oxygen transport from air to the tissues of the body
is deficient in a number of disease states including the anaemias and cardiovascular diseases which cause great human suffering, and some of these
conditions might at least in principle be amenable to haemoglobin therapy. The
study of haemoglobin as a model receptor system therefore seemed to offer
worthwhile opportunities, although it was clear that the limitations of the
model would have to be kept constantly in mind.

EXPERIMENTAL INTERPRETATION

When haemoglobin is first isolated from human erythrocytes there
are substantial quantities of bound phosphates which are not easily removed
from the protein. It appears that Barcroft and Roberts succeeded in 1908, but
the number of successful purifications between then and the studies of Chanutin
in the 1960's is open to doubt. In our own work prolonged sterile counterflow
dialysis was used (Paterson, Eagles, Young and Beddell, 1976) to remove the
natural ligand DPG, and subsequent experiments then consisted in the simultaneous equilibrium measurement of two variables under appropriate experimental
conditions (Goodford, Norrington, Paterson and Wootton, 1977). As shown in
figure 2a the proportion of oxyhaemoglobin (ordinate) was measured at different partial pressures of oxygen (abscissa) and a smooth "oxygen dissociation
curve" was then plotted through the experimental points.

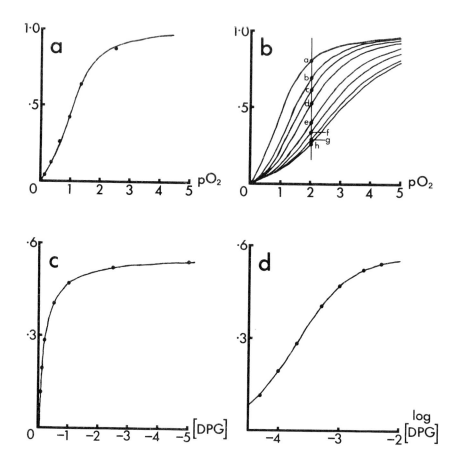

FIGURE 2   Shows how a pharmacological log dose:response curve (d) can be derived from a set of physiological oxygen dissociation curves (b). In this model system the first messenger or pharmacological agonist is the physiological effector 2,3-diphosphoglycerate, which displaces the oxygen dissociation curve of haemoglobin to the right. The pharmacological response is the liberation of oxygen from oxyhaemoglobin at a constant partial pressure of 2 kPa. See text.

An oxygen dissociation curve (Fig 2a) is the standard physiologists approach to the study of haemoglobin function, but it is not entirely appropriate for a pharmacological model. The next step was therefore to repeat the observations in the presence of different concentrations of the "first messenger" DPG. The individual experimental points are omitted in figure 2b which shows a collection of eight dissociation curves in the presence of 0, 0.05, 0.1, 0.2, 0.5, 1, 2.5 and 5 mM DPG. The top curve is the control and as stated above the compound works by diminishing the affinity of haemoglobin for the second messenger oxygen. This can be seen most simply by studying what happens (Fig 2b) at a fixed oxygen pressure represented by the vertical line. In the absence of DPG some 80% of the haemoglobin is oxygenated at point **a**, but the percentage falls successively until at the highest DPG concentrations less than 30% oxygenation is maintained (points **g, h**). The change is smooth and progressive as shown in figure 2c which illustrates the amount of oxygen displaced from the protein as a function of DPG concentration.

On this model the liberation of oxygen when the first messenger DPG reacts on haemoglobin is analogous to the liberation of Ca^{++} when the first messenger histamine reacts at H_1-receptors. The calcium becomes available for initiating a response and the oxygen becomes available for metabolism. The first message has been transformed into the second message, but it is only for the haemoglobin analogue system that the detailed mechanism of this transformation process has been extensively explored at the atomic and molecular level. The relevance of this analogue to pharmacology can now be studied as Clark first suggested in 1937.

However it is first necessary to transform the observations into traditional pharmacological format. This is shown in figure 2d where the release of second messenger oxygen is plotted against the logarithm of the first messenger concentration. As expected the response is sigmoidal tending to zero at low DPG concentrations and to a maximum release of second messenger at high DPG levels, but it is most important to distinguish between this sigmoid in figure 2d and the initial sigmoid curve of figure 2a. The initial sigmoid is the physiologists representation with linear axes. The final curve is only sigmoidal because of the pharmacological tradition calling for a logarithmic abscissa.

MECHANISM

Adair's four chemical reactions (3) can be used as a basic to explain the mechanism of DPG action by assuming that the compound changes the numerical values of the equilibrium constants of some or all of his equations. More elegant is the proposal of Monod, Wyman and Changeux (1965) that haemoglobin can exist in two different states. On their interpretation one state has a high affinity for oxygen and the other low, and in the complete absence of oxygen both are of course fully deoxygenated. Then if oxygen is slowly admitted it tends to combine preferentially with the high affinity state (R), while the low affinity state (T) does not react completely with oxygen until higher pressures are reached. There is moreover a dynamic equilibrium between the two states so that at any particular oxygen pressure both are present, and both are partly oxygenated according to the Law of Mass Action.

This model is attractive because it can account for the general shape of an oxygen dissociation curve on the basis of only two disposable parameters:-

(i) The equilibrium constant L between the R and T state in the complete absence of oxygen or any other compound.

(ii) The ratio **c** of the oxygen affinity of the T-state to the affinity of the R.

Adair's equations had four disposable constants and it is possible to refine the two-state model by the addition of more parameters (Goodford, St-Louis and Wootton, 1978), but the two constants L and **c** are adequate for many purposes and it is quite feasible to determine these by experiment directly.

Such experiments show that the effects of DPG could be mainly due to a change of L, and a physical explanation of this would be that DPG combines selectively with the T but not with the R state. This interpretation is moreover fully compatible with the X-ray structural observations which show an appropriate difference of conformation between the two states at that part of the haemoglobin molecule where DPG binds. Hence this simple Mass Action equilibrium description is physically plausible, and so is its extension if more parameters are brought into play. The simple model outlined above assumes that the reduced oxygen affinity of the T-state is the same whether the haemoglobin is combined with DPG or not. Haemoglobin can exist in only two states on this interpretation, and their oxygen affinities always differ by the factor **c**. It might not be unreasonable to concede, however, that combination of DPG with the T-state could in principle alter its oxygen affinity, and this assumption is favoured by the actual experimental results (Goodford et al, 1977). DPG apparently reduces the already low oxygen affinity of the T-state to a still lower value, and other ligands show this effect to a greater or lesser extent. Thus one can envisage two different mechanisms by which an effector might promote second messenger release in such a system:-

(a) By selective combination with one state (T) in preference to the other (R).

(b) By changing the affinity of the T-state for second messenger after effector combination.

The first of these proposals is now widely accepted for pharmacological receptors, but there is little evidence for the second in pharmacological systems. However, this may be simply because appropriate experiments have yet to be designed and carried out.

EFFECTOR DESIGN

If DPG can distinguish between the R and T states and alter the R:T balance thereby promoting oxygen liberation, it might also be possible to design novel molecules to do the same thing. This was demonstrated by Beddell, Goodford, Norrington, Wilkinson and Wootton (1976) who prepared three compounds which had been designed to fit the DPG site of the T-state. These workers focused their attention on two amino groups of the protein, and the compounds were aromatic dialdehydes in which the reactive aldehyde groups should be just the right distance apart to react with the amino groups in the T-state of the protein. This straightforward approach was apparently successful (Brown and Goodford, 1977), and the compounds were of comparable potency to DPG itself as oxygen liberators although they were of a completely different chemical type.

It must be emphasised that such compounds may have been novel effector ligands, but they were not necessarily novel drugs. It is not unreasonable to argue that compounds like these which promote the release of oxygen from blood, might also make that oxygen more readily available to the tissues of the body and therefore confer therapeutic benefits in some disease states. But a real therapeutic agent to cure sick people must also satisfy many more requirements besides interacting with a chosen macromolecule. It must

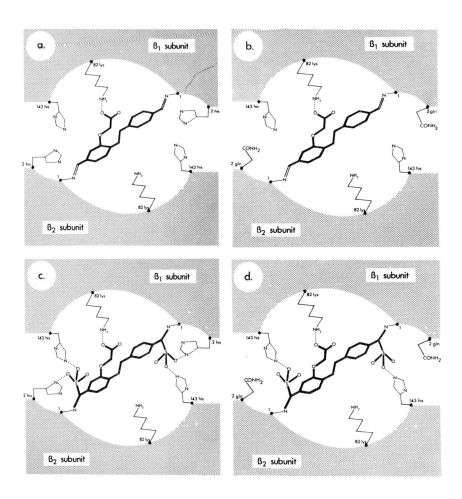

FIGURE 3   Schematic diagrams of the DPG receptor site. Figs 3a and 3b
show an aromatic dialdehyde which should bind similarly to human (a) and horse
(b) haemoglobins. Figs 3c and 3d show that the bisulphate addition complex of
the same compound should interact with more groups in human (c) than horse (d)
haemoglobin. The observed potencies of the compounds were compatible with
these predictions. See text.

be chemically stable; must be readily absorbed and distributed appropriately in the body; must be metabolised at an acceptable rate to give acceptable products; must be non-toxic; must not have any potential for abuse; must be excreted adequately and must not cause side-effects. Some of these factors (toxicity; abuse-potential; side-effects) can be gathered together as a general requirement for specificity, and this must now be considered.

Haematologists have discovered more than 200 different human haemoglobins, many of which are quite widespread while others only occur in a single family or even a single person. In addition there are still more haemoglobins in various animal species and Goodford (1978) decided to compare human and horse haemoglobins. These macromolecules differ at the DPG binding site where a human histidine residue is replaced by glutamine in the horse protein. The rationale adopted by Beddell et al (1976) when they designed their compounds suggested that only one of the novel ligands should be equipotent on human and horse haemoglobins, while the others should differ. This prediction was fulfilled (Fig 3) and was ultimately extended to cover three different ligands and six different haemoglobins giving 18 different combinations in all (Beddell, Goodford, Stammers and Wootton, 1979). From their models (Beddell et al, 1973) of the binding site and of the compounds, these authors were able to predict the number of specific ligand-protein interactions which should occur for each combination. They divided the interactions into two classes, covalent (e.g., Schiff's base) and ionic, and predicted that the total number might vary from only one in some cases to as many as seven. From their measurements of the oxygen dissociation curves in the presence and absence of ligands they were able to derive the corresponding free energies of binding which ranged from only -0.5 up to -36.0 $kJmol^{-1}$. It was then found that the observed free energies (G) were related to the predicted numbers of covalent (c) and ionic (i) interactions by the equation:-

$$G = -3.1i - 6.8c - 8.3 \qquad (6)$$

which is highly significant with a multiple correlation coefficient of 0.928 for 29 observations.

It must be stressed that this interpretation derives from an extremely crude model based on assumptions which were difficult or impossible to justify rigorously. Computer graphics were not available at the drug-design stage, nor were computer calculations used to assess the goodness of fit of the test-ligands to the protein. On the other hand a good deal of care went into the physical modelling of ligand to receptor, and Perutz (1970) and his co-workers had already developed a comprehensive interpretation of structure-function relationships for haemoglobin. The findings with the six different haemoglobins suggest that it may be possible in favourable cases to predict subtle differences of ligand binding in systems where structure and function are reasonably well-defined. Hence the problems of predicting drug toxicity, abuse potential and other side-effects may also be soluble given a wide enough data base, and these are some of the key problems in the search for new therapeutic agents today.

ACKNOWLEDGEMENTS

I am most grateful to the Wellcome Foundation, to the Laboratory of Molecular Biophysics at Oxford and to all my colleagues for their advice, encouragement and support. I am also indebted to Elsevier for permission to reproduce figures 1 and 2, and to Macmillans for permission to reproduce figure 3.

REFERENCES

Adair, G.S. (1925) J. biol. Chem., 63, 529-545.

Arnone, A. (1972) Nature, Lond., 237, 146-149.

Barcroft, J. and Roberts, Ff. (1909) J. Physiol., Lond., 39, 143-148.

Beddell, C. R., Goodford, P. J., Norrington, F. E. and Wilkinson, S. (1973) Br. J. Pharmac. Chemother., 48, 363P-364P.

Beddell, C. R., Goodford, P. J., Norrington, F. E., Wilkinson, S. and Wootton, R. (1976) Br. J. Pharmac. Chemother., 57, 201-209.

Beddell, C. R., Goodford, P. J., Stammers, D. K. and Wootton, R. (1979) Br. J. Pharmac. Chemother., 65, 535-543.

Bohr, C. (1909) In Handbuch der Physiologie des Menschen. (Ed: W. Nagel. Publ: Friedrich Vieweg und Sohn, Braunschweig.) I, 54-222.

Brown, F. F. and Goodford, P. J. (1977) Br. J. Pharmac. Chemother., **60**, 337-341.

Chanutin, A. and Curnish, R. R. (1967) Archs. Biochem. Biophys., **121**, 96-102.

Clark, A. J. (1926a) J. Physiol., Lond., **61**, 530-546.

Clark, A. J. (1926b) J. Physiol., Lond., **61**, 547-556.

Clark, A. J. (1937) Handb. exp. Pharmak., (Publ: Springer, Berlin) **4**, 1-228.

Cuatrecasas, P. (1970) J. biol. Chem., **245**, 3059-3065.

Douglas, C. G., Haldane, J. S. and Haldane, J. B. S. (1912) J. Physiol., Lond., **44**, 275-304.

Fischer, E. (1894) Ber. dt. chem. Ges., **27**, 2985-2993.

Goodford, P. J. (1978) Br. J. Pharmac. Chemother., **62**, 428P-429P.

Goodford, P.J., Norrington, F.E.N., Paterson, R.A. and Wootton, R. (1977) J. Physiol., Lond., **273**, 631-645.

Goodford, P. J., St-Louis, J. and Wootton, R. (1978) J. Physiol., Lond., **283**, 397-407.

Goodford, P. J., St-Louis, J. and Wootton, R. (1980) Br. J. Pharmac. Chemother., **68**, 741-748.

Greenwald, I. (1925) J. biol. Chem., **63**, 339-349.

Hill, A. V. (1910) J. Physiol., Lond., **40**, iv-vii.

Langley, J. N. (1878) J. Physiol., Lond., **1**, 339-369.

Langley, J. N. (1905) J. Physiol., Lond., **33**, 374-413.

Monod, J., Wyman, J. and Changeux, J-P. (1965) J. molec. Biol., **12**, 88-118.

Ostwald, W. (1908a) Z. Chemie. Ind. Kolloide., **2**, 264-272.

Ostwald, W. (1908b) Z. Chemie. Ind. Kolloide., **2**, 294-301.

Paterson, R.A., Eagles, P.A.M., Young, D.A.B. and Beddell, C.R. (1976) Int. J. Biochem., **7**, 117-118.

Perutz, M. (1970) Nature, Lond., **228**, 726-739.

Rubsamen, H., Montgomery, M., Hess, G.P., Eldefrawdi, A.T. and Eldefrawdi, M.E. (1976) Biochem. biophys. Res. Commun., **70**, 1020-1027.

Scatchard, G. (1949) Ann. N.Y.Acad. Sci., **51**, 660-672.

Schild, H. O. (1947) Br. J. Pharmac. Chemother., **2**, 189-206.

Sugita, Y. and Chanutin, A. (1963) Proc. Soc. exp. Biol. Med., **112**, 72-75.

8 Structural aspects of drug–nucleic acid interactions

Stephen Neidle

I. INTRODUCTION

The past decade has witnessed a remarkable increase in the power of chemotherapy to combat human cancers. Indeed, it is now possible in certain
favoured instances to effect long-term survival, or even permanent cure, on
the basis of drug treatment alone. However, many of the more prevalent
cancers, such as lung or breast, remain at the present time relatively
resistant to the twenty or so drugs in current clinical use. It is a principal task for the workers in this field to develop new drugs that will not
only have a higher potency to tumour cells yet lower toxicity to normal
tissue than those currently available, but will also exhibit a wide spectrum of activity to a variety of tumour types (Pratt and Ruddon, 1979.
Calman, Smyth and Tattersall, 1980).

It is remarkable that many of the drugs both in current usage and in
trial phases, are ultimately active at the DNA level. On the one hand,
compounds such as methotrexate or 5-fluoro-uracil are involved in alterations of DNA metabolism. Other compounds are believed to interact directly with the DNA molecule, many by non-covalent physical binding, and
thereby interfering with the genetic apparatus of cells, especially rapidly
proliferating ones. This review is concerned with molecular structural
aspects of drugs and drug-receptor interactions in this latter category;
these have been stimulated by the unique body of information on DNA structure that has been amassed over the past thirty years.

At this point, we introduce representative examples of these drugs.
Structurally simple compounds such as proflavine (1) and ethidium (2) are
of mainly historic medical interest, although they are still the primary
investigative tools for those interested in drug-nucleic acid interactions.
Several of them are clinically effective against neoplasms, and are natur-
ally-occuring antibiotics (Goldin et al., 1981); daunomycin (4) is active
against acute lymphocytic leukemia, ellipticine (3) is finding use in the
treatment of colon cancer, and actinomycin (6) is employed against Hodgkin's
lymphoma and several solid tumours. Adriamycin (a derivative of daumomycin
with -COMe replaced by -CH$_2$OH), is one of the most active of all anti-

cancer agents, and has an exceptionally wide spectrum of activity which is however accompanied by severe cumulative cardiotoxicity. The synthetic acridine m-AMSA (5) is showing promise in the treatment of diffuse neoplasms.

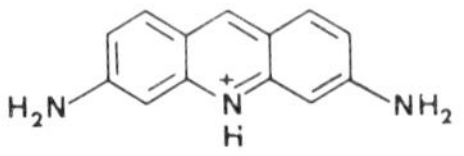

(1)

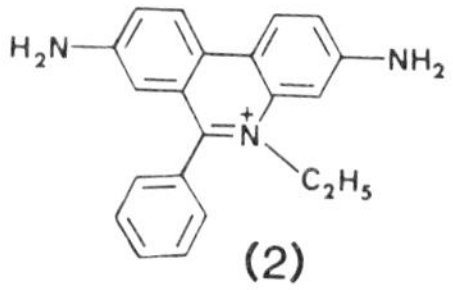

(2)

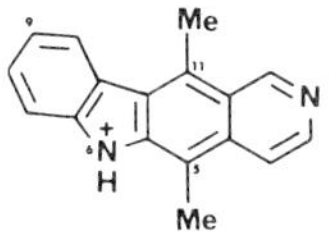

(3)

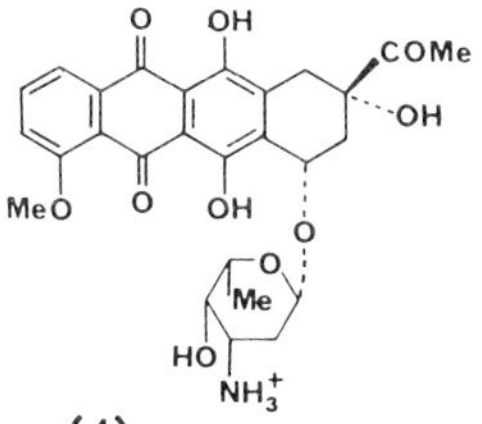

(4)

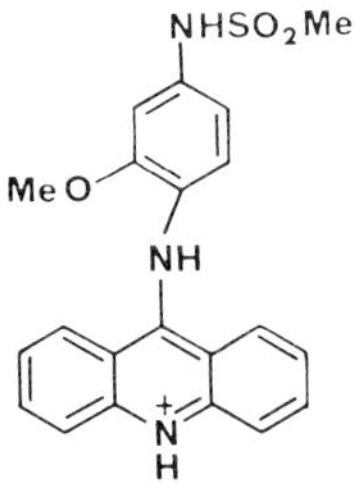

(5)

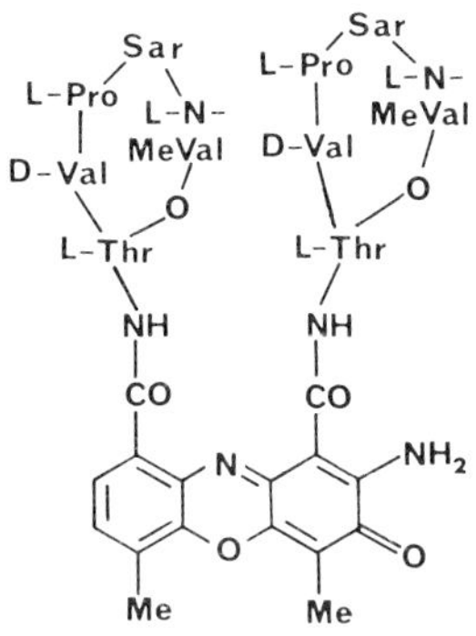

(6)

All of these drugs share a common structural feature with simpler DNA-binding agents exemplified by proflavine (1) and ethidium (2). This feature, the possession of a polarisable planar aromatic chromophore with a minimum of three fused rings, is the key to the interactions of these drugs with nucleic acids. It is thus unsurprising that there have been very many studies of these binding processes using proflavine and ethidium as models for the more complex drugs (see, inter alia, Berman and Young,

1981; Gale et al., 1981, Neidle, 1979; Waring, 1981; Wilson and Jones,
1981). It should be borne in mind that the in vivo interactions of these
drugs are invariably more complex than the in vitro model studies suggest,
primarily because of metabolic activations. Nonetheless, such models have,
and continue to be, powerful tools for probing structure-activity relationships.

2. FUNDAMENTAL FEATURES OF NUCLEIC ACID STRUCTURE

Nucleic acids comprise linear chains of repeating units. Each consists of
one of four heterocyclic bases each of which is attached to ribose sugar
(in the case of RNAs) or a deoxyribose sugar (for DNA). The residues are
connected to one another by phosphate groups between the 3' and 5' positions of successive sugars. The Watson-Crick model for DNA postulates
specific hydrogen-bonded base-pairing between two strands of polynucleotide,
which are inter-twined together in an anti-parallel manner to form a double
helical arrangement. This model was completely consistent with the X-ray
fibre diffraction data obtained by Franklin and Wilkins, and was very soon
after its inception recognised as being of profound significance in biological terms. Indeed, it is now generally agreed that the determination
of the structure of DNA stands as the landmark in the development of modern
biological science.

Fibres of DNA can display considerable structural polymorphism, which
is dependent on ionic strength and relative humidity (see Arnott, 1981).
The B form (Fig. 1) has been considered to be the physiologically most

FIG. 1. The molecular structure
of classical B-DNA (Courtesy of
M.H.F. Wilkins).

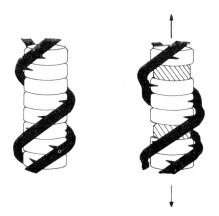

FIG. 2. Schematic representation of the Lerman intercalation model.
Native DNA is shown on the left. On the right is DNA with bound drug
molecules (as shaded discs).

3.1. Physico-chemical effects of intercalation

The Lerman model satisfactorily accounts for the retention of the
characteristic 3.4Å spacing seen in X-ray fibre diffraction patterns of
drug-DNA complexes (although because of disorder problems the patterns
cannot be analysed in detail). A wide range of other physical methods
are now employed to diagnose and probe the intercalation process.

Spectroscopic techniques are conviently employed for monitoring and
quantifying drug-DNA interactions. For example, the systematic shift in a
drug's absorption spectrum to longer wavelengths upon binding to DNA can be
analysed by Scatchard plot approximations, to yield an association constant
(typically $1-5 \times 10^6$ moles) for intercalative binding, and an indication of
maximal level of binding at c.a. one drug molecule per two base pairs. NMR
methods are useful at several levels. On the one hand, the systematic
shifts observed for the drug's aromatic protons on binding, can be indicative of ring current effects caused by intercalation (Patel, 1979; Patel,
Pardi and Itakura, 1982). On the other, analysis of ^{31}P signals from the
nucleic acid itself can demonstrate, at least in a qualitative sense,
changes that occur in the backbone conformation. Studies of drug binding
to short-length oligonucleotides have been especially fruitful. Thus, it
has been shown that ethidium and proflavine show sequence-preference interaction upon binding to dinucleosides and tetranucleosides (Reinhardt and
Krugh, 1978; Patel and Canuel, 1977) - both drugs preferentially bind to
pyrimidine-3'5'-purine sequences. At a more detailed level, analysis of

relevant type. It is characterised by having ten residues per turn of
helix, with the base pairs almost exactly perpendicular to the helix axis.
The A form exists in more hydrophobic environments, and has an eleven-fold
helix with the base pairs inclined to the helix axis. These distinctions
between the two forms are further revealed by differences in the size of
major and minor grooves - these are the entry points of interacting molecules, so their geometric features are of importance. A-DNA has a very
deep yet narrow major groove and a shallow wide minor groove. B-DNA has
both grooves equally deep, with the major groove being roughly twice as
wide as the minor.

The concept of DNA being a monotonously repetitive polymer, with all
residues being conformationaly equivalent, has been profoundly modified by
sngle-crystal studies on short lengths of oligonucleotide. Although these
have shown that the overall features of the fibre-derived A- and B-DNA
models are correct, they have revealed the existence of sequence-dependent
structural effects (Dickerson and Drew, 1981; Dickerson et al., 1982;
Shakked et al., 1981), and thus that nucleic acids possess much more complex and subtle structures than hitherto realised. As yet, unifying rules
for describing such features cannot be formulated; they are clearly of
major potential importance in the recognition of DNA by foreign molecules
such as drugs.

3. THE BINDING MODEL
The advent of the original double-helical model for DNA provided a firm
basis on which to propose a structural model for the interaction of drugs
such as ethidium and the acridines (c.f. proflavine) with DNA. This model
(Lerman, 1961) provides a ready rationalisation for the large body of
physico-chemical data that had, and continues to be produced on these interactions.

Lerman suggested that since the dimensions of a base pair are very
similar to those of the planar chromophores of the drugs, the latter may
become inserted inbetween adjacent base pairs. This process, termed intercalation (Fig. 2), involves an increase in DNA length of 3.4Å for each drug
molecule bound, as well as some unwinding of the double helix around each
binding site. The insertion of a drug molecule then naturally results in
an alteration of the base sequence read during gene transcription, and
thus produces frameshift mutagenesis.

proton NMR spectra from a daunomycin- poly (dA-dT) complex has enabled some details of the base-pair and chromophore geometry at this drug's intercalation site to be deduced (Patel, Kozlowski and Rice, 1981). However, to date no complete conformational study of a drug-oligonucleotide complex has been reported.

The intercalation process induces several characteristic changes in gross nucleic acid structure. Their detection and monitoring are essential steps in the evaluation of a drug in structure/activity terms, and so a brief survey of them is given since it is relevant to subsequent discussions in this review (see Gale et al., 1981 for further detail). Mention has already been made of the length increase in a DNA molecule produced by intercalation. This may be directly monitored by various techniques, such as electron microscopy or autoradiography, which have indicated a maximum extension corresponding to one drug molecule binding at every other possible site. Measurements of viscosity enhancement produced in sonicated rod-like DNA fragments, are especially easy to perform and analyse in terms of length increases.

Intercalation can only take place with concomitent local unwinding of the double helix. This can be most readily monitored using topologically-constrained closed-circular superhelical DNA. In general, an intercalative agent initially removes and then reverses the initially negatively-super-coiled state of these DNAs. Using, for example, changes in sedimentation coefficient or viscosity, it is possible to determine an unwinding angle per intercalated drug molecule. The angle is normally given relative to one of 26° for ethidium (Table 1). This unwinding parameter is not readily

TABLE 1.

Unwinding angles for various intercalating
drugs (from Waring, 1981), in degrees

Proflavine (1)	17	Actinomycin (6)	26
Ethidium (2)	26	Echinomycin (7)	48
Ellipticine (3)	17	Diacridine with	
Daunomycin (4)		$R=-(CH_2)_6-$ (8)	33
m-AMSA (5)			

related to other parameters of intercalation. It does not, for example, correlate with strength of binding to DNA, or indeed with biological

properties. It is most likely that the value of an unwinding angle reflects the stereochemical properties of the drug in relation to the geometry of the binding site.

The extent to which the DNA double helix is stabilised by drug binding, although not in itself fully indicative of intercalation, is nonetheless a simply-applied and widely-used method for comparing closely-related
analogues (see, for example, Brown, 1983).

4. SPECIFICITY OF INTERCALATION

The Lerman model implies that all sites in DNA are equally accessible to
intercalating drugs, and that the strength of interaction will be the same
throughout. There are several lines of evidence suggesting that this picture is an over-simplification for even the simplest of intercalators,
such as proflavine, and is quite incorrect for the more complex ones. The
preference for pyrimidine-3',5'-purine sites (such as -CpG-) for several
drugs has been mentioned above. Proflavine, ethidium and daunomycin
exhibit this preference. Evidence is now beginning to be found that even
as simple a compound as ethidium has specific high-affinity sites - a study
of the blockage of restriction enzyme sites specific for the sequence CGCG
shows non-random blockage by drugs (Coffman et al., 1982). Actinomycin D
has a near-absolute requirement for a 2-amino purine (guanine in the case
of DNA) at the 5' end of the intercalation site, and thus binds non-
randomly to natural DNA (Müller and Crothers, 1968). Echinomycin (7) binds
2-3 times more strongly to the synthetic polynucleotide poly(dG-dC) than to
poly (dA-dT), whereas its analogue triostin A, which differs by having an
extra sulphur atom in the central peptide cross-bridge, shows a reversal of
this order, and des-N-tetramethyl triostin A does not bind at all to the former
nucleic acid (Wakelin and Waring, 1976).

Echinomycin and its analogues are examples of drugs that interact
simultaneously at two DNA sites. Bis-intercalation has been extensively
explored with compounds such as the diacridine (8), with variable lengths
of spacer R. In general, bifunctional binding only occurs when R is of a
critical size such that the two chromophores occupy sites separated by an
empty site (Fig. 3a). This neighbour-exclusion binding is indeed a governing principle even for mono-intercalation. It is not difficult to envisage
that bis-intercalators would show enhanced sequence-selectivity or even
specificity compared to mono-ones (Malcolm et al., 1982), and it is thus
not surprising that a number of current programmes of DNA-binding design
are incorporating this concept.

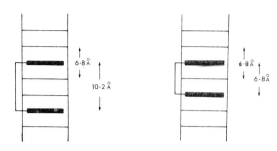

FIG. 3. Bifunctional
intercalation into DNA.
(a) neighbour exclusion
binding, and (b)
violation of this.

a   b

(7)   (8)

5. STEREOCHEMICAL BASIS OF INTERCALATION

It is now widely realised that the simple view of intercalation as a nonspecific event on classical Watson-Crick DNA, is an increasingly outmoded
one. In order to understand drug action at a molecule level, stereochemical information beyond that available from the classical model is
required. Such information is also needed for systematic drug-design
studies. We can then formulate the principal factors which structural
studies may be directed to evaluating:

- what is the precise conformation of the DNA at, and adjacent to
  the intercalation site for a given drug?
- what is the disposition of a drug in the intercalation site?
- how do these factors vary for different drugs?
- what is the effect of varying DNA sequence and what is the origin
  of sequence selectivity?
- what is the structural basis of unwinding and neighbour exclusion?

The classical solution methods used for studying drug-DNA interactions provide data averaged over all nucleotide sequences, and cannot
give the molecular-level insights that the fine-structure techniques of NMR
and X-ray crystallography can in principle provide. Rather, one would
conceive of these two types of approach as being complementary; full understanding of the intercalation process will come from a synthesis of kinetic,

thermodynamic, structural as well as biological data.

6. STRUCTURAL STUDIES ON INTERCALATION

X-ray fibre diffraction patterns of drug-DNA complexes typically indicate highly disordered structures, with little analysis being possible beyond the confirmation of the retention of the 3.4Å stacking separation present in the native DNA. The quality of a fibre pattern is dependent upon the regularity of the structure (which is high for DNA itself, at a relatively coarse level of resolution). Thus this finding is not without interest, since it suggests that drug intercalation produces rather greater structural abberations than those suggested by the Lerman model.

Single-crystal studies on drugs bound to nucleic acid fragments have and continue to be, the most fruitful and detailed source of structural information on intercalation. Since it is not feasible at present to analyse to an appropriate resolution even low molecule-weight DNAs (of typically several hundred nucleotide residues), even if they could be crystallised, the consensus approach has been to analyse very much smaller systems that may model the larger ones.

The analysis of an actinomycin D-deoxyguanosine complex (Sobell et al., 1971; Sobell and Jain, 1972) was undoubtedly a landmark in drug-receptor studies generally; this was the first atomic-level visualisation of a drug bound to a receptor model. Illuminating as this structure was at the time, in retrospect its deficiencies in terms of relevance to actino-mycin-DNA binding have become increasingly apparent. Deoxyguanosine cannot be expected to display the conformational features and constrainst of nucleosides within double-stranded nucleic acids, and therefore one has to look to more meaningful models.

6.1. The Structures

A dinucleoside monophosphate of appropriate sequence can in principle self-associate to form a dimer; such a structure has been shown to closely resemble a dinucleoside fragment of polynucleotide. Thus, the ribo-dinucleosides GpC and ApU have many features of double-stranded RNA, with anti-parallel Watson-Crick base pairs and RNA-like backbone torsion angles (Rosenberg et al., 1976; Seeman et al., 1976). Such models have therefore received much attention in the present context. The majority of drug-oligonucleoside crystal structures show Watson-Crick base-paired dimers although some have quite different structures, (for example Neidle et al., 1978; Takusagawa et al., 1982). We shall not discuss this latter category as it has only limited relevance to the theme of drug design pursued here.

To date, some dozen drug-duplex dinucleoside structures have been reported, all but two of which are with ribodinucleosides. Ethidium with 5 iodo-CpG (Jain, Tsai and Sobell, 1977) and with 5-iodo-UpA (Tsai, Jain and Sobell, 1977), forms two very similar structures. These have base pairs extended to a 6.8Å separation with ethidium intercalated (from the minor groove direction) in between them. The complexes as a whole have approximate two-fold symmetry. The base turn angle subtended by the C1'...C1' interstrand vectors, (the helical twist angles for helical nucleic acids are $36°$ for B-DNA and $32.7°$ for A-DNA and RNA), is $\sim5°$ in both cases. This has been interpreted (Sobell et al., 1976) as being due to the drug's unwinding of the dinucleoside upon intercalation, and there-fore as corresponding to an unwinding angle of $\sim28°$. The sugar puckers at the 3' and 5' ends of each dinucleoside strand are dissimilar, the con-sistent pattern being C3' endo-3',5'-C2'endo. Both complexes have a 2:2 drug:dinucleoside ratio, with one drug molecule intercalated and the other stacked external to the nucleic acid fragment dimer. Very similar struc-tures have been found by Sobell and co-workers for several other inter-calators. Ellipticine for example, (Jain, Bhandary and Sobell, 1979) has a $\sim11°$ base twist angle, again with alternating sugar puckers.

A number of proflavine-dinucleoside complex crystal structures have been analysed, and their features examined in considerable detail (Berman and Neidle, 1980; Neidle, 1981a; 1981; Neidle and Berman, 1982; 1983). The 3:2 proflavine:CpG complex (Fig. 4) (Neidle et al., 1977; Berman et al., 1979) has one proflavine molecule exactly symmetrically intercalated between C...G base pairs, with additional stabilisation provided by hydrogen bonds (of 3.00Å length) between phosphate-oxygen atoms of the dinucleoside backbone and the exocyclic amino groups of the drug. The other proflavine molecules of the complex are involved in non-intercalative interaction with the exterior of the nucleic acid dimer, via hydrogen bonds to phosphate-oxygen atoms and sugar hydroxyl groups. This outside binding is surprisingly not observed in any of the other intercalation crystal structures, even though in solution, outside binding to nucleic acids is an important (though relatively non-specific) component of the total binding process. Indeed, fast kinetic data (Li and Crothers, 1969) indicates that the initial recognition of DNA by proflavine is by outside binding, which is followed by opening-up of binding sites and consequent drug inter-calation:, this crystal structure may be said to show a frozen-out inter-mediate state. The proflavine-CpG structure has several other features

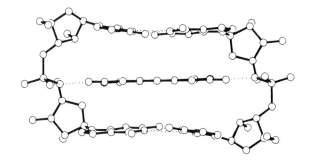

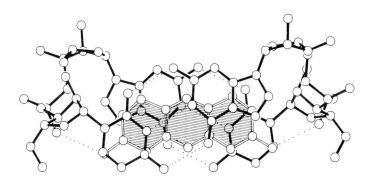

FIG. 4. Two views of the proflavine-CpG complex in the crystal structure.

that contrast remarkably with others such as the ethidium ones. Most
importantly, the base turn angle is ⋏32°, which is virtually identical to
the helical twist angle for double-stranded RNA (this is a ribo-dinucleo-
side complex). The sugar puckers do not alternate, and are all C3'_endo_.

The proflavine complex with the deoxydinucleoside analogue d(CpG)
(Shieh _et al_., 1989) is a 2:2 one with a 17° base turn angle and a complex
pattern of sugar puckers (Neidle, Berman and Shieh, 1980). The inter-

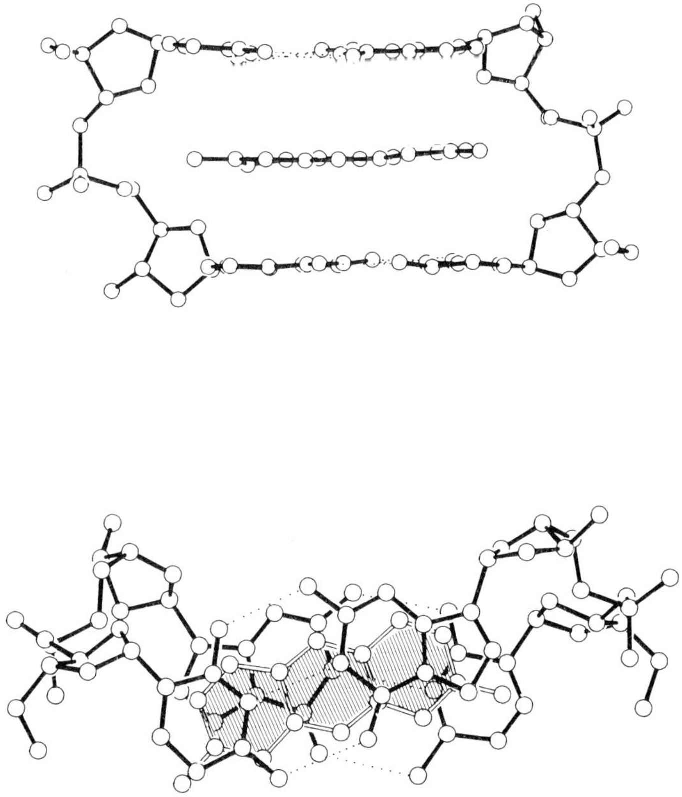

FIG. 5. Two views of the proflavine-d(CpG) complex in the crystal structure.

calated drug molecule is not symmetrically disposed, and does not hydrogen-
bond to the dinucleoside backbones (Fig. 5). The very recent analysis of a
ternary proflavine complex involving the two dinucleosides CpA and UpG
(Aggarwal et al., to be published), has revealed a structure with many
features that are intermediate between the first two proflavine ones. This
structure, which like the others has the two dinucleoside strands antiparallel, but now with one C...G and one A...U base pair, has the inter-

calated drug pseudo-symmetrically stacked between the base pairs and
hydrogen-bonded to one strand only. The base turn angle of 16° and the
distinctly larger inter-strand separation compared to the CpG complex,
have suggested that the differences in nucleotide sequence between the
structures are responsible for these distinctive features. All the proflavine complexes have the base pairs rolled open towards the major groove,
in accord with the long-held supposition that proflavine intercalates from
the major groove direction; the extent of roll is variable, and again
appears to depend upon base sequence.

    The crystal-structure analysis of a complex between daunomycin and
the hexanucleotide d(CpGpTpApCpG) is to date the only intercalation structure solved beyond the dinucleoside level (Quigley et al., 1980). In it,
a drug molecule is intercalated at both CpG sites of the hexamer duplex,
illustrating a sequence preference for this drug. Daunomycin intercalates
via the minor groove, with its chromophore aligned at right angles
to the base pairs, contrary to model-building predictions (summarised by
Neidle, 1978), although in agreement with NMR chemical shift data (Patel,
Kozlowski and Rice, 1981). Of particular relevance to the dinucleoside
structures is the observation of zero unwinding at the intercalation site
in the daunomycin complex. Instead, unwinding is distributed over neigh-
bouring residues. The relatively poor resolution and consequent uncertainties over conformational details, have made this structure less useful
than might be expected. Furthermore, as mentioned below, its relevance to
polynucleotide intercalation is no less clear than are the dinucleoside
complexes.

6.2. Overview of the dinucleoside intercalation complexes

    These structures appear at first sight to have many features of
dissimilarity, especially with respect to base turn, drug-base pair overlap and sugar conformation. However, surveys of the detailed backbone
conformations (Berman, Neidle, and Stodola, 1978; Berman and Neidle, 1980;
Neidle, 1981a) have shown that these are highly conserved in all the structures,
regardless of the above factors (Table 2). The spread of values for any
one backbone torsion angle can probably be ascribed to (a) sequence and
environmental effects (b) secondary effects of a particular drug, and (c)
experimental uncertainties in the structures themselves. This last factor
is almost certainly the most important, even though it is too frequently
under-estimated. Many of these structures suffer from problems arising
out of poor crystal quality, small crystal size, less than atomic re-

TABLE 2.

Average backbone torsion angles in dinucleoside-drug

intercalation complexes (in$^{\circ}$)

Bond:	5'-glycosidic	C3'-O3'	O3'-P	P-O5'	O5'-C5'	C5'-C4'	3'-glycosidic
Angle:	-167	-155	-68	-72	-136	57	-93

solution of the diffraction data, and consequent difficulties in refinement. It is a feature of oligonucleotide structures in general that they are very sensitive to the precise refinement procedures employed-individual torsion angles may easily vary by 10-15° from 'expected' values.

Thus, regardless of the nature of the drug, a common backbone structure always occurs. The effect of varying the sugar pucker at the 3' end is of quite secondary importance, although the C5'-C4' torsion angle is unsurprisingly somewhat correlated with it. The observations of widely-varying base-turn angles and base-pair tilt, roll and bend (Berman, Neidle and Stodla, 1978; Shieh et al., 1980) are not directly related to backbone conformation. Thus, computer simulations have shown that one can obtain a range of structures differing in base turn (and 3'end sugar pucker) yet with virtually identical backbones. It appears that base turn is a complex function of the variability in base pair geometry and flexibility, which in turn probably relates to the nature of the drug and the steric and electronic hindrance that it invokes.

It is surprising that this equivalence of conformation extends not only to all the ribodinucleoside but also to the deoxyribodinucleoside complexes- though there are as yet far fewer examples of this latter category. Examination of Table 2 shows that the torsion angles relate to those of double-helical RNA (which are almost identical to the values for the A' varient of A-DNA); only the O5'-C5' and the 3'-end glycosidic torsion angles are different from the polynucleotide values by ∿40° and ∿75° respectively. It is these two angles that might be responsible for opening up a dinucleoside so as to accommodate a drug. Further increases in these angles can be shown to increase the base-pair separation still further, from 6.8Å to beyond 8Å. The observation of A-DNA-like character for the deoxy complexes has been rationalised in terms of solvent structure and properties in relation to the relatively hydrophobic side of the inter-calated drug that is presented to the minor groove (Neidle, Berman and Shieh, 1980).

There has been numerous molecular-mechanics and allied modelling studied on drug-DNA intercalation most of which have not been relevant to the experimental findings. We mention only the recent thorough study by Kollman and his associates on proflavine intercalation into CpG and d(CpG), which have investigated the mixed sugar pucker phenomenon (Dearing, Weiner and Kollman, 1981). Studies in the author's laboratory (Islam and Neidle to be published), on the same systems, have shown that the crystallograph-ically-observed positions for the drug are close, though not identical, to the minimum-energy ones. The slight differences, of $\sim$0.5$\overset{o}{A}$, are mostly ascribable to the lack of explicit solvent contributions in the theoret-ical model. The overall success of these simulations has encouraged attempts at subsequent ones, as described in later sections.

6.3. Relationships to drug-DNA complexes

It is important not to forget that all of the crystallographic data on intercalation relates to model systems. Their relevance to polymeric or even larger-length oligomeric ones is unclear at the present time, in spite of considerable speculation.

The intercalated dinucleosides, unlike their uncomplexed precursors, do not have helical geometry, in the sense that they cannot be directly used to either generate helical intercalated polymers, or to have standard double-helical RNA or DNA residues attached at their ends. This is primarily due to the non-equivalence of the two ends of each strand in the complexes, as shown by the asymmetry of glycosidic angles. Two approaches have been used to generate polymer intercalated models using the known structural data as a starting-point. In one (Sobell et al., 1977), the observed backbone conformations have been relaxed sufficient for standard base-paired residues to be built on to both ends of the intercalated dinucleoside. The other approach (Berman and Neidle, 1980) rigidly con-strained the dinucleoside geometry to its crystallographically-observed values; the residues that were added to the ends had highly non-standard conformations and distorted base-pairs, although they were energetically feasible.

As yet, there is no structural information available to resolve this question. In a sense, the daunomycin-hexanucleotide structure (Quigley et al., 1980) is no more relevant to the polymer situation, since the fact that there are residues in this adjacent to just one side of each inter-calation site must produce pronounced asymmetry in the conformational forces restraining the geometry at and adjacent to the site.

7. AN APPROACH TO INTERCALATOR DESIGN

Even though one cannot as yet assess the full relevance of the known drug-dinucleoside structures to polynucleotide complexes, or indeed answer all the questions posed in the earlier section, it has now become possible to utilise the available data in a rational manner. Since the backbone geometries are relatively invariant, one can take, for example, the d(CpG)-proflavine structure as a standard with which to compare other hypothetical geometries. Such an approach is quite distinct from solution studies of drug-DNA interaction, since it is exclusively focussing attention on a high-affinity drug binding site (at least for most intercalators). Solution data in general is averaged over all sequence permutations and affinities.

At this point, it becomes pertinent to examine the often-stated proposition that DNA-binding ability correlates with biological activity, especially anti-tumour properties. The correlation between binding ability and in vitro inhibition of nucleic acid synthesis is often good, which is perhaps unsurprising since the latter is in effect a measure of the block-age of the DNA template by drug. It is somewhat more suprising that good qualitative correlations have often been found between DNA binding ability and/or helix stabilisation and in vivo anti-tumour activity in test systems. Notable cases are the anthracyclines (Table 3), derivatives of m-AMSA (Baguley et al., 1981), and actinomycin D analogues (Sobell, 1973). It would not be prudent to envisage that these correlations are always obeyed, or indeed that DNA binding is the sole event that such compounds undergo in a cell. Suffice it to say that for many of these drugs, DNA binding is a

TABLE 3.

Biological effects and in vitro binding behaviour of some daunomycin analogues

	Dose required for 50% inhib-ition of DNA synthesis $(Mx10^6)$	Average in-crease in survival time as % of controls for ascites tumour	Association constant $(M^{-1}x10^6)$	ΔTm in ^{o}C
Daunomycin	1.6	222	3.8	13.4
Adriamycin	3.4	227	3.0	14.8
N-Acetyl-daunomycin	>8.3	100	$1.8x10^2$	1.0
4-Demethoxy-daunomycin	1.8	264	2.4	21.0
2-Amino-2-deoxyglycosyl-daunomycinone	8.8	107	$7.1x10^2$	8.0
4-Epi-daunomycin	9	234	2.0	12.4

necessary, though by itself not a sole determinant of activity. Armed with
this encouraging data, several groups are now attempting to rationally
design compounds that are avid DNA intercalators, and thus have the potential of high and useful anti-neoplastic activity.

7.1. Anthraquinone studies

Tricyclic anthraquinones (Fig. 6) have recently shown promising anti-

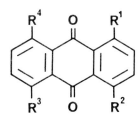

FIG. 6. The anthraquinone skeleton. The substituent group(s) X is
$-NH(CH_2)_2N(CH_2CH_3)_2$

In compound 1, $R^1 = X$, $R^2 = R^3 = R^4 = H$

2, $R^1 = R^2 = X$, $R^3 = R^4 = H$

3, $R^1 = R^3 = X$, $R^2 = R^4 = H$

4, $R^1 = R^4 = X$, $R^2 = R^3 = H$

cancer activity, and one $(R^1=R^2=NH(CH_2)_2NH(CH_2)_2OH$ and $R^3=R^4=OH)$ is now in
clinical trial. These compounds have the advantage over the related anthracyclines of far greater synthetic accessibility, reduced cost, and probable
significantly lower cardiotoxicity. The anthraquinones are therefore
suitable for detailed studies of DNA binding and activity correlations.
Our approach has been to systematically investigate the varying effects of
changes in substituent position (Fig. 6), with respect to geometry of DNA
interaction and parameters from solution binding experiments. Initial
studies have been with the substituent(s) $-NH(CH_2)_2N(CH_2CH_3)_2$ $(=X)$

In order to evaluate the geometry of intercalation, the standard
opened-up dinucleoside conformation detailed in earlier sections (Table 2)
has been taken as a template. Computer simulations of the interactions have
been performed, using a combined interactive computer graphics and empirical
molecular-mechanics approach to obtained approximate geometries, which
have been refined by a total energy-minimisation technique.

TABLE 4.

Characteristics of the DNA binding of anthraquinones (1-4),

and their interaction energies with a d(CpG) model

COMPOUND	ΔTm	K (assoc. constant)	Unwinding Angle in $^\circ$.	Total interaction energy in Kcal mole^{-1}	
1	8.8	1.48×10^6	10.6	(a)	-70.3
				(b)	-85.0
2	22.0	3.17×10^6	14.9		-118.5
3	25.1	3.97×10^6	14.2		-120.0
4	9.5	1.71×10^6	16.0		-89.3

Four compounds have been studied (Table 4). Crystal-structure
analyses of three of them (to be published) have been performed, which as
well as providing starting coordinates, have indicated likely ranges of
flexibility for the aliphatic side chains. Intramolecular hydrogen-
bonding between the quinoid oxygen atoms and adjacent N-H groups consistently restrict the conformational freedom of part of the side-chains
in all four compounds. The terminal groups on the chains are much more
flexible, their conformations in the crystal structure being sometimes
quite different from minimum-energy positions in the intercalated state.

The final intercalated structures (Figs. 7-9) show a variety of
chromophore orientations and minimum energies (Table 4), many of which
involve the side-chains in interactions with nucleoside groups. The 1-
substituted compound has several feasible binding modes; the perpendicular
one (Fig. 8) has reduced base-pair overlap overlap and less stabilisation
compared to the orientation in Fig. 7. The 1,5 compound (Fig. 9) is of
interest in that in order to either intercalate it into the dinucleoside,
or dissociate the resulting complexes the base pairs have to be broken.
Thus, the total energy of interaction, which may be thought of as the
stabilisation energy that is required to change from a complexed to a
separated state, includes an appropriate term for base-pair dissociation.
Table 3 gives the stabilisation energies for all four compounds. It is
gratifying that their order parallels both ΔTm and association constants,
and that the 1,4 and 1,5 compounds are consistently the strongest intercalator. The unwinding angles notably do not correlate with any of these

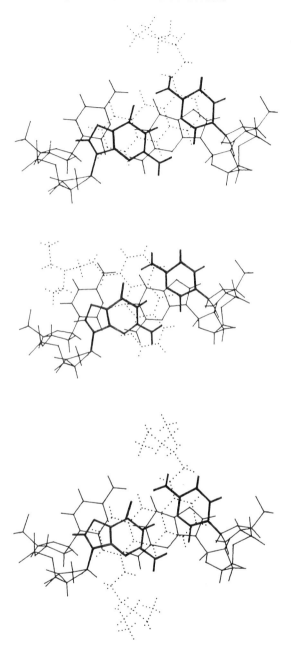

FIGS. 7-9. Interaction of anthraquinone 1 (top and middle) and
anthraquinone 3 (bottom) with d(CpG), as found from computer
simulation analyses.

parameters; extent of inhibition of cell growth in culture does follow the pattern of in vitro binding activity.

8. CONCLUSIONS

It is now apparent that drug intercalation is a much more complex family of phenomena than was hitherto imagined. Structural studies have started to reveal the extent of these subtleties; many more are needed to clarify questions of, in particular, sequence specificity and intercalation into long sequences of oligonucleotide. Such analyses will undoubtedly enable rational drug design to be conducted to higher levels of sophistication than at the present time.

ACKNOWLEDGEMENTS

I am grateful to colleagues and collaborators, past and present who have made major contributions to the work reported here, and to the Cancer Research Campaign for grant support and a Career Development Award.

REFERENCES

ARNOTT, S. (1981), In Topics in Nucleic Acid Structure, ed. Neidle, S., MacMillan Press, London.

BAGULEY, B.C., DENNY, W.A., ATWELL, G.J. and CAIN, B.F. (1981) J. Med. Chem., 24, 520.

BERMAN, H.M. and NEIDLE, S. (1980) In Nucleic Acid Geometry and Dynamics, ed. Sarma, R.H., Pergamon Press, New York.

BERMAN, H.M. NEIDLE, S. and STODOLA, R.K. (1978) Proc. Natl. Acad. Sci. U.S.A, 75, 828.

BERMAN, H.M., STALLINGS, W., CARRELL, H.L., GLUSKER, J.P., NEIDLE, S., TAYLOR, G. and ACHARI, A. (1979) Biopolymers, 18, 2405.

BERMAN, H.M. and YOUNG, P.R. (1981) Ann. Rev. Biophys. Bioeng., 10, 87.

BROWN, J.R. (1983) In Molecular Aspects of Anticancer Drug Action, eds. Neidle, S. and Waring, M.J., MacMillan Press, London.

CALMAN, K.C., SMYTH, J.F. and TATTERSALL, M.H.N. (1980) Basic Principles of Cancer Chemotherapy MacMillan Press, London.

COFFMAN, G.L., GAUBATZ, J.W., YIELDING, K.L. and YIELDING, L.W. (1982) J. Biol. Chem., 257, 13205.

DICKERSON, R.E. and DREW, H.R. (1981) J. Mol. Biol., 147, 761.

DICKERSON, R.E., DREW, H.R., CONNER, B.N., WING, R.M., FRATINI, A.V. and KOPKA, M.L. (1982) Science, 216, 475.

GALE, E.F., CUNDLIFFE, E., REYNOLDS, P.E., RICHMOND, M.H. and WARING, M.J.

(1981). The Molecular Basis of Antibiotic Action 2nd Ed., John Wiley,
London.

GOLDIN, A., VENDITTI, J.M., MACDONALD, J.S., MUGGIA, F.M., HENNEY, J.E. and
DEVITA, V.T. (1981) Eur. J. Cancer, 17, 129.

JAIN, S.C., BHANDARY, K.K. and SOBELL, H.M. (1979). J. Mol. Biol., 135,
813.

JAIN, S.C., TSAI, C.-C. and SOBELL, H.M. (1977) J. Mol. Biol., 114, 317.

LERMAN, L.S. (1961). J. Mol. Biol., 3, 18.

MALCOLM, A.D.B., MOFFATT, J.R., FOX, K.R. and WARING, M.J. (1982). Biochim.
Biophys. Acta, 699, 211.

MÜLLER, W. and CROTHERS, D.M. (1968). J. Mol. Biol., 35, 251.

NEIDLE, S. (1978) In Topics in Antibiotic Chemistry. vol. 2, ed. Sammes,
P.G., Ellis Horwood, Chichester.

NEIDLE, S. (1979) Prog. Med. Chem., 16, 151.

NEIDLE, S. (1981a) In Topics in Nucleic Acid Structure ed. Neidle, S.,
MacMillan Press, London.

NEIDLE, S. (1981b). Comments Mol. Cell. Biophys., 1, 171.

NEIDLE, S., ACHARI, A., TAYLOR, G.L., BERMAN, H.M. CARRELL, H.L., GLUSKER,
J.P. and STALLINGS, W.C. (1977) Nature, 269, 304.

NEIDLE, S. and BERMAN, H.M. (1982) In Molecular Structure and Biological
Activity eds. Duax, W.L. and Griffin, J.F., Elsevier, New York.

NEIDLE, S. and BERMAN, H.M. (1983) Prog. Biophys. Mol. Biol., 41, 43.

NEIDLE, S., BERMAN, H.M. and SHIEH, H.-S. (1980) Nature, 288, 129.

NEIDLE, S., TAYLOR, G., SANDERSON, M., SHIEH, H.-S. and BERMAN,H.M. (1978)
Nucl. Acids Res., 5, 4417.

PATEL, D.J. (1977) Acc. Chem. Res., 12, 118.

PATEL, D.J. and CANUEL, L.L. (1977) Proc. Natl. Acad. Sci. USA., 73, 2624.

PATEL, D.J., KOZLOWSKI, S.A. and RICE, J.A. (1981) Proc. Natl. Acad. Sci.
USA., 78, 3333.

PATEL, D.J., PARDI, A. and ITAKURA, K. (1982) Science, 216, 581.

PRATT, W.B. and RUDDON, R.W. (1979) The Anticancer Drugs, Oxford University
Press, New York.

QUIGLEY, G.J., WANG, A.H.-J. UGHETTO, G., VAN DER MAREL, VAN BOOM, J.H.
and RICH, A. (1980). Proc. Natl. Acad. Sci. USA., 77, 7204.

REINHARDT, C.G. and KRUGH, T.R. (1978) Biochem., 17, 4845.

ROSENBERG, J.M., SEEMAN, N.C., DAY, R.O. and RICH, A. (1976) J. Mol. Biol.,
104, 145.

SEEMAN, N.C., ROSENBERG, J.M., SUDDATH, F.L., KIM, J.J.P. and RICH, A.
(1976) J. Mol. Biol., 104, 109.

SHAKKED, Z., RABINOVICH, D., CRUSE, W.B.T., EGERT, E., KENNARD, O., SALA, G., SALISBURY, S.A. and VISWAMITRA, M.A. (1981) Proc. Roy. Soc Lond., A213, 479.

SHIEH, H.-S., BERMAN, H.M., DABROW, M. and NEIDLE, S. (1980) Nucl. Acids Res., 8, 85.

SOBELL, H.M. (1973) Prog. Nucleic Acid Res. Mol. Biol., 13, 153.

SOBELL, H.M. and JAIN, S.C. (1972) J. Mol. Biol., 68, 21.

SOBELL, H.M., JAIN, S.C., SAKORE, T.D. and NORDMAN, C.E. (1971) Nature, 231, 200.

SOBELL, H.M., TSAI, C.-C., JAIN, S.C. and GILBERT, S.G. (1977) J. Mol. Biol., 114, 333.

TAKUSAGAWA, F., DABROW, M., NEIDLE, S. and BERMAN, H.M. (1982) Nature, 296, 466.

TSAI, C.-C., JAIN, S.C. and SOBELL, H.M. (1977). J. Mol. Biol., 114, 301.

WAKELIN, L.P.G. and WARING, M.J. (1976) Biochem. J., 157, 721.

WARING, M.J. (1981). Ann. Rev. Biochem. 50, 159.

WILSON, W.D. and JONES, R.L. (1981) Adv. Pharmacol. Chemoth., 18, 177.

9 The α-helix dipole and the binding of phosphate groups of coenzymes and substrates by proteins

Wim G.J. Hol and Rik K. Wierenga

1. INTRODUCTION

Electrostatic interactions play a key role in many processes in the living cell (1,2). Consequently, these interactions are undoubtedly important for the action of many existing drugs and for the design of new drugs. Electrostatic interactions often involve only charged groups, monopoles, such as in salt bridges. They also occur between two small dipoles, such as between the NH and C=O groups in hydrogen bonds. The present paper focusses on the hitherto largely neglected macro-dipole of the α-helix. This macro-dipole is of considerable strength and appears to be involved in catalysis and in the binding of substrates or coenzymes by numerous enzymes. For the design of new drugs on the basis of the three-dimensional structures of potential target enzymes it is important to be aware of the crucial role helix dipoles may play in the mechanism of action of proteins.

2. THE ORIGIN OF THE α-HELIX DIPOLE

The α-helix dipole originates from the regular arrangement of the peptide units in the α-helical spiral. Each peptide unit has a dipole moment of approximately 3.5 Debye, or 0.68 e.Å, which corresponds roughly with the charge distribution given in Figure 1 (3,4). In the α-helix, these peptide units point all virtually in the same direction, i.e. parallel to the helix axis (4), as shown in Figure 2. The resulting dipole moment of the α-helix is an addition of the dipole moments of the individual peptide units. For a helix of for example 20 residues this amounts to about 60-65 Debye. For comparison, it may be mentioned that the well-known, and important, dipole moment of a water molecule is 1.8 Debye. The N-terminus of the α-helix is the positive end of the helix-dipole, the C-terminus the negative end.

FIG. 1. The partial charges   FIG. 2. Peptide dipoles in the
on the peptide atoms of the   α-helix. The resultant macro-
peptide unit, giving rise to   dipole of the helix corresponds to
a dipole of 3.5 D in the   half a positive unit charge at the
direction as indicated.   N-terminus, and half a negative
                                   unit charge at the C-terminus.

A good approximation of the strength of the helix-dipole is obtained
by placing half unit charges of opposite sign at the helix termini (3).
Extensive ab-initio quantum-mechanical calculations have indicated
that the peptide dipoles in the α-helix increase considerably due to
mutual polarisation (5). This is in agreement with a dipole moment per
peptide unit of about 5 Debye as measured by Applequist & Mahr (6) and
as considered likely by Wada (4) on theoretical grounds. Therefore,
the helix dipole may be significantly larger than is obtained from a
simple summation of static dipoles. It may be recalled here that a
similar situation occurs for the water dipole. For ice, various investigators have reported dipoles per water molecule of up to 2.7 Debye
(7,8,9), which is 50% larger than the value for a single water molecule.

3. THE α-HELIX DIPOLE AND ENZYMATIC CATALYSIS

     Enzymes are often extremely efficient catalysts with usually a
precise specificity. Many energetic and entropic factors play a role
in the rate enhancement by these specialized macro-molecules. In a
number of instances, the electric field generated by the α-helix

dipole is important in increasing the reactivity of active site side
chains or of coenzyme moieties. A summary of enzymes in which an
"active site helix" occurs is given in Table 1. The "classical" example
of an α-helix being important for catalysis is the sulfhydryl protease
papain. Based on the three-dimensional structure of this enzyme, Drenth
and co-workers have proposed a reaction mechanism in which the essential Cys-25 forms an ion pair with His-159 in the native protein (ref.
10; Fig. 3). The thiolate ion thus obtained is the crucial nucleophile
in the initial catalytic step. In ab-initio calculations without incorporation of the helix dipole, the existence of the ion pair was doubtful. However, taking into account the helix field, changed the picture

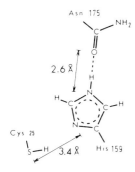

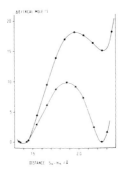

FIG. 3. The catalytic triad
in the sulfhydryl protease
papain. Cys-25 is the essential nucleophile and has a
low pK. The dipole of helix
24-43 facilitates proton
transfer from Cys-25 to
His-159.

FIG. 4. The (relative) energy
of the active site residues in
papain as a function of the position
of the proton shared between Cys-25
and His-159. Left: the proton
resides on the SH group; right:
on the imidazole ring. The upper-
line is the energy without including the helix dipole. The
lowerline with the inclusion of
this dipole.

completely: the energy of the ion pair became equal to that of the
non-zwitterionic situation (ref. 11; Fig. 4). Consequently, the existence of an ion pair in the active site of papain is certainly possible. Other unusually reactive sulfhydryl groups situated at the N-
terminus of an α-helix occur in the sulfur-transferring enzyme rhoda-
nese (12) and in glyceraldehyde-3-phosphate dehydrogenase (13,14). An
example of the α-helix field influencing the reactivity of coenzyme
moieties is probably observed in the FAD-binding enzymes glutathione-
reductase and p-hydroxybenzoate hydroxylase (15,16,17; Fig. 6). Here,
in two enzymes catalyzing quite different reactions, the N1-atom of

Table 1   Enzymes with a helix dipole pointing to the active site

Enzyme	Active site residue	Helix	Reference
Papain	Cys-25	24-43	46
Subtilisin	Ser-221	220-238	47,48
GAPDH	Cys-149	148-165	13,14
Rhodanese	Cys-247	251-264	12,49
Triose phosphate -isomerase	His-95	95-101	50
Glutathione peroxidase	Seleno Cys-35	36-50	51
Glutathione reductase	Cys-58 – Cys-63	63-80	15
p-Hydroxybenzoate hydroxylase	isoalloxazine of FAD	298-318	16,17
Glutathione reductase	isoalloxazine of FAD	338-354	15,17

the isoalloxazine-ring is hydrogen bonded, in precisely the same way,
by the N-terminus of an α-helix. The negative charge of the isoalloxa-
zine radical and anion is localised in the N1-O2 region of the iso-
alloxazine (18), which is stabilised by the helix-dipole (17).

As some future drugs may depend on an optimal exploitation of
unusual reactivities of side chains near active centres of potential
target enzymes, it seems appropriate to be aware of the influence the
helix field may exert on these reactivities.

4. THE α-HELIX DIPOLE AND BINDING OF CHARGED GROUPS BY PROTEINS

A surprisingly large number of phosphate groups of coenzymes appears to be bound near the N-terminus of an α-helix (Table 2). This is the positive end of the helix dipole and therefore a favourable electrostatic interaction exists. Of course, other modes of binding coenzyme-phosphates occur (see e.g. hexokinase (19)) just as nature has also devised several quite different ways for hydrolysing the peptide bond.

Limiting ourselves to nucleotides, it is remarkable that these compounds often bind to the well-known "βαβ-unit". This folding pattern is probably quite stable for a number of reasons and appears to provide an α-helix dipole, and binding crevices (20), which makes it very suitable for binding dinucleotides as will be discussed in some detail in section 6 below. Well-known examples of enzymes with structurally related coenzyme binding domains are: lactate dehydrogenase, malate dehydrogenase, alcohol dehydrogenase and glyceraldehyde-3-phosphate dehydrogenase(21). One example is given in Figure 5. From an analysis of these and other enzymes, it has become apparent that the crucial interactions occur between the βαβ-unit of the enzyme and the ADP-moiety of the coenzyme(22). This "ADP-binding βαβ-fold" has also been observed in the two FAD-binding proteins glutathione-reductase (human) and p-hydroxybenzoate hydroxylase (bacterial) (17) which have been

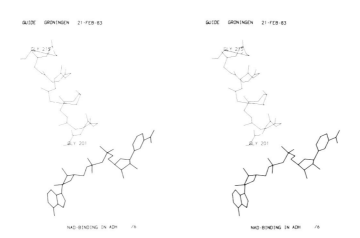

FIG. 5. The pyrophosphate moiety of NAD interacting with an α-helix dipole in horse liver alcohol dehydrogenase (54).

Table 2 Helices involved in binding negatively charged groups

Protein	Bound Ligand	Charged Moiety	Helix or Helices	Reference
Lactate dehydrogenase	NAD	pyrophosphate	30-42	21,52
Malate dehydrogenase	NAD	pyrophosphate	15-26	21,53
Alcohol dehydrogenase	NAD	pyrophosphate	201-215	21,54
G-3-P dehydrogenase	NAD	pyrophosphate	9-22	13,14,21
Dihydrofolate reductase	NADP	pyrophosphate	42-48 & 99-106	25,26
Glutathione reductase	NADP	pyrophosphate	196-210	15,17,23
Ferredoxin reductase	NADP	pyrophosphate	3 helices	55
p-OH-benzoate hydroxylase	FAD	pyrophosphate	11-25	16,17
Glutathione reductase	FAD	pyrophosphate	29-42	15,17,23
Flavodoxin	FMN	phosphate	10-26	29,30,31
Phosphorylase	Pyridoxalphosphate	phosphate	675-682	27
Aspartate transaminase	Pyridoxalphosphate	phosphate	107-124	28
Phosphorylase	Glucose-1-phosphate	phosphate	134-140	27
Triose-phosphate isomerase	Dihydroxy-acetone-P	phosphate	232-237	50
Phospho fructokinase	ADP, ATP	pyrophosphate	'helix 5'	56
Phosphoglycerate kinase	ATP	γ-phosphate	372-380	57
Phosphoglycerate kinase	Phosphoglycerate	phosphate	393-401	57
Adenylate kinase	AMP	phosphate	23-30	58
Asp. carb. transferase	CTP	pyrophosphate	52-65 & 136-149	59,60
Myoglobin	SO_4	SO_4	α_E and α_D	61
D-alanyl-transpeptidase	Penicillin	Carboxylates	several	33

discussed above in relation to the influence of a, different, helix
dipole on catalysis. Both of these helices interacting with FAD in
both enzymes are shown in Figure 6. It may be worthwhile to point
out that the similarity in structure and function of this βαβ-unit

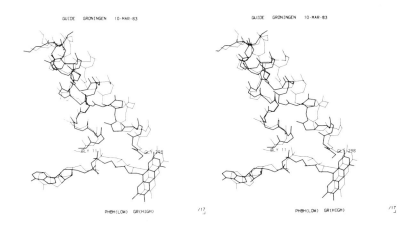

FIG. 6. The pyrophosphate-moiety of FAD and the N-1 atom of the isoalloxazine-ring interacting with helix-dipoles in p-hydroxybenzoate
hydroxylase (thin lines) and in glutathione-reductase (heavy lines).

is reflected in a significant, but not immediately obvious, amino
acid sequence similarity: only 7 out of 31 amino acids are identical
(Table 3; ref. 17). The structurally related complete FAD-domains,
which include the βαβ-units, also have a rather different amino acid
sequence: only 10 out of 69 residues at analogous positions are identical (17).

Glutathione reductase makes ample use of the helix dipole as
inspection of Tables 1 and 2 shows. In addition to the two helices
discussed above, this enzyme contains a third helix involved in
binding the pyrophosphate moiety of the coenzyme NADP at its N-terminus and a fourth helix which may be important for the unique properties of the essential disulphide bridge which is positioned precisely
at the N-terminus of this helix (15,23). As glutathione reductase
from the malaria parasite is under consideration as a potential target

enzyme for pharmaceutical compounds (24), some of these helix dipole effects described may one day turn out to be important for the action of a new drug.

A well known target enzyme for a variety of existing drugs is dihydrofolate reductase. The inhibitor methotrexate is used as a carcinostatic, the folate analog pyremethamine is a malaria drug and trimethoprim is used as an antibacterial agent. In the ternary complex of dihydrofolate reductase, methotrexate and NADPH (25,26), it appeared that N-termini from two helices point to the pyrophosphate moiety of this coenzyme (Fig. 7). As the three-dimensional structure

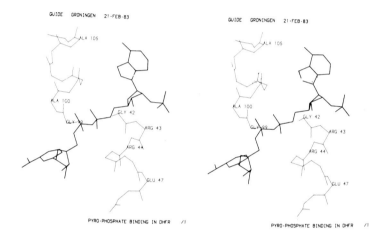

FIG. 7. The pyrophosphate moiety of NADP interacting with two helix dipoles in dihydrofolate reductase from *L. casei* (26).

of this, and other, ternary complexes are currently being used for the development of new drugs, some of which may be covalent substrate-coenzyme analogs, it is important to be aware of the helix dipoles interacting with the negatively charged pyrophoshate group.

Not only the phosphates from dinucleotide coenzymes bind to N-termini from helices. This is also the case in the two known structures of proteins which bind the coenzyme pyridoxal-phosphate (27,28). The functions and structures of these enzymes are totally unrelated, and even the use of the pyridoxal-phosphate during catalysis is distinctly different, so that one may state that the only common feature of these two proteins is the interaction of a helix dipole with a coenzyme phosphate group.

The flavodoxins bind an FMN moiety with an α-helix and a special loop of the backbone (29,30,31; Fig. 8). It is intriguing that no

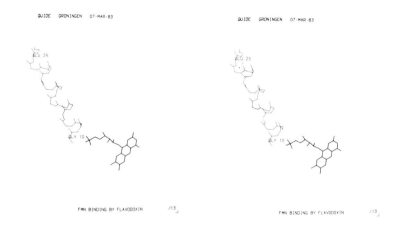

FIG. 8. The phosphate-group of FMN interacting with the helix dipole in flavodoxin (29).

positive charge interacts with the phosphate group and that, instead, the region of the protein near this negatively charged phosphate-moiety contains a large excess of negatively charged carboxylate groups (32).

The characteristic feature of a phosphate moiety bound to a helix-dipole is the presence of one hydrogen bond with one NH group of the helix. There may be one or more positively charged side chains nearby but these positive charges never occur in between the helix and the phosphate group. Further, the helix is shielded from solvent by the nucleotide or substrate, while the charged side chains are often fully exposed to solvent. In some instances, no charged side chains occur at all (29,31) and only the helix dipole remains to interact with the negatively charged phosphate group. A schematic view of a characteristic phosphate-binding helix is given in Figure 9.

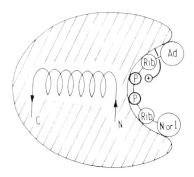

FIG. 9. Typical mode of di-nucleotide binding by proteins: the pyrophosphate moiety interacts with a helix-dipole.

It seems appropriate to finish this section with one of the
most important drugs discovered in the twentieth century: penicillin.
This antibiotic may provide an example where carboxyl groups instead
of phosphate groups are interacting with helix dipoles. In a preliminary report of the three-dimensional structure of penicillin bound
to its target enzyme D-Ala-transpeptidase-carboxypeptidase (33), it
was stated that this drug is bound near the N-termini of two helices.
More detailed information is awaited with great interest to see how
helix dipoles are involved in binding this important drug.

5. CANCER

Among the fascinating discoveries which are currently being
made with regard to the molecular biology of cancer, one which has
attracted great attention is the demonstration that a single point
mutation is responsible for the acquisition of transforming properties by the EJ and T24 human bladder oncogene (34,35,36). The point
mutation consists of the conversion of guanidine into thymidine,
which results in the replacement of a glycine by a valine at position
12 of the p21 protein encoded by the EJ and T24 oncogenes. Moreover,

comparison of the 37 N-terminal residues of the normal p21 protein
with their retroviral analogues shows perfect identity of all residues
except at position 12, where, instead of glycine, an arginine or
serine is observed (34-38). In the normal rat homologue of the human
p21 protein, also a glycine is observed at this position (M. Ruta
as cited in refs. 34 and 35). These results indicate that the glycine
residue at position 12 is crucial for a proper functioning of the
p21 protein. The normal function of normal p21 is still unknown.

Comparison of the sequence of the 37 residues encoded by the
first exon of human p21 protein with the N-terminal sequence of the
bacterial enzyme p-hydroxybenzoate hydroxylase (34,35,36,39) and
several other, structurally related, proteins, leads to the surprising
result that an important sequence homology exists (40; Table 3).
In the known structures, this sequence folds into a "βαβ-unit" as
schematically depicted in Figure 10. As can be seen from Table 3,

FIG. 10. Schematic view of the proposed βαβ-unit of p21 protein,
binding a (di)nucleotide. The arrow indicates the steric clash which
would occur if Gly-12 were changed into a residue containing a side
chain.

it is very likely that the first 37 residues of the p21 protein also
forms a nucleotide binding βαβ-unit. The "fingerprint" of this
nucleotide binding βαβ-unit is:

Table 3

Enzyme	nucleotide	Sequence (positions 1 – 30)	end res.	Number of residues identical with p21	Refs.
LDH	NAD	K I T V V G V G A V G M A C A I S I L M K D L A D E V A L V D V M	54	10	52,62
ADH	NAD	T C A V F G L G G V G L S V I M G C K A A G A A – R I I G V D I N	225	8	54,63
GPD	NAD	K I G I D G F G R I G R L V L R A A L S C G A Q – V V A V N D P F	33	9	13,64
GR	FAD	D Y L V I C G G G L A S A R R A A E L G A – – R A A V V E S H	52	4	15,65
PHBH	FAD	Q V A I I G A G P S G L L L G Q L L H K A G I – – D N V I L E R Q	34	4	16,39
p21	?	K L V V V G A [G] G V G K S A L T I Q L I Q N H – – F V D E Y D P T	35	–	34–36

fingerprint: (fingerprint symbols ●, ■, ▲, Θ as discussed in text)

sec struct.: β β β β β   α α α α α α α α α α α α α α   β β β β β

Alignment of the sequence of the first exon of normal human p21 protein with the sequences of a number of nucleotide binding βαβ-units from proteins of which the three-dimensional structure has been determined. The fingerprint symbols are discussed in the text. LDH = dogfish lactate dehydrogenase. ADH = horse liver alcohol dehydrogenase. GPD = lobster glyceraldehyde-3-phosphate dehydrogenase. GR = human erythrocyte glutathione reductase. PHBH = p-hydroxybenzoate hydroxylase from *Ps. fluorescens*. Hyphens indicate deletions. The boxed glycine-12 of the p21 protein is mutated into a valine in bladder carcinoma cells (see text).

(i)   the sequence Gly-X-Gly-X-X-Gly in the region joining the first
       β—strand and the N-terminus of the α—helix. The function of
       the second of these three essential glycines is particularly
       important in view of the mutation seen in the p21 protein:
       this glycine provides space for a close approach to the pyro-
       phosphate moiety to the N-terminus of the α—helix. This allows
       the formation of several hydrogen bonds with the protein, in
       addition to a favourable interaction between the negative
       charge of the pyrophosphate group and the α—helix dipole.

(ii)   the occurrence of several neutral, often hydrophobic residues,
       forming the core of the βαβ—unit (Fig. 10,11). These residues
       are indicated with ■ in Table 3.

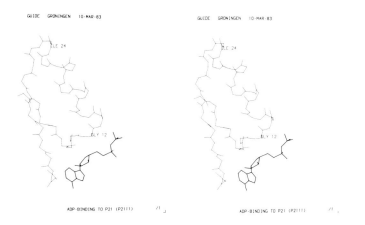

FIG. 11. The βαβ—unit in the p21 protein binding ADP, as derived from
model building based upon the homologous βαβ—unit in p-hydroxybenzoate
hydroxylase.

(iii) the presence of a negatively charged residue at the C-terminus
       of the second β—strand. The carboxylate moiety of this side
       chain forms hydrogen bonds with the 2'-hydroxyl group of the
       adenine ribose (Fig. 10,11).

(iv)   the presence of a hydrophylic residue at the beginning of the
       first β—strand, indicated with ▲ in Table 3. The function of
       this residue is as yet unknown.

Residues 5-33 of the p21 protein follow these four characteristics almost perfectly (Table 3). With computer graphics techniques, a model of the p21 βαβ-unit was constructed, based upon the known three-dimensional structure of p-hydroxybenzoate hydroxylase (16). After the appropriate amino acid substitutions were made, only minor alterations of a few atoms were required to remove a small number of short contacts. Residues Leu-6, Val-8, Leu-19, Gln-22 and Val-29 form a nice hydrophobic core in the centre of the βαβ-unit. The glutamic acid at position 31 can form a salt bridge with Lys-16. The proposed model implies that the binding of the ribose-pyrophosphate moiety of the nucleotide would be very similar to that in p-hydroxybenzoate hydroxylase. Several hydrogen bonds can be formed, while Asp-33 of the p21 protein forms a good hydrogen bond with the 2'-hydroxyl group of the ribose ring. This latter interaction implies that the nucleotide bound by the normal p21 protein is probably not NADP as this would result in a repulsion between the negative charges of Asp-33 and the 2'-phosphate group of the coenzyme. Fig. 11 shows a model of the p21 protein with bound ADP, as derived from the structure of a flavin enzyme. It should be stressed, however, that no evidence exists that FAD is bound by normal p21. The bound nucleotide could as well be NAD, ATP, GTP or other coenzymes. GTP-binding by the p21 protein has been demonstrated by various investigators (41-43).

The proposed structure and function of residues 5-33 explains why the presence of a glycine residue at position 12 is essential for a proper functioning of normal p21 protein: the side chain of another residue at this position would interfere with the binding of the pyrophosphate moiety of the nucleotide to the N-terminus of the α-helix. Gay & Walker (44) have drawn attention to the sequence homology between the p21 protein and mitochondrial ATP-synthase. The homology is quite striking, but it does leave one crucial question unanswered: why is replacement of Gly-12 in normal p21 by another residue so important? In the family of proteins related to ATP-synthase (45), great variability is permitted for the chain at position 12. Evidently, further investigations are required to establish the function, structure and nucleotide binding properties of normal p21 protein.

Nevertheless, in case dinucleotides are indeed bound by the

first exon of p21 in a manner as depicted in Figure 11, then our model might form the starting point for the development of new cancer drugs. Certain dinucleotide analogues with suitable replacements for the pyrophosphate-ribose moiety may be able to overcome the steric hindrance occurring in mutated oncogenic p21 for binding the natural dinucleotide. Oncogenic p21 protein binding such a potential cancer drug may function quite similarly to normal p21 binding its natural nucleotide.

6. CONCLUSION

It is certain that three-dimensional structures of potential target proteins will become an increasingly important starting point for the development of future drugs. Numerous interactions will then have to be taken into account, including the electrostatic effect of the α-helix dipole.

Acknowledgements

We are indebted to Drs. H.J.C. Berendsen, P.Th. van Duijnen and J. Drenth for stimulating discussions. We also like to thank Ms. R.A. Hogenkamp for excellent assistance in the preparation of the manuscript. This research was supported, in part, by the Dutch Foundation for Chemical Research (SON) with financial aid from the Dutch Organisation for the Advancement of Pure Research (ZWO).

REFERENCES

1. Perutz, M.F., Science 201, 1187-1191 (1978).
2. Warshel, A. and Levitt, M., J. Mol. Biol. 103, 227-249 (1976).
3. Hol, W.G.J., Van Duijnen, P.Th. and Berendsen, H.J.C., Nature 273, 443-446 (1978).
4. Wada, A., Adv. Biophys. 9, 1-63 (1976).
5. Van Duijnen, P.Th. and Thole, B.T., Biopolymers 21, 1749-1761 (1982).
6. Applequist, J. and Mahr, T.G., J. Am. Chem. Soc. 88, 5419 (1966).
7. Coulson, C.A. and Eisenberg, D., Proc. Roy. Soc. A291, 445-453 (1966).
8. Barnes, P., Bliss, D.V., Finney, J.L. and Quinn, J.E., Faraday Disc. Roy. Soc. Chemistry 69, 210-220 (1980).
9. Adams, D.J., Nature 293, 447-448 (1981).

10. Drenth, J., Swen, H.M., Hoogenstraaten, W. and Sluyterman, L.A.Æ., Kon. Nederl. Akad. Wetensch. 78, 104-110 (1974).

11. Van Duijnen, P.Th., Thole, B.T. and Hol, W.G.J., Biophysical Chemistry 9, ?73-280 (1979).

12. Ploegman, J.H., Drent, G., Kalk, K.H. and Hol, W.G.J., J. Mol. Biol. 127, 149-162 (1979).

13. Moras, D., Olsen, K.W., Sabesan, M.N., Buehner, M., Ford, G.C. and Rossmann, M.G., J. Biol. Chem. 250, 9137-9162 (1975).

14. Biesecker, G., Ieuan Harris, J., Thierry, J.C., Walker, J.E. and Wonacott, A.J., Nature 266, 328-333 (1977).

15. Schulz, G.E., Schirmer, R.H. and Pai, E.F., J. Mol. Biol. 160, 287-308 (1982).

16. Wierenga, R.K., De Jong, R.J., Kalk, K.H., Hol, W.G.J. and Drenth, J., J. Mol. Biol. 131, 55-73 (1979).

17. Wierenga, R.K., Drenth, J. and Schulz, G.E., J. Mol. Biol. (1983) in press.

18. Müller, F., Hemmerich, P., Ehrenberg, A., Palmer, G. and Massey, V., Eur. J. Biochem. 14, 185-196 (1970).

19. Shoham, M. and Steitz, T.A., J. Mol. Biol. 140, 1-14 (1980).

20. Brändén, C.I., Quart. Revs. Biophys. 13, 317 (1980).

21. Rossmann, M.G., Liljas, A., Brändén, C.I. and Banaszak, L.J. in "The Enzymes" (P.D. Boyer, Ed.) Vol. 11, pp. 61-102 (1975).

22. Wierenga, R.K., Hol, W.G.J. and De Maeyer, M.C.H., in preparation.

23. Thieme, R., Pai, E.F., Schirmer, R.H. and Schulz, G.E., J. Mol. Biol. 152, 763-782 (1981).

24. Schirmer, R.H., personal communication.

25. Volz, K.W., Matthews, D.A., Alden, R.A., Freer, S.T., Hansch, C., Keufman, B.T. and Kraut, J., J. Biol. Chem. 257, 2528-2536 (1982).

26. Filman, D.J., Bolin, J.T., Matthews, D.A. and Kraut, J., J. Biol. Chem. 257, 13663-13672 (1982).

27. Johnson, L.N., Jenkins, J.A., Wilson, K.S., Stura, E.A. and Zanotti, G., J. Mol. Biol. 140, 565-580 (1980).

28. Ford, G.C., Eichele, G. and Jansonius, J.N., Proc. Natl. Acad. Sci. USA 77, 2559-2563 (1980).

29. Smith, W.W., Burnett, R.M., Darling, G.D. and Ludwig, M.L., J. Mo. Biol. 117, 195-225 (1977).

30. Watenpaugh, K.D., Sieker, L.C. and Jensen, L.H., Proc. Natl. Acad. Sci. USA 70, 3857-3860 (1973).

31. Ludwig, M.L., Pattridge, K.A., Smith, W.W., Jensen, L.H. and
    Watenpaugh, K.D. in "Flavins and Flavoproteins" (V. Massey &
    C.H. Williams, Jr., Eds.) Elsevier North Holland pp. 19-27 (1982).

32. Halie, L. and Hol, W.G.J., unpublished results.

33. Kelly, J.A., Moews, P.C., Knox, J.R., Frère, J.M. and Ghuysen,
    J.M., Science 218, 479-481 (1982).

34. Tabin, C.J., Bradley, S.M., Bargmann, C.I., Weinberg, R.A.,
    Papageorge, A.G., Scolnick, E.M., Dhar, R., Lowy, D.R. and Chang,
    E.H., Nature 300, 143-149 (1982).

35. Reddy, E.P., Reynolds, R.K., Santos, E. and Barabacid, M., Nature
    300, 149-152 (1982).

36. Taparowsky, E., Suard, Y., Fasano, O., Shimizu, K., Goldfarb,
    M. and Wigler, M., Nature 300, 762-765 (1982).

37. Dhar, R., Ellis, R.W., Shih, T.Y., Oroszlan, S., Shapiro, B.,
    Maizel, J., Lowy, D. and Scolnick, E., Science 217, 934-937
    (1982).

38. Tsuchida, N., Ryder, T. and Ohtsubo, E., Science 217, 937-939
    (1982).

39. Weijer, W.J., Hofsteenge, J., Vereijken, J.M., Jekel, P.A. and
    Beintema, J.J., Biochim. Biophys. Acta 704, 385-388 (1982).

40. Wierenga, R.K. and Hol, W.G.J., accepted by Nature.

41. Scolnick, E.M., Papageorge, A.G. and Shih, T.Y., Proc. Natl.
    Acad. Sci. USA 76, 5355-5359 (1979).

42. Shih, T.Y., Papageorge, A.G., Stokes, P.E., Weeks, M.O. and
    Scolnick, E.M., Nature 287, 686-691 (1980).

43. Papageorge, A.G., Lowy, D. and Scolnick, E.M., J. Virol. 44,
    509-519 (1982).

44. Gay, N.J. and Walker, J.E., Nature 301, 262-264 (1983).

45. Walker, J.E., Saraste, M., Runswick, M.J. and Gay, N.J., EMBO J.
    1, 945-951 (1982).

46. Drenth, J., Jansonius, J.N., Koekoek, R. and Wolthers, B.G. in
    "The Enzymes" (P.D. Boyer, Ed.) Academic Press, New York, Vol.
    3, pp. 485-499 (1971).

47. Wright, C.S., Alden, R.A. and Kraut, J., Nature 221, 235-242
    (1969).

48. Drenth, J., Hol, W.G.J., Jansonius, J.N. and Koekoek, R., Cold
    Spring Harbor Symp. Quant. Biol. 26, 107-116 (1971).

49. Ploegman, J.H., Drent, G., Kalk, K.H. and Hol, W.G.J., J. Mol.

Biol. <u>123</u>, 557-594 (1978).

50. Phillips, D.C., Rivers, P.S., Sternberg, M.J.E., Thornton, J.M. and Wilson, I.A., Biochem. Soc. Trans. <u>5</u>, 642-647 (1977).

51. Ladenstein, R., Epp, O., Bartels, K., Jones, A., Huber, R. and Wendel, A., J. Mol. Biol. <u>134</u>, 199-218 (1979).

52. Rossmann, M.G., Moras, D. and Olsen, K.W., Nature <u>250</u>, 194-199 (1974).

53. Hill, E., Tsernoglay, D., Webb, L. and Banaszak, L.J., J. Mol. Biol. <u>72</u>, 577-591 (1972).

54. Eklund, H., Samama, J.P., Wallén, L., Bränden, C.I., Åkeson, A. and Jones, T.A., J. Mol. Biol. <u>146</u>, 561-567 (1981).

55. Sheriff, S. and Herriott, J.R., J. Mol. Biol. <u>145</u>, 441-451 (1981).

56. Evans, P.R. and Hudson, P.J., Nature <u>279</u>, 500-504 (1979).

57. Watson, H.C., Walker, N.P.C., Shaw, P.J., Bryant, T.N., Wendell, P., Fothergill, L.A., Perkins, R.E., Conroy, S.C., Dobson, M.J., Tuite, M.F., Kingsman, A.J. and Kingsman, S.M., EMBO J. <u>1</u>, 1635-1640 (1982).

58. Pai, E.F., Sachsenheimer, W., Schirmer, R.H. and Schulz, G.E., J. Mol. Biol. <u>114</u>, 37-45 (1977).

59. Honzatko, R.B., Crawford, J.L., Monaco, H.L., Ladner, J.E., Ewards, B.F., Evans, D.R., Warren, S.G., Wiley, D.C., Ladner, R.C. and Lipscomb, W.N., J. Mol. Biol. <u>160</u>, 219-263 (1982).

60. Honzatko, R.B. and Lipscomb, W.N., J. Mol. Biol. <u>160</u>, 265-286 (1982).

61. Phillips, S.E.V., J. Mol. Biol. <u>142</u>, 531-554 (1980).

62. Taylor, S.S., J. Biol. Chem. <u>252</u>, 1799-1806 (1977).

63. Jörnvall, H., Eur. J. Biochem. <u>14</u>, 521-534 (1970).

64. Davidson, B.E., Sajgò, M., Noller, H.F. and Ieuan Harris, J., Nature <u>216</u>, 1181-1185 (1967).

65. Kraut-Sieghel, R.L., Blatterspiel, R., Saleh, M., Schiltz, E. and Schirmer, R.H., Eur. J. Biochem. <u>121</u>, 259-267 (1982).

10 Dihydrofolate reductase: its structure, function, and binding properties

C.R. Beddell

In recent years, dihydrofolate reductase (DHFR) has become one of the most intensively studied enzymes. A wide range of physicochemical techniques has been applied in an attempt to gain knowledge and understanding of the molecular properties of DHFR isolated from many different sources. Much of this interest must stem from the importance of the enzyme as a therapeutic target. The anti-cancer properties of methotrexate, the anti-bacterial properties of trimethoprim and the anti-malarial properties of pyrimethamine are related to inhibition of DHFR. That this enzyme is a therapeutic target in such a diversity of disease states reflects its important metabolic role and wide distribution. As would be expected in an area of such activity, there have been many reviews. That of Blakley (1969) deals extensively with the development of the subject and specific aspects have been reviewed quite recently (Gready, 1980; Blakley, 1981; Jardetzky and Roberts, 1981; Roth and Cheng, 1982). The present brief review, in common with that of Hitchings and Smith (1980) can only high-light how our understanding of DHFR has developed over several decades.

BIOCHEMISTRY

Some fifty years ago, Wills (1931) reported that extracts of autolysed yeast or liver were effective in the treatment of tropical macrocytic anaemia and later it was shown that a diet similar to that associated with the disease in man caused in monkeys a blood condition which responded to the administration of yeast or liver extracts (Wills and Stewart, 1935; Day *et al*, 1938). Similarly it was shown that in yeast, alfalfa and wheat bran there was a factor which stimulated the growth of chicks maintained on highly purified diets (Stokstad and Manning, 1938) and that the diet-induced macrocytic anaemia in chicks could be prevented by feeding a factor from liver (Hogan and Parrott, 1940; Pfiffner *et al*, 1943). In other studies it was found that liver yeast and spinach could provide factors essential to the growth of certain lactic acid bacteria (Snell and Peterson, 1940; Mitchell *et al*, 1941, 1944). Subsequent isolation work established that the responsible factors were folic acid *1* and derivatives containing additional residues of glutamic acid.

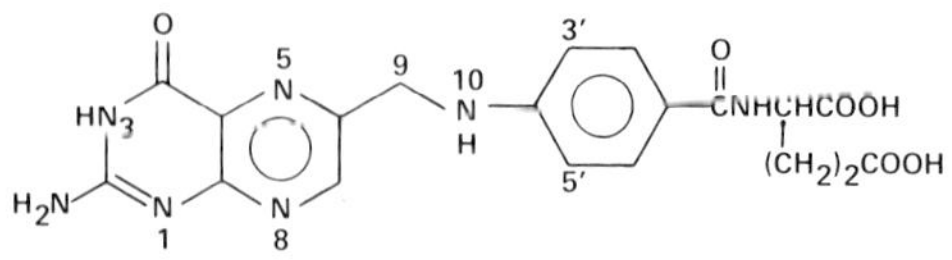

1

Within a year of the establishment of the structure of folic acid by
Angier *et al*, (1946) a number of 2,4-diamino-pteridine derivatives had
been shown to act as folate antagonists (Daniel and Norris, 1947;   Daniel
et al, 1947;   Franklin *et al*, 1948;   Oleson *et al*, 1948).   Effects reported for these antagonists included growth inhibition of various
bacteria and of chicks and rats, and anaemia, all reversed by folic acid
provided the antagonist dose was sufficiently low.   Hitchings and his
colleagues demonstrated that 2,4-diamino-pyrimidines were also inhibitory
to bacterial growth (Hitchings *et al*, 1945) and antagonistic to folic acid
(Hitchings *et al*, 1948).   Since antagonists inhibited the growth of
bacteria which could not synthesise folic acid, as well as those that
could, it appeared that the antagonist was preventing folate use rather
than folate synthesis (Daniel and Norris, 1947).   However, folic acid did
not reverse such antagonism in folic acid synthesising bacteria.   In 1948,
Farber reported that temporary remission of acute leukaemia in children
could be obtained with one of these folate antagonists, aminopterin *2a*.
The synthesis of methotrexate *2b* was reported shortly thereafter (Seeger
et al, 1949).   Methotrexate remains today the single antifolate in
wide use for cancer treatment.

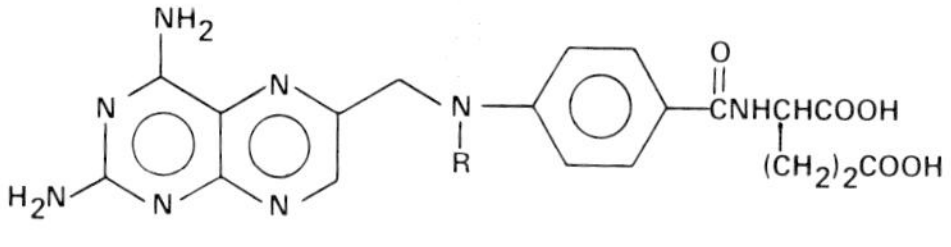

(a)   R = H
(b)   R = CH$_3$

2

It was in the succeeding decade that the mechanism by which folate is
used came to be understood.   Nichol and Welch (1950) showed that aminopterin inhibited conversion of folic acid to 'citrovorum factor' (CF)

(so named because it supported the growth of *Leuconostoc citrovorum*, an organism unable to effectively use folic acid itself) in rat liver and in animals. Hitchings (1952) showed that the 2,4-diaminopyrimidine pyrimethamine *3* prevented this conversion in the rat and in *Streptococcus faecalis*.

3

The synthesis of CF by Shive *et al*, (1950) and Brockman *et al*, (1950) soon led to the proposed structure, 5-formyltetrahydrofolate. Natural and synthetic CF are similar, but the latter is racemic at C6. With the demonstration of an enzyme in liver (DHFR) which reduced folic acid (Futterman and Silverman, 1957) the mapping of folate metabolic inter-relationships entered a new phase.

It is now evident that folate related cofactors are involved in many processes. A simplified scheme is shown in Figure 1. Compared with di-hydrofolate (FH_2), folic acid is a poor substrate for bacterial DHFR and for vertebrate DHFR, and the *de novo* synthesis of folate cofactors proceeds through FH_2 as intermediate and not through folic acid. Indeed the major dietary form of folate is 5-methyltetrahydrofolate (5-methyl-FH_4). Reduced folates lacking 5-substitution are readily oxidised and one role of DHFR is to salvage oxidised cofactor. The ineffectiveness of folate at reversing the effects of the folate antagonists acting at DHFR in organisms which synthesise folates is seen to be due to the small role of folic acid itself in the synthesis of reduced folate cofactors. Where the *de novo* route for FH_2 synthesis is absent, folate can more effectively compete with antagonist at DHFR in the regeneration of reduced folate derivatives.

But why should cells be so sensitive to the inhibition of DHFR? Of the many reactions which involve FH_4 as a carrier of Cl units which in turn become incorporated into purines, into some amino acids and into thymidylate, only the last process diminishes the pool of fully reduced folate derivatives by converting FH_4 to FH_2. Since thymidylate is a constituent of DNA, DNA synthesis can be sustained only through thymine salvage routes, if at all, once FH_4 is exhausted. DHFR is the primary

means of restoring FH4 levels, which otherwise can be replenished only
from exogeneous 5-substituted FH4. Thus the inhibition of DHFR leads to
such effects of folate insufficiency as cessation of DNA synthesis and
also interference with amino acid metabolism.

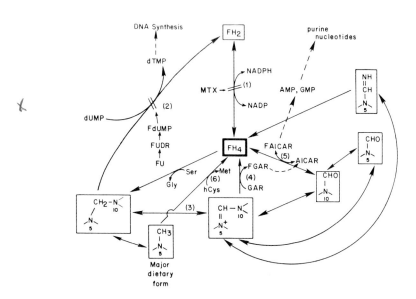

FIGURE 1. Some important folate metabolic pathways, illustrating the
role of dihydrofolate reductase (1). (1) Dihydrofolate reductase
(tetrahydrofolate dehydrogenase, EC 1.5.1.3; (2) thymidylate synthetase
(EC 2.1.1.45); (3) methylene tetrahydrofolate dehydrogenase (EC 1.5.1.5);
(4) phosphoribosylglycinamide formyltransferase (EC 2.1.2.2); (5) phos-
phoribosylaminoimidazolecarboxamide formyltransferase (EC 2.1.2.3); (6)
tetrahydropteroylglutamate methyltransferase (EC 2.1.1.13); FH2 = 7,8-
dihydrofolate; FH4 = 5,6,7,8-tetrahydrofolate; (5>N-CH2-N<) = 5,10-
methylenetetrahydrofolate; (5>N-CH3)=5-methyltetrahydrofolate;
(5>N+=CH-N<) = 5,10-methenyltetrahydrofolate; (5>N-CHO) = 5-formyl-
tetrahydrofolate (citrovorum factor, CF); (5>N-CH=N) = 5-formimino-
tetrahydrofolate; (10>N-CHO) = 10-formyltetrahydrofolate; hCys =
homocysteine, FGAR = formylglycinamide ribonucleotide; GAR = glycina-
mide ribonucleotide; AICAR = aminoimidazolecarboxamide ribonucleotide;
FAICAR = 5-formamidoimidazolecarboxamide ribonucleotide; dUMP = deoxyuridylate; AMP = adenylate; GMP = guanylate; dTMP = thymidylate;
FU = 5-fluorouracil; FUDR = 5-fluorodeoxyuridine; FdUMP = 5-fluoro-
deoxyuridylate.

PROPERTIES OF DIHYDROFOLATE REDUCTASES

Dihydrofolate reductase is widely distributed. Tables 1 and 2 (modified from Hitchings and Smith, 1980) show respectively how various inhibitors differentially inhibit enzymes from different sources and how some general properties of different DHFRs compare.

TABLE 1

Comparative IC_{50} Values for
Various Enzymes and Inhibitors
Values X $10^{-8}M$

Source of Enzyme	*E.coli* (1)	*S.aureus* (1)	Rat Liver(1)	*P.berghei* (2)	*T.equiperdum* (3)
COMPOUND					
MTX*	0.1	0.2	0.21	0.07	0.02
PYR	250	300	70	0.05	20
TMP	0.5	1.5	26000	7.0	100
BuPP	50	4	46	1.7	50
MeBuPP	2	7	26	1.0	–
BuDHT	65000	50000	14	0.8	2000

*MTX, 2,4-diamino-methylpteroylglutamate (methotrexate); PYR, 2,4-diamino-5-p-chlorophenyl-6-ethylpyrimidine (pyrimethamine); TMP, 2,4-diamino-5(3',4',5'-trimethoxybenzyl)pyrimidine(trimethoprim); BuPP, 2,4-diamino-6-butylpyrido[2,3-d]pyrimidine; MeBuPP, 2,4-diamino-5-methyl-6-butylpyrido[2,3-d]pyrimidine; BuDHT, 1-(p-butylphenyl)-1,2-dihydro-2,2-dimethyl-4,6-diamino-s-triazine.
(1)   Burchall and Hitchings (1965);   (2)   Ferone *et al* (1969);
(3)   Jaffe (1972).

It is now evident that structural differences between enzymes are responsible for the highly selective inhibition of *E.coli* DHFR, relative to vertebrate DHFR, exhibited by the anti-bacterial drug trimethoprim and the particular selectivity of pyrimethamine for DHFR of the malarial protozoan *Plasmodium berghei*. The Crithidia 'DHFR' appears also to be a thymidylate synthetase (Ferone and Roland, 1980). Comparison of the amino acid sequence of the various DHFRs reveals further the extent of the structural differences between them. Figure 2 shows the amino acid sequence of chicken DHFR and emphasises how this sequence compares with completed amino acid sequences for DHFR from other sources. The chicken sequence is closely homologous (72-89% identity) with those of other vertebrate DHFRs

TABLE 2

Properties of Purified Dihydrofolate Reductases

Enzyme Source	M.W.*	No.of subunits	Turn over No.	pI	pH optimum	Activated by salt	Kinetic Constants(μM)		Ref.
							NADPH	FAH$_2$	
E.coli (TMP resistant)	18250	1	1800	4.6	4,7		4.4	8.9	(1)
E.coli (MTX resistant)	18250	1	600		6.8		6.45	0.44	(2)
S.faecium var.Durans	19600	1	900		5.8	+	5.0	10.1	(3)
L.casei	18300	1	175	6.25	7.3	+	0.78	0.36	(4)
T4 phage	44500	2	6675		7		18	2.3	(5)
R-plasmid (R67)	33780	4	36				6.4	4.1	(6)
L1210	21460	1	230 / 1050	8.1	4,7.5	+	1.36 / 5.0	0.3 / 0.4	(7)
Porcine Liver	21450	1	810	8.4	5,7.5	+	3.2	0.74	(8)
Bovine Liver	21450	1	700 / 2180	7.15 / 6.8	5	+	22	7.5	(9)
Chicken Liver	21450	1	315	8.4	4.0,7.4	+	0.63	0.18	(10)
Crithidia fasciculata	107000	2	1280				6.7	<2	(11)

*Molecular weight calculated from amino acid sequence except for T4 phage and Crithidia which have not been sequenced.

(1) Baccanari et al.(1977); (2) Bennett et al.(1978), Poe et al.(1973); (3) Nixon and Blakley (1968), Gleisner et al.(1974), Blakley et al.(1971); (4) Gundersen et al.(1972), Freisheim et al.(1978), Dann et al.(1976); (5) Erickson and Mathews (1971, 1973), Purohit et al.(1981); (6) Pattishall et al.(1977), Smith et al.(1979), Stone and Smith (1979); (7) Perkirs and Bertino (1966), Reyes and Huennekens (1967), Huennekens et al.(1976), Stone et al.(1979), McCullough et al.(1971); (8) Smith et al.(1979); (9) Lai et al.(1979), Bauman and Wilson (1975), Peterson et al.(1975), Kaufman and Kemerer (1976), Rowe and Russel (1973); (10) Barbehenn and Kaufman (1982); (11) Ferone and Roland (1980).

and also with human DHFR (J. Freisheim, personal communication). DHFR
from different species of bacteria show only roughly 25-35% residue
identity with one another or with vertebrate DHFR and this also appears
to be true for *Neisseria gonorrhoea* DHFR (D. Stone, personal communication).
Of the 17 fully conserved residues shown in Figure 2, only 5 occur beyond
residue 75 (chicken numbering) and therefore it would be anticipated that
the first half of the amino acid sequence will contribute more markedly
than the latter half to the construction of the enzyme active site.

The crystallographic investigation of dihydrofolate reductase has
been conducted by groups at La Jolla, USA and Beckenham, UK. The crystal
structure of the binary complex of methotrexate with the *E.coli* enzyme
(Matthews *et al.*,1977) was followed shortly thereafter by that of the
ternary complex of methotrexate, reduced nicotinamide adenine dinucleotide phosphate (NADPH) with *L.casei* dihydrofolate reductase (Matthews
et al., 1978, 1979; Bolin *et al.*, 1982; Filman *et al.*, 1982). Further
crystallographic work with the *E.coli* enzyme has shown how the anti-
bacterial drug trimethoprim (Baker *et al.*, 1981a) and certain analogues
(Baker *et al.*, 1981b) interact in the binary complex. This group has
work in progress on the trimethoprim -NADPH-*E.coli* dihydrofolate
reductase ternary complex. Crystallographic studies on the vertebrate
enzymes were initiated after those on the bacterial enzymes. The structure
of the ternary complex between chicken liver dihydrofolate reductase,
NADPH and a dihydrotriazine inhibitor (Volz *et al.*, 1982) or trimethoprim
(Matthews and Volz, 1982) are reported and crystallographic work is

---

FIGURE 2. Comparison of dihydrofolate reductase sequences.
A, chicken liver (Kumar *et al.*, 1980); B, mouse L1210 (Stone *et al.*,
1979); C, porcine liver (Smith *et al.*, 1979); D, bovine liver (Lai *et al*
1982); E, *Escherichia coli* RT500 (Stone *et al.*, 1977, Baccanari *et al.*,
1981); F, *Escherichia coli* MB1428 (Bennett *et al.*, 1978); G, *Streptococcus
faecium* (Gleisner *et al.*, 1974); H, *Lactobacillus casei* (Bitar *et al.*,
1977). The residue numbering shown is for chicken liver (top), *E. coli*
(middle) and *L.casei* enzymes and the alignment of the sequences, together
with insertion of gaps where appropriate, follows that of Volz *et al.*
(1982) and is based on three-dimensional structural superposition of
the chicken, *E.coli* and *L.casei* enzymes. Identity with the corresponding
residue in the chicken liver enzyme is shown by means of a dot. The
high degree of homology between vertebrate enzymes (A-D) is clearly
evident and also the conservation for all enzymes shown at positions
9, 16-17, 23-24, 27, 34, 38, 53, 56, 67, 75, 116-117, 136, 143 and
145 (chicken numbering).

```
     1   2   3   4   5   6   7   8   9  10  11  12  13  14  15  16  17  18  19  20  21  22  23  24  25  26  27  28  29  30
A  VAL ARG SER LEU ASN SER ILE VAL ALA VAL CYS GLN ASN MET GLY ILE GLY LYS ASP GLY ASN LEU PRO TRP PRO LEU ARG ASN GLU
B   .   .  PRO   .   .  CYS   .   .   .   .   .  SER   .   .   .   .   .   .  ASP   .   .   .   .   .   .   .   .   .   .
C   .   .  PRO   .   .  CYS   .   .   .   .   .  SER   .   .   .   .   .   .  ASN   .  ASP   .   .   .   .   .   .   .   .
D   .   .  PRO   .   .   .   .   .   .   .   .   .   .   .   .   .   .   .  ASP   .   .   .   .   .   .   .   .   .   .   .
     1   2   3   4   5   6   7   8   9  10  11  12  13  14  15  16  17  18  19  20  21  22      23  24  25  26  27
E  -   -  MET ILE SER LEU   .  ALA   .  LEU ALA VAL ASP ARG VAL   .  MET GLU ASN ALA MET   .   .   .  ASN   .  PRO ALA ASP
F  -   -  MET ILE SER LEU   .  ALA   .  LEU ALA VAL ASP ARG VAL   .  MET GLU ASN ALA MET   .   .   .  ASN   .  PRO ALA ASP
G  -   -  MET PHE ILE   .  MET TRP   .  GLN ASP LYS   .  GLY LEU   .   .   .   .  LEU   .   .   .  ARG   .  PRO   .  ASP
     1   2   3   4   5   6   7   8   9  10  11  12  13  14  15  16  17  18  19  20  21      22  23  24  25  26
H  -   -   -  THR ALA PHE LEU TRP   .  GLN ASN ARG ASP GLY LEU   .   .   .   .  HIS   .   .   .  -  HIS   .  PRO ASP ASP
```

```
    31  32  33  34  35  36  37  38  39  40  41  42  43  44  45  46  47  48  49  50  51  52  53  54  55  56  57  58  59  60
A  TYR LYS TYR PHE GLN ARG MET THR SER THR SER HIS VAL GLU GLY LYS GLN ASN ALA VAL ILE MET GLY LYS LYS THR TRP PHE SER ILE
B  PHE   .   .   .   .   .  THR   .  SER   .   .   .   .   .   .   .  LEU   .   .   .  ARG   .   .   .   .   .   .   .   .   .
C   .   .   .   .   .   .  THR   .  SER   .   .   .   .   .   .   .  LEU   .   .   .  ARG   .   .   .   .   .   .   .   .   .
D  PHE GLN   .   .   .   .  THR VAL   .  SER   .   .   .   .   .   .  LEU   .   .   .  ARG   .   .   .   .   .   .   .   .   .
    28  29  30  31  32  33  34  35  36                      37  38  39  40  41  42  43  44  45  46  47  48  49  50
E  LEU ALA TRP   .  LYS   .  ASN   .  LEU  -   -   -   -   -   .  ASN LYS PRO   .   .   .  ARG HIS   .  GLU   .   .
F  LEU ALA TRP   .  LYS   .  ASN   .  LEU  -   -   -   -   -   .  ASN LYS PRO   .   .   .  ARG HIS   .  GLU   .   .
G  MET ARG PHE   .  ARG GLU HIS   .  MET  -   -   -   -   -   .  ASP LYS ILE LEU VAL   .  ARG   .  TYR GLU GLY MET
    27  28  29  30  31  32  33  34  35                      36  37  38  39  40  41  42  43  44  45  46  47  48  49
H  LEU HIS   .  ARG ALA GLN   .  VAL  -   -   -   -   -   .  GLY LYS ILE MET VAL VAL   .  ARG ARG   .  TYR GLU   .  PHE
```

```
    61  62  63  64  65  66  67  68  69  70  71  72  73  74  75  76  77  78  79  80  81  82  83  84  85  86  87  88  89  90
A  PRO GLU LYS ASN ARG PRO LEU LYS ASP ARG ILE ASN ILE VAL LEU SER ARG GLU LEU LYS GLU ALA PRO LYS GLY ALA HIS TYR LEU SER
B   .   .   .   .   .   .   .   .   .   .   .   .   .   .   .   .   .   .   .  PRO   .  ARG   .   .   .   .  PHE   .  ALA
C   .   .   .   .   .   .   .   .   .   .   .   .   .   .   .   .   .   .   .  PRO   .  GLN   .   .   .   .  PHE   .  ALA
D   .   .   .   .   .   .   .   .   .   .   .   .   .   .   .   .   .   .   .  PRO   .   .   .   .   .   .  PHE   .  ALA
    51              52  53  54  55  56  57  58  59  60  61  62  63  64  65  66      67  68  69  70  71      72  73  74  75
E  GLY  -   -   -   .  PRO GLY   .  LYS   .   .  ILE   .   .   .  SER GLN PRO  -  GLY THR ASP ASP ARG   .  VAL THR TRP VAL
F  GLY  -   -   -   .  PRO GLY   .  LYS   .   .  ILE   .   .   .  SER GLN PRO  -  GLY THR ASP ASP ARG   .  VAL THR TRP VAL
G  GLY  -   -  LYS LEU SER   .  PRO TYR   .  HIS ILE   .   .  THR THR GLN LYS ASP PHE LYS VAL GLU GLY LYS ASN ALA GLU VAL
    50          51  52  53  54  55  56  57  58  59  60  61  62  63  64  65  66  67  68  69  70  71  72      73  74  75  76
H   .   .   -   -  LYS   .   .   .  PRO GLU   .  THR   .  VAL   .   .  THR HIS GLN GLU ASP TYR GLN ALA GLN   -  ALA VAL VAL VAL
```

```
    91  92  93  94  95  96  97  98  99 100 101 102 103 104 105 106 107 108 109 110 111 112 113 114 115 116 117 118 119 120
A  LYS SER   .  LEU ASP ASP ALA LEU ALA LEU LEU ASP SER PRO GLU LEU LYS SER LYS VAL ASP MET VAL TRP ILE VAL GLY GLY THR ALA
B   .   .   .   .   .   .   .  ARG   .  ILE GLU GLN   .   .   .  ALA   .   .   .   .   .   .   .   .   .   .   .  SER SER   .
C   .   .   .   .   .   .   .  LYS   .  THR GLU GLN   .   .   .  LYS ASP   .   .   .   .   .   .   .   .   .   .   .  SER SER   .
D   .   .   .   .   .   .   .  GLU   .  ILE GLN ASN   .   .   .  THR ASN   .   .   .   .   .   .   .   .   .   .   .  SER SER   .
    76  77  78  79  80  81  82  83  84  85                      86  87  88  89  90  91  92  93  94  95  96  97  98  99
E   .   .  VAL ASP GLU ALA ILE ALA   .  CYS  -   -   -   -   -  GLY THR   .  PRO GLU ILE MET VAL ILE   .  GLY ARG   .
F   .   .  VAL ASP GLU ALA ILE ALA   .  CYS  -   -   -   -   -  GLY ASN   .  PRO GLU ILE MET VAL ILE   .  GLY ARG   .
G  HIS   .  ILE ASP VAL GLU GLY ALA TYR ALA LYS GLN  -   -   -  ASP ILE PRO GLU ASP ILE TYR VAL SER   .  SER ARG ILE
    77  78  79  80  81  82  83  84  85  86  87  88                      89  90  91  92  93  94  95  96  97  98  99 100 101 102
H  HIS ASP VAL ALA ALA VAL PHE ALA TYR ALA LYS GLN  -   -   -   -  HIS LEU ASP GLN GLU LEU VAL   .  ALA   .   .  ALA GLN ILE
```

```
   121 122 123 124 125 126 127 128 129 130 131 132 133 134 135 136 137 138 139 140 141 142 143 144 145 146 147 148 149 150
A  TYR LYS ALA ALA MET ASP LYS PRO ILE ASN HIS ARG LEU PHE VAL THR ARG ILE LEU HIS GLU PHE GLU SER ASP THR PHE PHE PRO GLU
B   .  GLU GLN   .  ASN GLU   .  GLY HIS LEU   .   .   .   .   .   .   .  MET GLN   .   .   .   .   .   .   .   .   .   .   .
C   .  GLU   .  ASN   .  GLY HIS ILE   .   .   .   .   .   .   .   .   .  MET LYS   .   .   .   .   .   .   .   .   .   .   .
D   .  GLU   .  ASN   .  GLY HIS VAL   .   .   .   .   .   .   .   .   .  MET GLN   .   .   .   .  ALA   .   .   .   .   .   .
   100 101 102 103 104         105 106 107 108 109 110 111 112 113 114 115 116 117 118 119 120 121 122 123 124 125 126 127
E   .  GLU GLN PHE LEU  -   -  LYS ALA GLN LYS   .  TYR LEU   .  HIS   .  ASP ALA GLN VAL   .  GLY   .   .  HIS   .   .  ASP
F   .  GLU GLN PHE LEU  -   -  LYS ALA GLN LYS   .  TYR LEU   .  HIS   .  ASP ALA   .  VAL   .  GLY   .   .  HIS   .   .  ASP
G  PHE GLN   .  LEU LEU  -   -  GLU THR LYS ILE ILE TRP ARG   .  LEU   .  ASP ALA   .   .  GLY   .   .   .  ILE GLY   .
   103 104 105 106 107         108 109 110 111 112 113 114 115 116 117 118 119 120 121 122 123 124 125 126 127 128 129 130
H  PHE THR   .  PHE LYS   .   .  ASP ASP VAL ASP THR   .  LEU   .   .   .  LEU ALA GLY SER   .   .  GLY   .   .  LYS MET ILE PRO
```

```
   151 152 153 154 155 156 157 158 159 160 161 162 163 164 165 166 167 168 169 170 171 172 173 174         175 176 177
A  ILE ASP PHE LYS ASP PHE LYS LEU LEU THR GLU TYR PRO GLY VAL PRO ALA ASP ILE GLN GLU GLU ASP ASP  -   -  ILE GLN TYR
B   .   .  LEU GLY LYS TYR   .   .   .  PRO   .   .   .  LEU GLU GLU VAL ALA   .   .   .   .   .   .   -   -  LYS   .
C   .   .  LEU GLU LYS TYR   .   .   .  SER   .  LYS SER   .   .  SER   .  VAL   .   .   .  LYS   .   -   -  LYS   .
D   .   .  PHE GLU LYS TYR   .   .   .  PRO   .   .   .  LEU   .  VAL   .   .   .   .  LYS   .   .   .   -   -  LYS   .
   128 129 130 131 132 133 134 135 136 137 138                     139 140 141 142 143 144 145 146 147 148 149 150 151
E  TYR GLU PRO ASP   .  TRP GLU SER VAL PHE SER  -   -   -   -  GLU PHE HIS ASP   .  ALA ILE ASN SER HIS SER GLU   .
F  TYR GLU PRO ASP   .  TRP GLU SER VAL PHE SER  -   -   -   -  GLU PHE HIS ASN ALA   .  ALA GLN ASN SER HIS SER   .
G   .  PHE THR HIS   .  GLU   .  VAL   .  HIS GLU   .  ILE   .   .  VAL ASN   .  ASN GLN TYR   .  PRO HIS ARG PHE GLN
   131 132 133 134 135 136 137 138 139 140 141                     142 143 144 145 146 147 148     149 150 151 152 153
H  LEU ASN TRP ASP   .   .  THR LYS VAL SER SER  -   -   -   -  ARG THR VAL   .  ASP THR ASN   .  PRO ALA LEU THR HIS
```

```
   178 179 180 181 182 183 184 185 186 187 188 189
A  LYS PHE GLU VAL TYR GLN LYS SER VAL LEU ALA GLN
B   .   .   .   .   .   .  GLU   .  LYS ASP
C   .   .   .   .   .   .  GLU   .  ASN ASN
D   .   .   .   .   .   .  GLU   .  ASN ASN
   152 153 154 155 156 157 158 159
E  CYS   .  GLU ILE LEU GLU ARG ARG
F  LYS   .  LYS ILE LEU GLU ARG ARG
G   .  TRP GLN LYS MET SER   .  VAL
   154 155 156 157 158 159 160 161 162
H  THR TYR   .   .  TRP   .   .  LYS ALA
```

FIGURE 2

in progress on the mouse enzyme in the UK and on the R67 plasmid enzyme
in the USA. The crystallographic studies of the enzyme architecture
have revealed a conservation of three dimensional structure far more
pronounced than the extent of residue identity between amino acid sequences (indeed these sequences were aligned in Figure 2 using the three
dimensional alignment of Volz *et al.*, 1982).

The architecture of the chicken and *E.coli* dihydrofolate reductase is
shown in Figure 3. The chain participates in the formation of an eight-
stranded β-sheet, with one strand, H, inserted between F and G and
antiparallel. Four α-helices, denoted B, C, E and F, are involved in
the interstrand connections which lead to strands, B, C, E and F respectively. There is a crescent-shaped cleft backed by strands A, B, C, E
and F and the carboxyl end of αF, and flanked on one side by the B helix
and the preceding connecting strand from βA and on the other side by the
C helix and subsequent strand connecting to βC. The cofactor occupies
one half of this cleft, in an extended conformation, with the adenine
outermost, and reduced nicotinamide near the middle of the cleft. The
diamino inhibitors which have been studied are found to bind close to the
nicotinamide in the other half of the cleft.

The close similarity of these structures is evident. This is emphasised by the topology diagram (Figure 4) for the structured DHFRs.
Associated with the non-planarity (twist) of the β-sheet it has been
observed that interconnections between strands take the shortest (right-
handed) route (for review see Richardson, 1981) and DHFR conforms to this
rule. Of the seven inserted segments and two deletions for vertebrate
enzyme relative to bacterial all but one occur in loops connecting to
α-helices or β-strands and are at the surface of the enzyme. Extra segments are thereby accommodated as outgrowths without undue interference
with the overall folding. This is even true for the large insertion in
the G strand (a "β-blow out")

ENZYME BINDING: METHOTREXATE

Figure 5 shows how methotrexate binds to *E.coli* DHFR. X-ray crystallographic studies of proteins have seldom provided information on the positions
of labile hydrogen and therefore on ionisation states and protonation positions. For methotrexate in solution at neutral pH, the carboxyl groups are
anionic and the pteridine moiety largely unprotonated (See Table 3). Cocco
et al. (1981a,b) and Blakley *et al.* (1982) have shown that the pKa at
N1 for methotrexate is >10 when complexed with *S. faecium*, *L. casei*

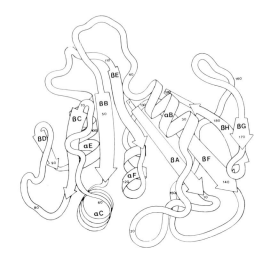

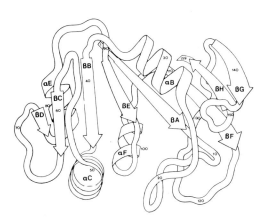

FIGURE 3. Schematic illustration of the folding of chicken liver (top) and of *E.coli* dihydrofolate reductase. The overall folding of the polypeptide chain is dominated by an eight-stranded β-sheet composed of seven parallel strands and a single antiparallel strand at the C-terminus. The sheet, viewed along the strands, shows the usual right-handed twist. The molecule contains four helical regions. Cofactor and inhibitors bind in a cleft some 15Å deep which cuts across one face of the molecule and gives the structure a bi-lobed appearance.

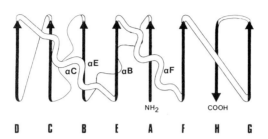

FIGURE 4. Typical topology for the peptide chain of DHFR. Strands in β-sheet are denoted by arrows and helices by waves. The topology of the β-strand interconnections is shown by a crossing above or below the topological plane

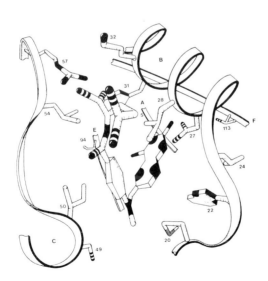

FIGURE 5

FIGURE 5. Schematic illustration of the active site of *E.coli* dihydrofolate reductase (view as in Figure 3) with bound methotrexate. Selected
atoms are highlighted, oxygen by stripes, nitrogen in black and sulphur
by hatching. The proximity of N1 and the 2-amino group of methotrexate
to the carboxyl group of residue 27 and of the α-carboxyl group to the
guanidinium group of residue 57 are examples of charge-assisted hydrogen-
bonded interactions. The benzene ring in methotrexate is in hydrophobic
contact with the side chains of residue 50 (isoleucine) and 28 (leucine).

TABLE 3

pKa Estimates for Titrating Atoms in Folic Acid,
Dihydrofolate, Tetrahydrofolate, Methotrexate
and Trimethoprim in Solution

Atom COMPOUND	N1	N3	N5	N10	α carboxyl	γ carboxyl	Ref
Folic Acid	2.35	8.38	<-1.5	0.2			1
Dihydro-folate	1.38	9.54	3.84	0.28			1
Tetrahydro-folate	1.24	10.5	4.82	-1.25	3.5	4.8	1
Methotrexate	5.71 5.73		<-1.5	0.5	3.36	4.7	1 2
Trimethoprim	7.41						2

(1) Poe (1977) and references therein;   (2) Blakley *et al.* (1982)

---

and bovine liver enzymes. This implies that the protonated compound binds
at least 20000 times more tightly to DHFR than does the unprotonated.
This would correspond to a difference in binding energy for the two
species of compound of at least 24 kJ/mol. At pH 7.4, when allowance is
made for the low proportion of compound protonated at N1 in solution the
corresponding differential is at least 14 kJ/mol. The pKa shift presumably reflects the proximity of the active site carboxylate to N1.
In *L.casei* DHFR, histidine 28 imidazole interacts with the γ-carboxylate.
The pKa of this histidine, 6.5, rises when methotrexate is bound.
Birdsall *et al.* (1978) found that two fragments of methotrexate, 2,4-
diaminopyrimidine (DAP) and p-aminobenzoylglutamate (PAB) show about
50-fold mutual cooperativity in binding to *L.casei* DHFR. Either DAP or
PAB bind more strongly to the enzyme when PAB or DAP respectively is
already binding to the enzyme. The corresponding binding energies for
methotrexate, DAP and PAB were -50, -18, -17 kJ/mol and the combined

binding energy for DAP and PAB, -44 kJ/mol was about 9 kJ/mol greater
than the individual fragment sum but still 6 kJ/mol less than that of
the complete methotrexate molecule. Cooperativity may arise through
direct interactions between fragments, through induced conformational
changes in the protein or the favouring of one of a number of protein
conformers, and through local changes in the vibration amplitudes of
atoms in the binding site and in the ligands. At present it is not possible to relate such events directly to binding energy.

ENZYME BINDING: TRIMETHOPRIM

The antibacterial drug trimethoprim *4* binds selectively to bacterial

4

enzyme. Baker *et al.*, (1981a) have studied crystallographically how this
drug binds to *E.coli* DHFR in the absence of cofactor and Matthews and Volz
(1982) how it binds to chicken DHFR in the presence of reduced cofactor.
The compound binds in the two enzymes with the 2,4-diaminopyrimidine ring
making interactions which are at least approximately analogous with those
made by the 2,4-diaminopteridine ring of methotrexate. However, the torsion angles between the rings are different in the two instances, providing different conformations for trimethoprim in the two complexes. The
two complexes differ in DHFR type and in presence of cofactor; either
difference might in principle be responsible for the lack of a single
trimethoprim conformation. Studies on the *E.coli* DHFR-trimethoprim-
NADPH complex in progress (J.N. Champness, personal communication) may
establish whether cofactor alone modifies the conformation of the bound
drug. Figure 6 shows schematically the active site of *E.coli* DHFR with
trimethoprim bound.

For trimethoprim [13]C n.m.r. studies (Blakley *et al.*, 1982) have indicated that the pKa of trimethoprim at N1 is 7.4 in solution and at least
10.0 in the complex with *S.faecium* or *L.casei* DHFR. With the latter
enzyme, G. Roberts and his colleagues (personal communication) have

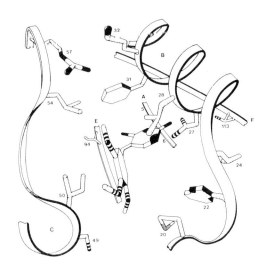

FIGURE 6. Schematic illustration of the active site of *E.coli* dihydrofolate reductase (view as in Figure 5) with bound trimethoprim. Like
methotrexate, N1 and the 2-amino group interact with aspartate-27. As
with the benzene ring in methotrexate, the benzene ring in trimethoprim
makes hydrophobic interaction with isoleucine-50 and leucine-28.

established directly by ^{15}N n.m.r. that N1 is protonated.

The relationship between structure and inhibitory potency has been
extensively reviewed (Roth and Cheng, 1982). Li *et al.* (1982) have
related the inhibitory potency to properties in the ligand molecule.
Empirically it was observed that the following equations were statistically significant:

E.coli DHFR

$$\log 1/K_{i(app)} = 1.36 \ MR'_{3',5'} + 0.88 \ MR'_{4'} + 0.75\pi_{3',4',5'}$$
$$-1.07 \log (\beta.10^{\pi_{3',4',5'}} + 1) + 6.20$$

n = 43;   r = 0.903;   s = 0.290; π_O = 0.25;   log β = -0.12

Bovine liver DHFR

$$\log 1/K_{i(app)} = 0.48\pi_{3',5'} - 1.25 \log(\beta.10^{\pi_{3',5'}} + 1) + 0.13MR_{3'}$$
$$+ 0.24 \ \Sigma\sigma + 5.43$$

n = 42;   r = 0.875;   s = 0.227; π_O = 1.52;   log β = -1.98

For trimethoprim binding to $E.coli$ DHFR the most important property is a specially selected parameter MR' which represents substituent bulk for small groups, but is truncated at a bulk approximately equal to that of methoxyl, for groups larger than this threshold size. Furthermore the coefficient applying to the $3',5'$ (i.e. meta) substituents is more influential than that applying to the $4'$ (i.e. para) substituent. By contrast, the most important property in relation to inhibition of bovine enzyme is π, a measure of the ability of a substituent to influence partitioning between water and n-octanol. The relationships in π are bilinear and π_0 is the optimum value for π. Sometimes π has been taken to be a measure of hydrophobicity but since compounds can hydrogen bond with octanol and with the water dissolved in it, such an interpretation is probably naive. However, it may be that the hydrophobicity which is optimum for inhibition of bovine enzyme ($\pi_0 = 1.52$) is higher than that for $E.coli$ enzyme ($\pi_0 = 0.25$). These observations are in good accord with the crystallographic observations. For the complex of trimethoprim with $E.coli$ DHFR, small pockets exist between the benzyl group of the ligand and, on one edge, the side chain of phenylalanine-31 and, on the other edge, the nicotinamide of the reduced cofactor, the position for which may be estimated from that observed with the $L.casei$ enzyme. These pockets could accommodate a group the size of methoxyl but larger flexible groups would spill out into solution and into loose contact with the enzyme. In contrast to the pockets for meta substituents, the para substituent is between leucine-28 and isoleucine-50 and substituents longer than methoxyl, will extend beyond this hydrophobic 'sandwich' into solution. In the corresponding complex for vertebrate DHFR, it appears that the benzyl ring of the ligand is in a substantial hydrophobic pocket and this would require a different binding mode for the inhibitor, perhaps involving a different conformation.

An attempt to use the structure of the $E.coli$ enzyme to design an inhibitor more potent than trimethoprim has been described (Kuyper et $al.$ 1982). In place of the methoxy substituent in one meta position various carboxyalkoxy substituents were incorporated to enable an anionic carboxylate to interact with arginine-57 in the $E.coli$ enzyme. As predicted, the inhibitory potency of these compounds was improved and optimum length for the substituent was that expected to provide optimum charge-assisted hydrogen-bonding   (See Table 4).

TABLE 4

Relative affinities of some analogues of trimethoprim for E.coli DHFR

R	Affinity relative to trimethoprim	
	Binary Complex	Ternary Complex
CH_3 (trimethoprim)	1.0	1.0
CH_2CO_2H	0.8	0.5
$(CH_2)_2CO_2H$	3.4	3.5
$(CH_2)_3CO_2H$	6.7	37.
$(CH_2)_4CO_2H$	7.7	20.
$(CH_2)_5CO_2H$	16.	54.
$(CH_2)_6CO_2H$	7.7	26.
$CH_2CO_2CH_3$		0.1
$(CH_2)_3CO_2CH_2CH_3$		2.8
$(CH_2)_4CO_2CH_3$		1.7
$(CH_2)_5CO_2CH_3$		1.5
$(CH_2)_6CO_2CH_3$		0.7

To test whether the carboxylate needed to be anionic, esters were prepared. These did not show the marked increase in inhibitory potency. To
establish whether the compounds did in actuality bind as predicted, two
were studied crystallographically (see Figure 7). The most potent inhibitor was found to make the expected charge-reinforced pair of hydrogen
bonds with the guanidinium group. The other inhibitor studied was the
smallest to show an increase in inhibitory potency. It was found to be
unable to place the carboxylate as close to the guanidinium group, although a single charge-reinforced hydrogen-bond was present. Evidently
this was the shortest carboxyalkoxy group to be able to hydrogen-bond with
arginine-57.

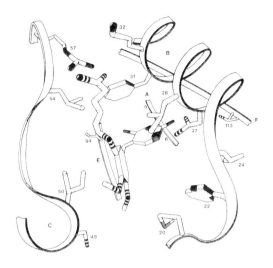

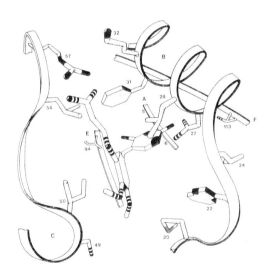

FIGURE 7. Schematic illustration of the active site of *E.coli* dihydrofolate reductase (view as in Figure 5) with two analogues of trimethoprim.
Each bears a carboxylate group, but the longer linkage between this group
and the benzene ring in one (top) permits a closer interaction with the
guanidinium group of arginine-57 for this compound than for the other
(bottom).

ENZYME BINDING: COFACTOR

There is considerable evidence about how cofactors interact with DHFR
and about their relationship to the binding of inhibitors and substrates.
Table 5, from Filman *et al.* (1982), shows the main interactions between
reduced cofactor and enzyme. Feeney *et al.* (1975) using ^{31}P have concluded that the 2'-phosphate group binds in dianionic form. This would
be compatible with the existence of protonated groups in the protein
interacting with it and also with the local effect of the dipole of
helix C, which terminates quite near to the phosphate group. There is
evidence that the interaction between NADPH and inhibitors is cooperative.
Birdsall *et al.* (1980a,b; 1981) have shown with *L.casei* enzyme NADPH
cooperativity factors of 670 with methotrexate, 130 with trimethoprim
and 0.002 with 5-formyltetrahydrofolate. For $NADP^+$ the corresponding
factors are 12, 2 and 0.33. Baccanari, Daluge and King (1982) have shown
that for 2,4-diamino-5-benzyl-pyrimidine and certain mono-, di- and trisubstituted analogues cooperativity increases with extent of substitution.
Since cooperativity is much less with the vertebrate DHFR, the selectivity
which trimethoprim itself manifests for bacterial DHFR relative to vertebrate DHFR is most marked (e.g. 2850) when NADPH is bound and considerably
less (e.g. 93) when NADPH is absent. Whether these cooperative effects
are due to direct interaction between cofactor and compound or due to
ligand-induced effects upon the enzyme or are due to both is not yet established, but some support for the first hypothesis comes from the observation that the nicotinamide ring is near the inhibitor (methotrexate) in
the *L.casei* DHFR complex.

ENZYME BINDING: SUBSTRATE

Less is known about substrate binding. The studies of Hood and Roberts
(1978) and of Cocco *et al.* (1981a,b) fail to indicate that the pteridine
in folate is protonated when bound to DHFR. It is also clear that folate
and dihydrofolate bind with the pteridine ring plane orientation opposite
to that seen for methotrexate. The enzyme transfers hydrogen from the
4-pro-R hydrogen of NADPH which is on the A side of the ring (Pastore
and Friedkin, 1962; Blakley *et al.*, 1963; Pastore and Williamson, 1968;
Poe and Hoogsteen, 1974; Gupta *et al.*, 1977) to the S position at C6
and C7 (Charlton *et al.*, 1979; Fontecilla-Camps *et al.*, 1979). The
observed cofactor orientation is compatible with transfer of hydride from
the A side of the nicotinamide ring. Only by inverting the ring plane

TABLE 5 (from Filman *et al.* 1982)

Component of NADPH		Contact in *L.casei* DHFR		Sequence Conservation
		Hydrophobic/Van der Waals		
Adenine		Leu-62	sidechain	conserved
		His-64	sidechain	
		Thr-63	mainchain	
		Gln-101	Cα, Cβ, Cγ	
		Ile-102	sidechain	
AMN ribose		Gly-42	mainchain	conserved
		Arg-43	mainchain	ex.chicken(Lys)
		Gln-101	sidechain	
		Ile-102	sidechain	
NMN ribose		Ile-13	sidechain	conserved
		Gly-14	mainchain	conserved
		Gly-99	mainchain	conserved
Carboxamide O7		Trp-5	Cα, Cβ, Cδ1	
		Ala-6	Cα, Cβ	conserved
		MTX	C4A, C8A, N8	
Carboxamide N7		Ile-13	sidechain	conserved
		Leu-19	Cδ1	conserved
		Trp-21	Cξ2	conserved
Nicotinamide (pyridine)		Ile-13	sidechain	conserved
		Leu-19	sidechain	conserved
		Thr-45	sidechain	conserved
		Phe-103	sidechain	Tyr or Phe
		MTX	pyrazine	
		Gly-98	carbonyl C	conserved Gly-Gly
		Gly-99	amido N, Cα	and *cis*-peptide
	C2	Ile-13	carbonyl O	conserved
	C6	Thr-45	Oγ1	conserved
	C4, C5	Ala-97	carbonyl O	conserved
		Hydrogen Bond/Ionic		
AMN ribose 2'-phosphate	O1R	Arg-43	Nε	ex.chicken(Lys)
	O3R	Arg-43	Nη2	ex.chicken(Lys)
	O1R	Thr-63	Oγ1	Ser or Thr
	O2R	His-64	peptide N	
	O1R	Gln-65	Nε2	Gln(bacterial DHFRs) Glu(vertebrate DHFRs)
AMN 5'-phosphate	O5'(a)	Arg-44	peptide N	Lys, Arg or His
	O1P(a)	Ile-102	peptide N	
	O2P(a)	Thr-45	peptide N	conserved
	O2P(a)	Thr-45	Oγ1	conserved
	O2P(a)	Gly-99	peptide N	conserved
NMN 5'-phosphate	O2P(n)	Arg-44	Nη1	Lys, Arg or His
	O1P(n)	Gln-101	peptide N	
NMN ribose	O2'(n)	His-18	carbonyl O	
	O3'(n)	His-18	carbonyl O	
Nicotinamide carboxamide	O7	Ala-6	peptide N	conserved
	N7	Ala-6	carbonyl O	conserved
	N7	Ile-13	carbonyl O	conserved

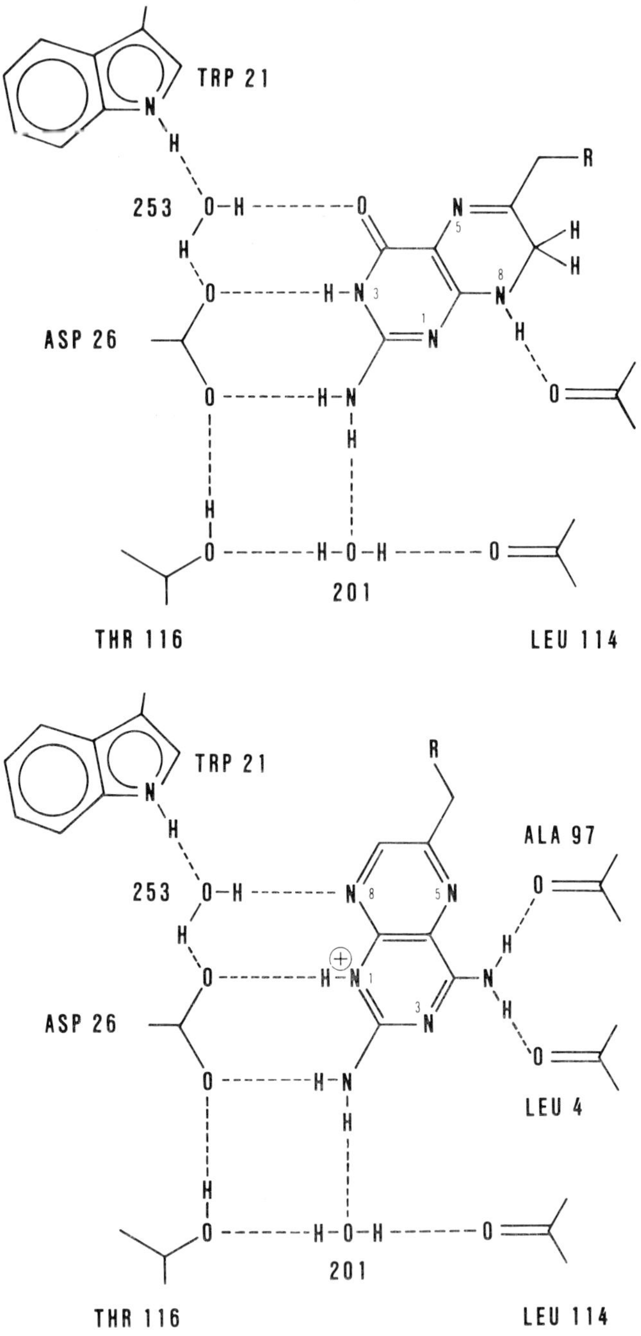

FIGURE 8

FIGURE 8. Schematic representation of hydrogen bonding between dihydrofolate reductase and the pteridine portions of (top) 7,8-dihydrofolate
(hypothetical) and (bottom) methotrexate, from Bolin *et al.* (1982).
The relative orientations of the pteridine rings differ by a 180° rotation
about the C2-amino-N bond.

relative to methotrexate can hydrogen from the nicotinamide ring be
directly transferred to the positions observed for the pteridine. Bolin
et al. (1982) have proposed a model for the binding of substrate (see
Figure 8) in which the pteridine rings of substrate and methotrexate are
related by a 180° rotation about the C-N bond of the 2-amino group with
compensating rotations about other bonds to allow the substrate to retain
those interactions between the PABG moiety and protein which are achieved
by methotrexate. In this pteridine ring orientation N3 rather than N1 is
near the active site carboxyl and N1 and N3 are shown neutral. Substrate
binds more weakly than methotrexate, perhaps because it binds with one
less ionic and one less hydrogen bond. In substrates N3 is amide-like
and the 'amide' pKa (Table 3), for the neutral amide acid rather than the
protonated amide acid, favours the neutral form in solution at pH7.
However this pKa is apparently not different in the complex (Cocco *et al.*,
1981a). If N3 were near the active site carboxyl, a shift in pKa might
be expected. Thus the major binding mode may not be that illustrated;
alternatively the amide may be protonated in the complex, in which case
the observed pKa in the complex might be the shifted pKa of the amide
cation. Huennekens and Scrimgeour (1964) proposed a reduction mechanism
involving initial protonation at nitrogen (i.e. N5 or N8 of the C = N
bond to be reduced) followed by nucleophilic attack by hydride on the
adjacent carbon atom. This mechanism is consistent with the general reduction properties of folates in solution and with mechanisms proposed
for other dehydrogenases (Dalziel 1975). However, the source of any such
proton is unclear and in the reduction of folate, intermediate hydride
transfer to C6 cannot be ruled out.

Acknowledgements: I am indebted to Dr. G. Hitchings, Dr. S. Smith and to
Pergamon Press Ltd. for permission to reproduce material for Tables 1 and
2; to Dr. D. Matthews for Table 5 and Figure 8; to Miss D. Baker for
Figures 3, 5, 6 and 7; to Dr. B. Roth for Figure 1 and to Mrs. R. Castle
for Figure 2.

REFERENCES

Angier, R.B., Booth, J.H., Hutchings, B.L., Mowat, J.H., Semb, J.,
    Stokstad, E.L.R., SubbaRow, Y., Waller, C.W., Cosulich, D.B.,
    Fahrenbach, M.J., Hultquiot, M.E., Kuh, E., Northey, E.H., Seeger,
    D.R., Sickels, J.P. and Smith, J.M. Jr. (1946) Science, 103, 667.
Baccanari, D.P., Averett, D., Briggs, C. and Burchall J. (1977) Bio-
    chemistry, 16, 3566.
Baccanari, D.P., Daluge, S. and King, R.W. (1982) Biochemistry, 21, 5068.
Baccanari, D.P., Stone, D. and Kuyper, L. (1981) J.Biol.Chem., 256, 1738.
Baker, D.J., Beddell, C.R., Champness, J.N., Goodford, P.J., Norrington,
    F.E.A., Roth, B. and Stammers, D.K. (1981b) 12th Gen. Assembly, Int.
    Union of Crystallography, Ottawa, C58.
Baker, D.J., Beddell, C.R., Champness, J.N., Goodford, P.J., Norrington,
    F.E.A., Smith, D.R. and Stammers, D.K. (1981a) FEBS Lett., 126, 49.
Barbehenn, E.K. and Kaufman, B.T. (1982) Archs.Biochem.Biophys., 219, 236.
Bauman, H. and Wilson, K.J. (1975) Europ.J.Biochem., 60, 9.
Bennett, C.D., Rodkey, J.A., Sondey, J.M. and Hirschmann, R. (1978)
    Biochemistry, 17, 1328.
Birdsall, B., Burgen, A.S.V., Hyde, E.I., Roberts, G.C.K., Feeney, J.
    (1981) Biochemistry, 20, 7186.
Birdsall, B., Burgen, A.S.V., Roberts, G.C.K. (1980a) Biochemistry, 19,
    3723.
Birdsall, B., Burgen, A.S.V., Roberts, G.C.K. (1980b) Biochemistry, 19,
    3732.
Birdsall, B., Burgen, A.S.V., Rodriquez de Miranda, J. and Roberts, G.C.K.
    (1978) Biochemistry, 17, 2102.
Bitar, K.C., Blankenship, D.T., Walsh, K.A., Dunlap, R.B., Reddy, A.V.
    and Freisheim, J.H. (1977) FEBS Lett., 80, 119.
Blakley, R.L. (1969) *The Biochemistry of Folic Acid and Related Pteridines*,
    (Frontiers in Biology, 13), North Holland Publ.Co., Amsterdam.
Blakley, R.L. (1981) *Molecular Actions and Targets for Cancer Chemo-
    therapeutic Agents*, Academic Press, New York, 303.
Blakley, R.L., Cocco, L., Montgomery, J.A., Temple, C. Jr., Roth, B.,
    Daluge, S. and London, R.E. (1982) Seventh International Symposium,
    Pteridine and Folic Acid Derivatives, Abstract O.18.
Blakley, R.L., Ramasastri, B.V. and McDougall, B.M. (1963) J.Biol.Chem.,
    238, 3075.
Blakley, R.L., Schrock, M., Sommer, K. and Nixon, P.F. (1971) Ann.N.Y.
    Acad.Sci. (Bertino, J.R., ed), 119.
Bolin, J.T., Filman, D.J., Matthews, D.A., Hamlin, R.C. and Kraut, J.
    (1982) J.Biol.Chem., 257, 13650.
Brockman, J.A. Jr., Roth, B., Broquist, H.P., Hultquist, M.E., Smith, J.M.
    Jr., Fahrenbach, M.J., Cosulich, D.B., Parker, R.P., Stokstad, E.L.R.
    and Jukes, T.H. (1950) J.Am.Chem.Soc., 72, 4325.
Burchall, J.J. and Hitchings, G.H. (1965) Molec.Pharmacol., 1, 126.
Charlton, P.A., Young, D.W., Birdsall, B., Feeney, J. and Roberts, G.C.K.
    (1979) J.Chem.Soc.Chem.Comm., 922.
Cocco, L., Groff, J.P., Temple, C. Jr., Montgomery, J.A., London, R.E.,
    Matwiyoff, N.A. and Blakley, R.L. (1981a) Biochemistry, 20, 3972.
Cocco, L., Temple, C., Montgomery, J.A., London, R.E., Blakley, R.L.
    (1981b) Biochem.Biophys.Res.Commun., 100, 413.
Dalziel, K. (1975) *The Enzymes*, (Boyer, P.D., ed) Academic Press,
    New York, XI, 1.
Daniel, L.J. and Norris, L.C. (1947) J.Biol.Chem., 170, 747.
Daniel, L.J., Norris, L.C., Scott, M.L. and Heuser, G.F. (1947) J.Biol.
    Chem., 169, 689.

Dann, J.G., Ostler, G., Bjur, R.A., King, R.W., Scudder, P., Turner,
  P.C., Roberts, G.C.K., Burgen, A.S.V. and Harding, N.G.L. (1976)
  Biochem.J., 157, 559.
Day, P.L., Langston, W.C. and Darby, W.K. (1938) Proc.Soc.Exp.Biol.Med.,
  38, 860.
Erickson, J.S. and Mathews, C.K. (1971) Biochem.Biophys.Res.Commun.,
  43, 1164.
Erickson, J.S. and Mathews, C.K. (1973) Biochemistry, 12, 372.
Farber, S., Diamond, L.K., Mercer, R.D., Sylvester, R.F. Jr. and
  Wolff, J.A. (1948) New Engl.J.Med., 238, 787.
Feeney, J., Birdsall, B., Roberts, G.C.K. and Burgen, A.S.V. (1975)
  Nature (London), 257, 564.
Ferone, R., Burchall, J.J. and Hitchings, G.H. (1969) Molec.Pharmacol.,
  5, 49.
Ferone, R. and Roland, S. (1980) Proc.Natn.Acad.Sci., USA, 77, 5802.
Filman, D.J., Bolin, J.T., Matthews, D.A. and Kraut, J. (1982) J.Biol.
  Chem., 257, 13663.
Fontecilla-Camps, J.C., Bugg, C.E., Temple, C., Rose, J.D., Montgomery,
  J.A. and Kisliuk, R.L. (1979) J.Am.Chem.Soc., 101, 6114.
Franklin, A.L., Stokstad, E.L.R. and Jukes, T.H. (1948) Proc.Soc.Exp.
  Biol.Med., 67, 398.
Freisheim, J.H., Bitar, K.G., Reddy, A.V. and Blankenship, D.T. (1978)
  J.Biol.Chem., 253, 6437.
Futterman, S. and Silverman, M. (1957) J.Biol.Chem., 224, 31.
Gleisner, J.M., Peterson, D.L. and Blakley, R.L. (1974) Proc.Natn.Acad.
  Sci., 71, 3001.
Gready, J.E. (1980) Advances in Pharmacology and Chemotherapy, (Garattini,
  S., Goldin, A., Hawking, F. and Kopin, I.J., eds) Academic Press, New
  York, 17, 37.
Gunderson, L.E., Dunlap, R.B., Harding, N.G.L., Freisheim, J.H., Otting,
  F. and Huennekens, F.M. (1972) Biochemistry, 11, 1018.
Gupta, S.V., Greenfield, N.J., Poe, M., Makulu, D.R., Williams, M.N.,
  Moroson, B.A. and Bertino, J.R. (1977) Biochemistry, 16, 3073.
Hitchings, G.H. (1952) Trans.Roy.Soc.Trop.Med.Hyg., 46, 467.
Hitchings, G.H., Elion, G.B., Vanderwerff, H. and Falco, E.A. (1948)
  J.Biol.Chem., 174, 765.
Hitchings, G.H., Falco, E.A. and Sherwood, M.B. (1945) Science, 102, 251.
Hitchings, G.H. and Smith, S.L. (1980) Adv.Enzyme.Regul., 18, 349.
Hogan, A.G. and Parrott, E.M. (1940) J.Biol.Chem., 132, 507.
Hood, K. and Roberts, G.C.K. (1978) Biochem.J., 171, 357.
Huennekens, F.M. and Scrimgeour, K.G. (1964) Pteridine Chemistry
  (Pfleiderer, W. and Taylor, E.C., eds) Pergamon, New York, 355.
Huennekens, F.M., Vitols, K.S., Whiteley, J.M. and Neef, V.G. (1976)
  Methods in Cancer Research (Busch, H., ed) XIII, 199.
Jaffe, J.J. (1972) Comparative Biochemistry of Parasites (Van den
  Bossche, H. ed) Academic Press, New York, 219.
Jardetzky, O. and Roberts, G.C.K. (1981) NMR in Molecular Biology,
  Academic Press, New York.
Kaufman, B.T. and Kemerer, V.F. (1976) Archs.Biochem.Biophys., 172, 289.
Kumar, A.A., Blankenship, D.T., Kaufman, B.T. and Freisheim, J.H. (1980)
  Biochemistry, 19, 667.
Kuyper, L.F., Roth, B., Baccanari, D.P., Ferone, R., Beddell, C.R.,
  Champness, J.N., Stammers, D.K., Dann, J.G., Norrington, F.E.A., Baker,
  D.J. and Goodford, P.J. (1982) J.Med.Chem., 25, 1120.
Lai, P-H., Pan, Y-C., Gleisner, J.M., Peterson, D.L. and Blakley, R.L.
  (1979) Chemistry and Biology of Pteridines (Kisliuk, R.L. and Brown,
  G.M., eds) Elsevier/North Holland Publ.Co., Amsterdam, 437.

Lai, P-H., Pan, Y-C., Gleisner, J.M., Peterson, D.L., Williams, K.R. and Blakley, R.L. (1982) Biochemistry, 21, 3284.

Li, R., Hansch, C., Matthews, D., Blaney, J.M., Langridge, R., Delcamp, T.J., Susten, S.S. and Freisheim, J.H. (1982) Quant.Struct.Act.Relat., 1, 1.

Matthews, D.A., Alden, R.A., Bolin, J.T., Freer, S.T. Hamlin, R., Xuong, N.H., Kraut, J., Poe, M., Williams, M. and Hootsteen, K. (1977) Science, 197, 452.

Matthews, D.A., Alden, R.A., Bolin, J.T., Filman, D.J., Freer, S.T., Hamlin, R., Hol, W.G.J., Kisliuk, R.L., Pastore, E.J., Plante, L.T., Xuong, N.H. and Kraut, J. (1978) J.Biol.Chem., 253, 6946.

Matthews, D.A., Alden, R.A., Freer, S.T., Xuong, N.H., Kraut, J. (1979) J.Biol.Chem., 254, 4144.

Matthews, D.A. and Volz, K. (1982) Molecular Structure and Biological Activity, (Griffin, J.F. and Duax, W.L., eds) Elsevier, New York, 13.

McCullough, J.L., Nixon, P.F. and Bertino, J.R. (1971) Ann.N.Y.Acad.Sci., 186, 131.

Mitchell, H.K., Snell, E.E. and Williams, R.J. (1941) J.Am.Chem.Soc., 63, 2284.

Mitchell, H.K., Snell, E.E. and Williams, R.J. (1944) J.Am.Chem.Soc., 66, 267.

Nichol, C.A. and Welch, A.D. (1950) Exptl.Biol.Med., 74, 403.

Nixon, P.F. and Blakley, R.L. (1968) J.Biol.Chem., 243, 4722.

Oleson, J.J., Hutchings, B.L. and SubbaRow, Y. (1948) J.Biol.Chem., 175, 359.

Pastore, E.J. and Friedkin, M. (1962) J.Biol.Chem., 237, 3802.

Pastore, E.J. and Williamson, K.L. (1968) Fed.Proc., 27, 764.

Pattishall, K.H., Acar, J., Burchall, J.J., Goldstein, F.W. and Harvey, R.J. (1977) J.Biol.Chem., 252, 2319.

Perkins, J.P. and Bertino, J.R. (1966) Biochemistry, 5, 1005.

Peterson, D.L., Gleisner, J.M. and Blakley, R.L. (1975) Biochemistry, 14, 5261.

Pfiffner, J.J., Binkley, S.B., Bloom, E.S., Brown, R.A., Bird, O.D., Emmett, A.D., Hogan, A.G. and O'Dell, B.L. (1943) Science, 97, 404.

Poe, M. (1977) J.Biol.Chem., 252, 3724.

Poe, M., Greenfield, N.J., Hirshfield, J.M., Williams, M.W. and Hoogsteen, K. (1973) Biochemistry, 11, 1023.

Poe, M. and Hoogsteen, K. (1974) Fed.Proc., 33, 1382.

Purohit, S., Bestwick, R.K., Lasser, G.W., Rogers, C.M. and Mathews, C.K. (1981) J.Biol.Chem., 256, 9121.

Reyes, P. and Huennekens, F.M. (1967) Biochemistry, 6, 3519.

Richardson, J.S. (1981) Advances in Protein Chemistry, (Anfinsen, C.B., Edsall, J.T. and Richards, F.M., eds) Academic Press, New York, 34, 167.

Roth, B. and Cheng, C.C. (1982) Progress in Medicinal Chemistry (Ellis, G.P. and West, G.B., eds) Elsevier/North Holland, Biomedical Press, Amsterdam, 19, 269.

Rowe, P.B. and Russel, A.J. (1973) J.Biol.Chem., 248, 984.

Shive, W., Bardos, T.J., Bond, T.J. and Rogers, L.L. (1950) J.Amer.Chem. Soc., 72, 2817.

Seeger, D.R. Cosulich, D.B., Smith, J.M. Jr. and Hultquist, M.E. (1949) J.Amer.Chem.Soc., 71, 1753.

Smith, S.L., Patrick, P., Stone, D., Phillips, A.W. and Burchall, J.J. (1979) J.Biol.Chem., 254, 11475.

Snell, E.E. and Peterson, W.H. (1940) J.Bact., 39, 273.

Stokstad, E.L.R. and Manning, P.D.V. (1938) J.Biol.Chem., 125, 687.

Stone, D., Paterson, S.J.,Raper, J.H. and Phillips, A.W. (1979) J.Biol. Chem., 254, 480.

Stone, D., Phillips, A.W. and Burchall, J.J. (1977) Europ.J.Biochem.,
  72, 613.
Stone, D. and Smith, S.L. (1979) J.Biol.Chem., 254, 10857.
Volz, K.W., Matthews, D.A., Alden, R.A., Freer, S.T., Hansch, C.T.,
  Kaufman, B.T. and Kraut, J. (1982) J.Biol.Chem., 257, 2528.
Wills, L. (1931) Br.Med.J., i, 1059.
Wills, L. and Stewart, A. (1935) Br.J.Exp.Path., 16, 444.

11 Folic acid and folic acid antagonists: crystal structures of substrates and inhibitors help to explain enzyme action

Arthur Camerman, Donald Mastropaolo, and Norman Camerman

INTRODUCTION

The goal of cancer chemotherapy is to use drugs to selectively kill cancer cells with minimal damage to normal tissue. This has proven to be an extremely difficult task. Bacterial and fungal infections can be treated with drugs which act by exploiting biochemical differences between mammalian cells and the foreign invaders. Cancer cells, however, are not foreign material very different from normal mammalian cells. They derive from normal host material and do not differ markedly from non-malignant cells in their structure and chemical organization. Fortunately, there are indications that small evolutionary or developmental changes in malignant cells result in altered uptake and transport characteristics for various drugs and alterations in enzyme-substrate affinities. A major emphasis in medicinal chemistry is to develop compounds which can exploit these small differences and be useful as cancer chemotherapeutic agents.

An enzyme system that has received a great deal of attention in cancer research is the dihydrofolate reductase -- thymidylate synthetase (DHFR--TS) chain involved in the synthesis of DNA components. In this system folic acid is reduced to dihydrofolate and then to tetrahydrofolate by DHFR. Tetrahydrofolate is converted to N^5,N^{10}-methylenetetrahydrofolate which is a necessary cofactor in the TS catalyzed synthesis of thymidylate from deoxyuridylate. Dihydrofolate is regenerated by this latter reaction and the process repeats itself. The cycle is illustrated in Figure 1. Thymidylate is essential for DNA synthesis so that blockade of DHFR or TS, halting thymidylate production, results in cell death.

Two of the most widely used anticancer drugs, methotrexate and 5-fluorouracil, are potent inhibitors of DHFR and TS respectively. Both of these compounds are very similar to the natural substrates of the enzymes. Fluorouracil is metabolized to 5-fluorodeoxyuridylate which then binds to TS in place of deoxyuridylate. Methotrexate, which differs from folic acid only by replacement of the 4-carbonyl oxygen atom with an amino group and methyl substitution at N(10) (Figure 2) binds very strongly to DHFR and inactivates the enzyme.

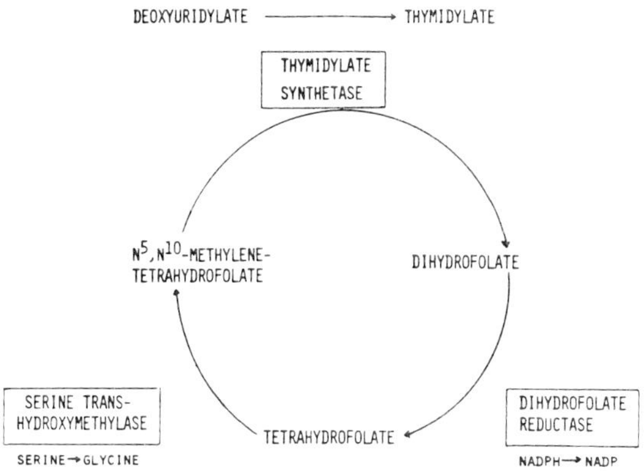

Figure 1. Dihydrofolate reductase and thymidylate synthetase are two important target enzymes in cancer chemotherapy. Inhibition of dihydrofolate reductase by methotrexate blocks regeneration of tetrahydrofolate.

Since folate derivatives are both substrates for DHFR and cofactors for TS, many potential folate antagonists have been synthesized in the quest for more selective drugs which could bind preferentially to tumor cells and inactivate their DHFR and TS. Such effective and selective compounds would be valuable anticancer agents. Synthetic efforts to this end have suffered from lack of stereochemical characterization of both the enzymes and their substrates or inhibitors. Very recently, however, crystallographic studies of DHFR and of folic acid and other folates have been performed. These studies have yielded valuable information on the mode of DHFR-folate binding and have enabled us to better understand the interaction of these compounds on the molecular level. They also have provided data which could serve as stereochemical bases for the design of new potential anticancer drugs.

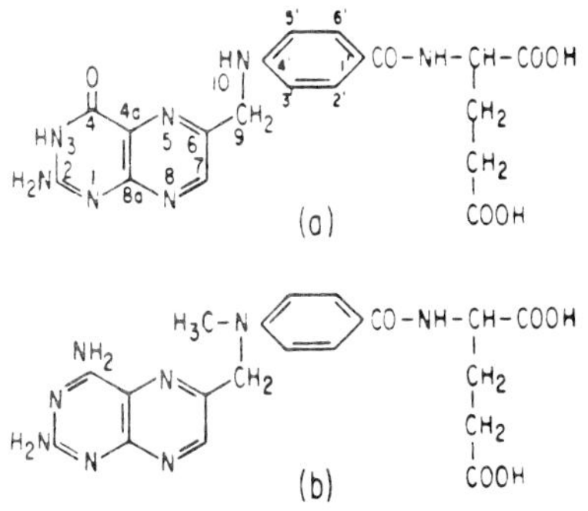

Figure 2. Chemical structure and atom numbering scheme for (a) folic acid and (b) methotrexate.

DIHYDROFOLATE REDUCTASE - METHOTREXATE COMPLEX

A major contribution to the field of folate-enzyme binding mechanisms was realized with the determination of the crystal structures of DHFR's from two bacterial sources, *E. coli* and *L. casei* (Matthews et al, 1977, 1978; Bolin et al, 1982). The *E. coli* DHFR was crystallized with methotrexate bound to the enzyme while the *L. casei* material is a ternary complex of DHFR, methotrexate and NADPH. Results of the two structure determinations are illustrated in Figures 3 and 4 for the

E. coli DHFR and in Figure 5 for the *L. casei* enzyme. The methotrexate molecules bound to the two bacterial DHFR enzymes are very
similar in conformation: both are bent in shape with the pteridine ring
nearly perpendicular to the phenyl ring of the p-aminobenzoyl group.
(The only significant difference between them is at the glutamic acid
end of the molecule which binds toward the enzyme surface rather than
at the active site). A very important conformational feature to note

is that in both enzyme
complexes the orientation
of the methotrexate pte-
ridine ring is such that
the N(5) and C(4)-amino
side is opposite in
direction from the N(10)-
methyl bond vector. This
can be easily visualized
by stereoscopic examina-
tion of Figure 4: the
N(10) methyl bond is
directed toward the viewer
while the C(4)-amino bond
points in the opposite
direction.

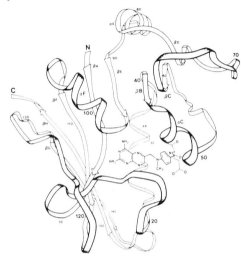

Figure 3. Backbone ribbon drawing of the
E. coli dihydrofolate reductase enzyme-
methotrexate binary complex. β strands
(represented by *arrows*) and α helices
are labeled. Methotrexate is shown
schematically. (Reprinted courtesy of
Dr. D. A. Matthews.)

Figure 4. Stereo drawing
of 159 α-carbon atoms in
E. coli dihydrofolate reductase with bound mole-
cule of methotrexate.
Nitrogen and oxygen
atoms are indicated by
blackening and striping
respectively. (Reprinted
courtesy of Dr. D. A.
Matthews.)

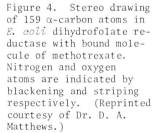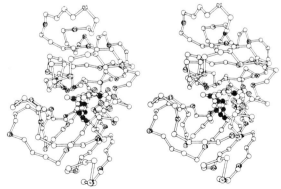

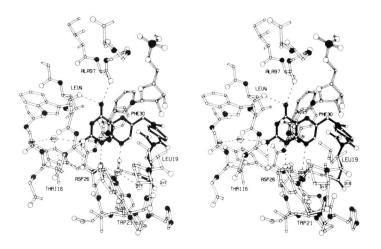

Figure 5.   Stereo drawing of the pteridine-binding site
of *L. Casei* dihydrofolate reductase at 1.7Å resolution.
Methotrexate is indicated by solid bonds, DHFR by open
bonds and a portion of the NADPH by striped bonds.
Carbon atoms are represented by smaller open circles,
oxygen atoms by larger open circles and nitrogen atoms
by blackened circles. Large numbered circles represent
fixed solvent molecules. (Reprinted courtesy of Dr.
D. A. Matthews.)

There have been numerous suggestions made, most of them having to do

with altered chemical properties of the pteridine ring due to replacement

of the C(4)-oxo atom with an amino group, to account for the dramatical-

ly increased binding constant of methotrexate over folate for DHFR. One

suggestion of a different sort, however, has been that methotrexate could

bind to the enzyme with its pteridine ring rotated 180° from the orien-

tation which would be adopted by folate substrates. Matthews et al

investigated this possibility and found that the DHFR active site area

could adequately accomodate a substrate with such an altered pteridine

orientation.

TETRAHYDROFOLATE

Strong evidence that dihydrofolate substrates do indeed bind to DHFR

with the pteridine oriented in a different manner has been recently obtained by Fontecilla-Camps et al (1979) who have determined the crystal
structure of two diastereoisomers of 5,10-methenyltetrahydrofolate
(Figure 6). Reduction of dihydrofolate to tetrahydrofolate produces an
asymmetric center at C(6), thus two isomers differing in configuration
at this atom are possible. Only the S or natural isomer is a substrate
for thymidylate synthetase, the R or unnatural isomer is inactive. Fontecilla-Camps et al have established, by structure analyses, the absolute
configurations of both isomers of their tetrahydrofolate derivative.
Since it is known that the DHFR mediated reduction of dihydrofolate involves hydride ion transfer from NADPH in a stereospecific mechanism,
examination of the conformational arrangement of methotrexate and NADPH
in the ternary complex with DHFR (Figure 5) indicates that if dihydrofolate binds to DHFR with the same pteridine orientation as methotrexate,
the tetrahydrofolate produced would have the unnatural C(6) R configuration. In order to obtain the natural S configuration at C(6), dihydrofolate would have to bind with its pteridine rotated approximately 180° from
the DHFR-bound methotrexate conformation.

Extension of these findings to
the manner of folic acid binding to
DHFR was investigated by Charlton
et al (1979) who used deuterium labelled NADPH and N.M.R. to follow
hydrogen transfer in the DHFR reduction of folic acid. Their results
indicate that there are no major
differences in the orientation of
the oxidized and reduced pteridine
rings i.e. between folic acid and
dihydrofolate, when bound to the
enzymes, and provide further support
for the suggestion that this orientation is substantially different
from that exhibited in the crystal
structures of the DHFR complexes.

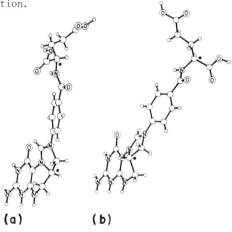

(a)   **(b)**

Figure 6. Structures of the
N(1) protonated derivatives of
(a) (+)-(5,10-methenyltetrahydro-
folate)+, the natural diastereomer,
and (b) (-)-(5,10-methenyltetra-
hydrofolate)+, the unnatural diastereomer. The asymmetric carbon
atoms are indicated by asterisks.
(Reprinted courtesy of the American
Chemical Society).

FOLIC ACID DIHYDRATE

We are reminded again from
Figure 2 that methotrexate, a

powerful DHFR inhibitor and widely used anticancer drug, differs only
slightly in chemical structure from folic acid, a DHFR substrate and an
essential vitamin. The primary difference between the two is replacement
of the 4-oxo atom of folic acid by an amino substituent (removal of the
N(10)-methyl does not alter methotrexate's binding properties). Many
laboratories have attempted for a considerable number of years to crystallize folic acid and methotrexate in order to determine and compare
their individual three dimensional conformations and perhaps visualize
on the molecular level a stereochemical explanation for their difference
in physiological action.

Mastropaolo, Camerman and Camerman (1980) have recently succeeded in
obtaining crystals of folic acid dihydrate and have elucidated the crystal
structure. As illustrated in Figure 7, folic acid adopts an extended conformation. This seems to be the preferred conformation for this class of
compounds, very similar linear molecular geometries are found in the crystal structures of both isomers of 5,10-methenyltetrahydrofolate, of two
non-classical (triazine) DHFR inhibitors (Camerman, Smith and Camerman,
1978) and of a quinazoline analogue
of methotrexate which will be discussed later.

The most significant feature
of the folic acid structure, in relation to the possible mode of enzyme substrate binding, is that the
C(4) oxygen and N(10) hydrogen atoms
are on the same side of the pteridine ring, stabilized in this orientation through hydrogen bonding to
a water molecule. Another way of
looking at this is to note that the
C(9)-N(10) bond is directed toward
N(5) (which is the conformation
necessary to form the N(5)-N(10)

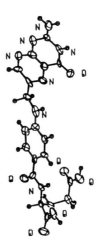

Figure 7. Molecular
structure of folic acid

methylene bridge after reduction to tetrahydrofolate). In contrast we
recall from figures 3 and 4 that the methotrexate C(4) amino group and
N(10) methyl are located on opposite sides of the pteridine ring, that is,
the C(9)-N(10) bond is directed away from N(5), when the inhibitor is
bound to DHFR. Thus, the orientation of the pteridine ring of folic acid
relative to the rest of the molecule differs from the orientation of the

REFERENCES

BOLIN, J.T., FILMAN, D.J., MATTHEWS, D.A., HAMLIN, R.C. and
KRAUT, J. (1982). J. Biol. Chem., 257, 13650-13662.

CAMERMAN, A., SMITH, H.W. and CAMERMAN, N. (1978). Biochem.
Biophys. Res. Comm., 83, 87-93.

CHARLTON, P.A., YOUNG, D.W., BIRDSALL, B., FEENEY, J. and
ROBERTS, G.C.K. (1979). J.C.S. Chem. Comm., 922-924.

FONTECILLA-CAMPS, J.C., BUGG, C.E., TEMPLE, C. JR., ROSE, J.D.,
MONTGOMERY, J.A. and KISLIUK, R.L. (1979). J. Amer. Chem. Soc.,
101, 6114-6115.

MASTROPAOLO, D., CAMERMAN, A. and CAMERMAN, N. (1980). Science,
210, 334-336.

MATTHEWS, D.A., ALDEN, R.A., BOLIN, J.T., FREER, S.T., HAMLIN, R.,
XUONG, N., KRAUT, J., POE, M., WILLIAMS, M. and HOOGSTEEN, K.
(1977). Science, 197, 452-455.

MATTHEWS, D.A., ALDEN, R.A., BOLIN, J.T., FILMAN, D.J., FREER, S.T.,
HAMLIN, R., HOL, W.G.J., KISLIUK, R.L., PASTORE, E.J., PLANTE, L.T.,
XUONG, N. and KRAUT, J. (1978). J. Biol. Chem., 253, 6946-6954.

12 The structure of human serum prealbumin and the nature of its interactions with the thyroid hormones

Stuart J. Oatley, Colin C.F. Blake, Jane M. Burridge, and Patricia de la Paz

1. INTRODUCTION

The thyroid hormones exhibit a wide range of biological activities. They play a fundamental role in the regulation of tissue development and differentiation, and exert considerable influence on metabolic processes. Receptors for the thyroid hormones are located in the nucleus of target cells and it appears that the hormones initiate their action at the nuclear level. The hormone synthesized in, and secreted from, the thyroid gland, under the control of thyroid stimulating hormone (TSH), is L-thyroxine (3,5,3´,5´-tetraiodothyronine, T_4), which is subsequently deiodinated in the peripheral tissues to yield 3,5,3´-triiodothyronine (T_3), biologically a more potent hormone than T_4. The chemical structures of the two forms of the hormone are shown in Fig. 1. It can be seen that the monodeiodination of T_4 produces two distinct conformers of T_3, since the bulky 3,5-iodine substituents of the α-ring (non-phenolic) form a barrier to rotation about the diphenyl ether linkage that prevents ready interconversion of the two forms. In the proximal conformation the 3´-iodine is positioned towards the α-ring, while in the distal conformation it is located away from the α-ring.

The solubilization of the thyroid hormone nuclear receptors has enabled some of their binding properties to be determined. Jorgensen and Andrea (1975) examined the nature of the hormone binding site by using a variety of T_3 and T_4 analogues to define the contribution to binding of the various substituent groups on the hormones. The generalized view of the T_3 binding site that emerged from this study suggested that the site contains a positively charged group that forms an ion-pair interaction with the carboxylate of the hormone, a hydrogen bonding site for the 4´-hydroxyl, and an intervening hydrophobic region which interacts highly specifically with aromatic rings and their iodine substituents. A further study by Macleod and Baxter (1975) showed that the T_3 receptor is capable of direct bonding to DNA.

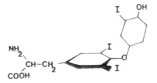

L - Thyroxine (T$_4$)

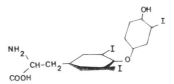

L - Triiodothyronine (proximal)

L - Triiodothyronine (distal)

Fig. 1. The molecular structures of thyroid hormones.

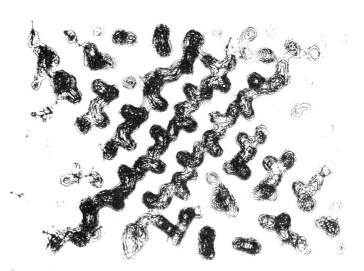

Fig. 2. A few sections of the prealbumin electron density map at
1.8Å resolution, showing the central region of β-sheet.

However, it is clear that the ability to examine at the molecular level the hormone binding site of the nuclear receptor protein, and the atomic interactions between protein and hormone, remains a distant prospect. In order to investigate the mode of interaction of the thyroid hormones with proteins at this level, we have carried out an extensive X-ray diffraction analysis of the molecular structure and ligand interaction of a thyroid hormone binding protein from human serum, prealbumin.

2.   PREALBUMIN

In human serum the thyroid hormones are largely present in the form of complexes with two specific binding proteins, thyroxine-binding globulin (TBG) and prealbumin. Prealbumin derives its name from the fact that it runs faster than albumin on electrophoresis of whole serum. Prealbumin has molecular weight 55,000 and is composed of four identical subunits of 127 amino acid residues (Kanda et al, 1974), arranged with tetrahedral symmetry (Blake et al, 1971). The protein has two binding sites for the thyroid hormones, whose association constants for T_4 are 1.05×10^8 M^{-1} and 9.55×10^5 M^{-1} ; T_3 binds at these same sites about one order of magnitude less strongly (Ferguson et al, 1975). Prealbumin also has binding sites for retinol-binding protein (RBP), the specific serum binding protein for vitamin A (Kanai et al, 1968) ; these sites are independent of those for the thyroid hormones. There are apparently four RBP sites on each prealbumin molecule, (van Jaarsveld et al, 1975), but their location and interactions are not at present known.

TBG is only a trace component of serum, but it has much greater affinity for the hormones ; Robbins (1976) has considered the relative importance of TBG and prealbumin in thyroid function and has concluded that the two proteins are of comparable importance. Table 1 summarizes the binding constants of these two proteins, and also includes those of the nuclear receptor

TABLE 1

Binding constants for T_3 and T_4 of thyroid hormone binding proteins

	prealbumin	TBG	nuclear receptor
molecular weight	55,000	54,000	∿60,000
number of subunits	4	1	?
$K_A(T_4)/M^{-1}$	1.1×10^8	6.0×10^9	5.0×10^8
	(9.6×10^5)		
$K_A(T_3)/M^{-1}$	1.0×10^7	5.0×10^8	6.4×10^9
	(1.0×10^5)		

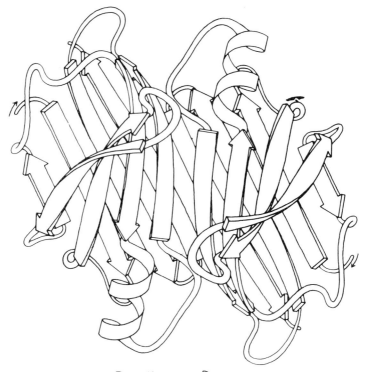

Prealbumin Dimer

Fig. 3. A schematic diagram of the structure of the prealbumin dimer. The arrows represent the β-strands. (Drawn by Jane Richardson and reproduced with her permission).

for the thyroid hormones. It can be seen that there is a major difference in that the two serum proteins show a higher affinity for T_4
while the nuclear receptor has a higher affinity for T_3.

Large crystals of prealbumin can be grown from 2M ammonium sulphate
at pH7 (Haupt and Heide, 1966). X-ray analysis of these crystals showed
that they belong to the orthorhombic system ; space group $P2_12_12_1$ with
cell dimensions a = 43.5Å, b = 85.7Å, c = 66.0Å, with two subunits
pseudosymmetrically related in the asymmetric unit (Blake et al., 1971).
The molecular structure of prealbumin was initially determined at 2.5Å
(Blake et al., 1974). The resolution of the X-ray data was subsequently
extended to 1.8Å (Blake and Oatley, 1977; Blake et al., 1978); the
crystallographic refinement of the structure against this data has
resulted in a very detailed and accurate model of the molecular structure
(S.J. Oatley, unpublished results), for which the R-factor[+] is now 0.18.
Figure 2 illustrates the quality of the present electron density map for
prealbumin.

The tertiary structure of prealbumin is shown as a schematic diagram
of the dimer in Fig. 3 and as a main chain drawing in Fig. 4. The
structure of the subunit is predominantly β-sheet; about 50% of the
amino acid residues are organized into eight β-strands (called strands
A-H according to their order in the sequence) which form two β-sheets,
DAGH and CBEF, each composed of four strands. All the hydrogen-bond
interactions are antiparallel except for those between strands A and G.
There is in addition a single short α-helix which lies at the end of
β-strand E. The monomer-monomer interactions in the dimer are simple,
consisting essentially of the edge strands of the two β-sheets, strands
F and H, forming antiparallel hydrogen-bond interactions with the
equivalent strands in the second monomer. As can be seen in Figs. 3
and 4, this results in a dimer composed of two eight-stranded β-sheets,
DAGHHGAD and CBEFFEBC. The twist of the β-sheets in the individual
monomer, coupled with the molecular symmetry, causes the eight-stranded
sheets to be markedly concave. The dimers assemble into tetramers by
opposing their sheets DAGHHGAD face-to-face about the crystallographic
two-fold axis, z. The concavity of the sheets results in their interaction occurring along their edges, and a cylindrical cavity being left
between them, which is about 8Å in diameter and runs for 55Å completely

---

[+] $$R = \frac{\sum |F_{obs}| - |F_{calc}|}{\sum |F_{obs}|} \quad \text{for all terms with d< 10Å}$$

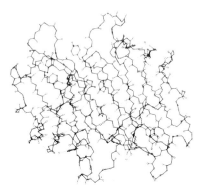

Fig. 4.   The main-chain of the prealbumin dimer, showing the hydrogen
bond arrangement in the β-sheets (in broken lines).

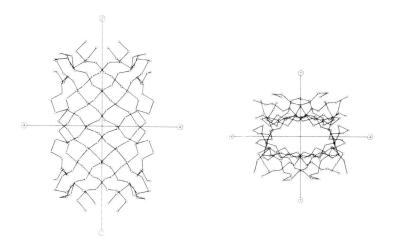

Fig. 5 (left).   Fig. 5 (right).

The organisation of the main chain around the central channel of
prealbumin. The circles represent the Cα positions. The β-sheet
structure may be clearly seen.

through the centre of the molecule. This channel is illustrated in Fig.
5. It will be appreciated that the subunits are very compactly arranged
in the tetramer, which is reflected in the unusual stability of the
prealbumin molecule.

3.   HORMONE BINDING

3.1   Thyroxine and 3,5,3´-triiodothyronine

The binding of the thyroid hormones T_3 and T_4 to prealbumin was analyzed
at 1.9Å and 1.8Å resolution respectively (Blake et al., 1979); the
complexes were prepared by soaking prealbumin crystals in solutions of
the hormones. The difference electron density maps calculated from
these data showed clearly that the hormones bind deeply within the
central channel of the prealbumin molecule, at two sites, each located
between two of the four subunits. The molecular 222 symmetry divides
the channel into two symmetry equivalent halves, each containing one
binding site. Therefore the explanation for the difference in binding
constants for two hormone molecules to prealbumin seems to be in terms
of a negative cooperative effect. The X-ray maps show an averaged
situation with occupancy of each site. Since these sites lie along the
crystallographic twofold axis z, each site itself has internal twofold
symmetry, and hence the asymmetric hormone must bind in two possible
orientations related by this axis.

As can be seen in Fig. 6, the major features in the difference maps
correspond to the iodine atoms, with little indication of the light-atom
hormone skeleton. Since full occupancy hormone binding corresponds to
one hormone molecule per prealbumin tetramer, the X-ray maps will show
half weight features in each binding site, as a result of the averageing
implicit in the method. Furthermore, the symmetry within each binding
site will cause the features to be reduced by a further factor of two.
Finally, since it is estimated that the crystal occupancy of hormone is
only two-thirds, the end result is that hormone atoms will appear at only
one-sixth weight, putting the features for the light atoms below the
error level of the maps. There will be further errors introduced as a
result of the hormone composition, since the intensity differences
measured will be dominated by scattering from the iodine atoms.

Given the limited conformational variability allowed in the thyronine
nucleus, it is possible to interpret these maps by fitting the iodine
atoms to the major features in the difference maps. The fit for T_4 is

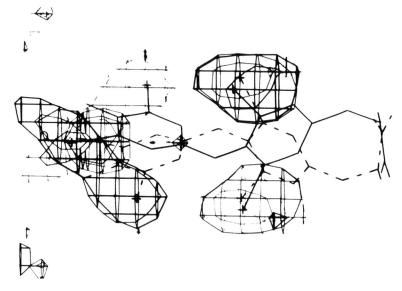

Fig. 6.   Interpretation of the difference electron density features for
          T₄ in one of the hormone binding sites. The two symmetry-
          related orientations of the hormone are shown in full and
          broken lines.

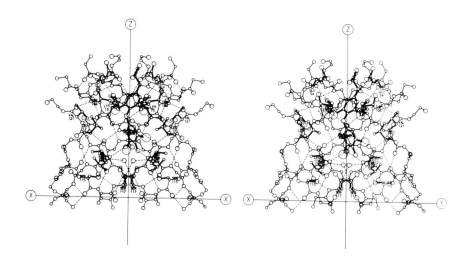

Fig. 7.   Stereo drawing of thyroxine within a single prealbumin binding
          site. The protein side chains involved in the binding are
          indicated in thick lines.

shown in Fig. 6. The iodine substituents of the hormone are the major
sites of interaction. The 3,5-iodines are seen to bind in positions
which are not quite symmetry-equivalent, and make very similar interactions with the methyl and methylene groups of Lys 15, Leu 17, Thr 106,
Ala 108 and Val 121. The interactions involving these iodines are both
less numerous and less close than those observed for the 3´,5´-iodines.
These iodines interact exclusively with one monomer in each case. The
3´-iodine lies in a groove formed between the parallel β-strands
composed of Ala 108, Ala 109 and Leu 110 on one side, and Lys 15 and
Leu 17 on the other, making a large number of close contacts with these
residues. In addition, a very close contact ($\sim$3.0Å) is observed between
this iodine and the carbonyl oxygen of Ala 109, of a type reported to
occur in other iodine and oxygen containing compounds (Murray-Rust and
Motherwell, 1979). The 5´-iodine is seen to be in a groove formed
between the antiparallel β-strands composed of residues Ala 108, Ala 109
and Leu 110 on one side and Ser 117 and Thr 119 on the other. The interactions observed for this iodine are not as close or as numerous as for
the 3´-iodine. All these interactions are shown in Fig. 8.

The amino and carboxyl groups of the hormone are positioned near
Glu 54 and Lys 15 and it seems likely that these residues are involved
in ion-pair interactions with the hormones, while the 4´-hydroxyl makes
largely hydrophobic contacts, not interacting directly with Ser 117 or
Thr 119. Fig. 7 shows the thyroxine molecules positioned within its
binding site.

In the case of T_3, with only a single outer ring iodine to accommodate within the protein, two different modes of binding are seen; these
involve binding sites for the β-ring iodine broadly similar to the two
binding sites seen in the case of T_4, although the overall hormone
conformations are somewhat different.

In the proximal form, the hormone may be oriented such that the 3´-
iodine fits into the groove between the strands carrying Ala 108 and
Leu 17, although it does not fit as closely as in the case of T_4. The
hormone is placed $\sim$ 0.8Å further from the molecular centre than in the
case of T_4. The interactions for the 3,5-iodines are rather similar to
those observed for T_4, however the interactions for the 3´-iodine are
less numerous and less close, and seem to be of an entirely hydrophobic
nature. The electron density also indicated that some T_3 was bound in
the distal conformation. In this case the hormone is seen to bind

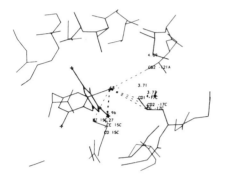

Fig. 8a.

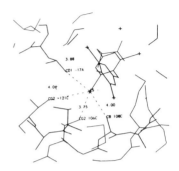

Fig. 8b.

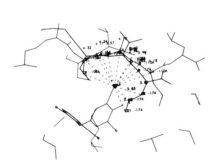

Fig. 8c.

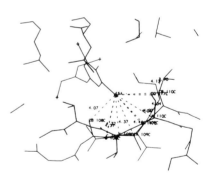

Fig. 8d.

The protein interactions observed for the four iodines of thyroxine. All contacts less than 4.5Å are shown in broken lines, for (a) the 3-iodine; (b) the 5-iodine; (c) the 3´iodine; (d) the 5´iodine.

∿ 0.5Å deeper within the molecule, placing its 3´-iodine optimally between
the strands carrying Ala 108 and Thr 119, making better contact than in
the case of T₄. This position may be attained since there is no proximal
iodine to accommodate. In this position its 4´-hydroxyl could interact
with Ser 117 and Thr 119 *via* a bound water molecule.

3.2 Hormone analogues

In all the above, it has been assumed that the hormone interacts in the
prealbumin binding site with its 4´-hydroxyl innermost in the binding site.
Initially, this orientation was chosen on the basis of steric consideration; the side chains near the molecular centre, in particular Ser 117
and Thr 119, would prevent entry of hormone with its amino acid moiety
innermost. This conclusion was apparently subsequently confirmed by a
2.0Å resolution study of the binding of 3´,5´-diisopropyl-3,5-diiodo-
thyronine, which clearly showed that the 3,5-iodines correspond to the
peaks furthest from the centre of the binding site (see Fig. 9), although
the peaks are positioned approximately 1Å closer to the molecular centre
than for T₄. An earlier experiment to test the orientation of binding,
with a two-dimensional projection difference map of the binding of 3,5-
diiodothyronine (3,5-T₂) had been inconclusive since the peaks corresponding to the two iodines fell between the iodine peaks of T₄, about 3Å
closer to the molecular centre. These two experiments indicated, though,
that the position of binding of the hormone along the channel seemed to
depend on the size of the substituents on the β-ring, which, as described
above, has closer interactions with the protein than the α-ring; replacing the iodine atoms with isopropyl groups allowed the hormone to move 1Å
closer to the centre, while their replacement by hydrogen allowed a
movement of 3Å, in which position the outer ring would be able to interact directly with Ser 117 and Thr 119.

   However, as can be seen in Fig. 9, when the analogues 3,3´,5´-
triiodothyronine (reverse-T₃) and 3´,5´-diiodothyronine were investigated,
the pattern of peaks did not fit with expectations. In the former case,
the peaks at the sites further from the centre are the stronger, for which
the most likely explanation appears to be an inverted mode of entry into
the binding channel, so that these outer sites are occupied by the 3´,5´-
pair of iodines. Similarly, for 3,´5´-T₂ peaks of roughly equal heights
are seen in the 'distal' sites and the outer sites, again suggesting an
inverted mode of binding. While for 3´,5´-T₂ the iodines occupy these
distal sites, for 3,5-T₂ (see above), they are presumably occupying the

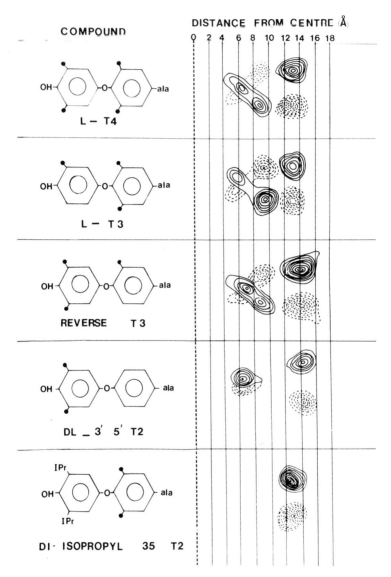

Fig. 9.   Composite drawings showing the major features from the
difference electron density maps for various hormones and
hormone analogues bound to prealbumin. Two sections are
shown: the contours drawn in full lines represent the section
closer to, and in broken lines that further from, the viewer.

proximal pockets.

Yet further complication has arisen with the difference map for $3,3'-T_2$, which would be expected to have equal sized features in the inner and outer sites. However, it is found that the peaks in the inner sites are stronger than those in the outer site. It appears possible therefore that yet another mode of binding exists, where the 3- and $3'$- iodines occupy the 'proximal' and 'distal' pockets simultaneously.

Detailed examination of the best available electron density maps for these analogues has shown that in fact a variety of conformational changes occur in the central region of the protein channel, and in partic- ular a rearrangement of a hydrogen bonding system involving the Ser 115, Ser 117 and Thr 119 residues from all four monomers and a number of water molecules. This rearrangement could allow the hormone to enter the channel with its amino acid moiety innermost, and could also be the means by which the negative cooperative effect is transmitted between one binding site and the other. It is however a difficult crystallographic problem to resolve the exact nature of these changes, as a result of the symmetry of the prealbumin crystal : the density observed for a particul- ar residue will show the summation of the conformations for that residue corresponding to the binding of the ligand in its two-symmetry related orientations, together with any transmitted effect resulting from ligand binding, again in two possible orientations, in the other binding site in the channel.

6.   CONCLUSIONS

The most striking result of these X-ray studies of the binding of thyroid hormones and hormone analogues to prealbumin is that the binding site is not 'monolithic', but appears to interact in a subtly different way with each of the ligands we have examined. It has proved extremely difficult to relate the different modes of binding observed with the structure of the ligand concerned, but it seems clear that the pattern of iodine substitution on the thyronine nucleus, the interchange of iodine for similarly-sized isopropyl moieties, the ionization state of the $4'$-hydroxyl group of the thyronine (which is strongly dependent on the nature of the $3'$ and $5'$ substituents), the conformational freedom about the diphenyl ether linkage (which is strongly dependent on the nature of the 3,5 substituents), and the presence of the amino-acid side-chain, are all important, and interacting, determinants of the mode of binding.

That the prealbumin molecule is able to detect, and respond differently
to, the subtle differences between the various iodinated thyronines that
make up the thyroid hormone complex is of considerable importance in
understanding the specificity of the hormone nuclear receptor. It is
also extremely relevant to those studies of protein-ligand interactions
that seek to provide the contributions of individual atoms, or chemical
groups, to the overall interaction, by studying the binding of a group
of closely related ligands. The underlying assumption of this approach
that closely related ligands will bind in a closely related manner is
certainly not supported by our work on prealbumin.

A further observation of interest is concerned with the response of
the prealbumin molecule to ligand binding, particularly with regard to
the known negative co-operativity of binding. Because prealbumin is
unusually stable, with very tight subunit interactions, and a binding
site composed of relatively 'inflexible' β-sheet structures, the negative
co-operativity of binding has seemed difficult to understand. However,
we can now see that the complex hydrogen-bonded cluster of 16 hydroxy
amino-acids and an equivalent number of water molecules that lies at the
very centre of the protein between the two binding sites, could provide
a means of communication between them. It is clear that certain ligands
cause alterations in the side-chain orientations and water positions
which result in rearrangements in the hydrogen-bond network that could
well account for the co-operativity of binding. This suggests that
transfer of information in protein molecules can take place without the
necessity of invoking gross conformational changes.

ACKNOWLEDGEMENTS

The financial support of the Medical Research Council is gratefully
acknowledged.   S.J.O. is a Mr. and Mrs. John Jaffé Donation Research
Fellow of the Royal Society.

REFERENCES

Blake, C.C.F. and Oatley, S.J. (1977). *Nature, London* 268, 115-120.

Blake, C.C.F., Burridge, J.M. and Oatley, S.J. (1979). *Biochem. Soc. Trans.* 6, 1114-1118.

Blake, C.C.F., Geisow, M.J., Oatley, S.J., Rerat, B. and Rerat, C. (1978). *J. Molec. Biol.* 121, 339-356.

Blake, C.C.F., Geisow, M.J., Swan, I.D.A., Rerat, C. and Rerat, B. (1974). *J. Molec. Biol.* 88, 1-12.

Blake, C.C.F., Swan, I.D.A., Rerat, C., Berthou, J., Laurent, A. and Rerat, B. (1971). *J. Molec. Biol.* 61, 217-224.

Ferguson, R.N., Edelhoch, H., Saroff, H.A. and Robbins, J. (1975). *Biochemistry* 14, 282-289.

Haupt, H. and Heide, K. (1966) *Experientia* 22, 449-451.

van Jaarsveld, P.P., Edelhoch, H., Goodman, DeW, S. and Robbins, J. (1975). *J. Biol. Chem.* 248, 4698-4705.

Jorgensen, E.C. and Andrea, T.A. (1975). In *Int. Thyroid Conf., Boston,* pp. 280-283. Academic Press, New York.

Kanai, M., Raz, A. and Goodman, DeW, S. (1974) *J. Clin. Invest.* 47, 2025-2031.

Kanda, Y., Goodman, DeW.S., Canfield, R.E. and Morgan, F.J. (1974) *J. Biol. Chem.* 249, 6796-6805.

Macleod, K.M. and Baxter, J.D. (1975). *Biochem. Biophys. Res. Commun.* 62, 577-583.

Murray-Rust, P. and Motherwell, W.D.S. (1979) *J. Amer. Chem. Soc.* 101, 4374-4376.

Robbins, J. (1976). In *Trace Components of Plasma : Isolation and Chemical Significance*, pp. 331-350. Alan R. Liss, New York.

13 The structure and function of the haemagglutinin glycoprotein of influenza virus

Ian A. Wilson, John J. Skehel, and Don C. Wiley

INTRODUCTION

The haemagglutinin is the major surface antigen of influenza virus. The glycoprotein has two important activities. Influenza virus binds to cells through a receptor on the haemagglutinin which recognises sialic acid containing proteins or lipids on the target cell surface. Fusion between the target membrane and the viral membrane must occur in order to initiate infection by permitting the viral nucleocapsid to enter the cytoplasm. In addition, recurrent epidemics of influenza are associated with changes in the antigenicity of the haemagglutinin. The structure of the haemagglutinin has been determined in order to establish mechanisms for these activities and to provide a model for recognition of viral antigens by the immune system.

The haemagglutinin is an integral membrane protein of 550 amino acids and contains 25% carbohydrate by weight. The glycoprotein consists of three major domains - the main component extracellular to the membrane, a 25 amino acid hydrophobic membrane-anchoring peptide and a small 10-14 residue hydrophilic peptide on the inside of the membrane. The extracellular component was cleaved from the viral membrane by bromelain a few residues before it enters the membrane and its structure was determined to 3Å resolution.

1. STRUCTURE DETERMINATION

The 1968 Hong Kong bromelain-released haemagglutinin was crystallised in 1.28-1.32M sodium citrate, pH 7.5, 0.1% sodium azide. Data were collected by film methods on a GX-6 Elliott rotating anode, 100 micron focus, Franks mirror optics (Harrison, 1968) at 4°C. Seventy-eight one degree oscillation photographs were recorded for the native data set and 62 for a mercury phenyl glyoxal derivative set. The space group is $P4_1$ with a=163.2Å and c=177.8Å. The asymmetric unit contains one glycoprotein trimer of

208,422 daltons with only 22% protein by weight in the crystals. The solution of the Patterson gave six heavy atom sites per trimer. The location
of the threefold axis was found from the heavy atom positions and used to
improve the quality of the single isomorphous replacement phases by non-
crystallographic symmetry averaging (Bricogne, 1976). The phases of 74,194
reflections were determined to 3Å resolution with agreement between calculated and observed amplitudes of 0.27 after 11 cycles of threefold averaging and solvent flattening. The amino acid sequence (Verhoeyen *et al.*,
1980, Ward and Dopheide, 1980) was fitted into the electron density map.
The location of all residues apart from the amino (1-8) and carboxyl termini (325-328) of HA1 were visible in the map although some of the external
loops and a large portion of the carbohydrate were associated with weak
density.

2. DESCRIPTION OF THE STRUCTURE

The haemagglutinin is synthesised as a single polypeptide chain, HA0, which
is subsequently cleaved into two chains HA1 (residues 1-328) and HA2 (1-
221) by excision of a single arginine residue. This cleavage activates
the fusion activity of the haemagglutinin and the new amino terminus of
HA2 is the most highly conserved sequence in the structure. This glycine-
rich, hydrophobic peptide has been implicated as the sequence responsible
for fusion by analogy with a similar sequence in the Sendai fusion protein.
The haemagglutinin trimer extends 135Å from the membrane and varies in
radius from 15-40Å. The monomer essentially forms two domains, a globular head distal to the membrane end and a fibrous stem proximal to the
membrane end (Figure 1). The amino terminus of HA1 is at the membrane
end and then the polypeptide extends some 80Å along the length of the molecule and forms the globular head with a Greek key, Swiss-roll type beta-
structure. The chain then extends back towards the membrane and ends some
35Å from the amino terminus of HA1. The amino terminus of HA2 is located
22Å from the carboxyl end of HA1 indicating a conformational change must
have accompanied the cleavage of the precursor HA0. The amino terminal
residues Leu 2 and Phe 3 form isologous three-fold contacts and are buried
in the trimer interface. The main part of HA2 then forms two antiparallel
helices to run along the long axis of the molecule. The second helix is
comprised of some 50 amino acids which form a triple coiled-coil with the
threefold-related helices which presumably stabilises the trimer. The
helices pack closely together along the top half of the helix with essen-

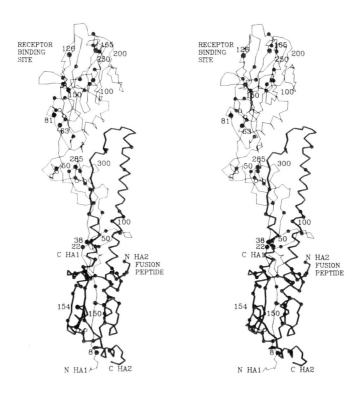

FIG. 1. Stereo drawing of the alpha-carbon tracing of an influenza haemagglutinin monomer. The HA1 polypeptide chain is shown by the thinner line
and HA2 by the thicker line. The location of invariant residues, as described in the text, of several A-viruses, a B-virus and two avian viruses
are shown by small filled-in circles. The location of the carbohydrate
attachment sites are shown by the larger circles and are numbered. The
open circles represent disulphide bonds. Sidechains are shown in thick
lines for the residues identified in the receptor binding site. Picture
produced by a program written by A.M. Lesk and K.D. Hardman (1982).

tially all hydrophobic residues making contact around the three-fold axis
(Ile 77, Leu 80, Val 84, Leu 91, 98, 99, 102). The triple coiled-coil
then begins to unravel close to where the amino terminus of HA2 inserts
itself among the helices. The helical residues then tend to be mainly
hydrophilic around the three-fold axis and the three-fold contacts may
involve, in addition, bound water molecules. HA2 then forms four strands
of an antiparallel five-stranded beta-sheet, the central strand being the
amino end of HA1, some 350 residues earlier in the sequence. This unusual
structure led us to suppose the amino terminus may have remained tethered
to the signal peptide during biosynthesis and folding before being removed
(Wilson *et al.*, 1981). The carboxyl terminus of HA2 is a short helix which
presumably would continue into the membrane to form the anchor had it not
been cleaved at Gly 175 by bromelain. The only interchain disulphide is
located in the five-stranded beta-sheet between residues 14 of HA1 and 137
of HA2. The location of the invariant residues in sequences of 14 human
A-viruses (WSN/33 (H1) - Hiti *et al.*, 1981; Pr/8/34 (H1) - Winter *et al.*,
1981; Jap/305/57 (H2) - Gething *et al.*, 1980; 11 H3 strains - see Laver
et al., 1980, for review; Bangkok/1/79 (H3) - Sleigh *et al.*, 1981; fowl
plague virus/34 - Porter *et al.*, 1979; Duck/Ukraine/63 - Fang *et al.*, 1981;
Ward and Dopheide, 1981) and a human B-virus (Lee/40 - Krystal *et al.*,
1982) show the fusion peptide and the receptor binding site to be the most
conserved (Figure 1).

3. CARBOHYDRATE ATTACHMENT

The 1968 Hong Kong haemagglutinin contains seven Asn-X-Ser/Thr sequences
which act as attachment sites for N-linked oligosaccharides. The carbohydrate is located all along the surface of the molecule (Figure 2). The
electron density for the oligosaccharide tends to be rather less well defined than that of the protein. The hexose core residues for the high
mannose moieties at 165 and 285 of HA1 are more clearly visible than those
of the complex oligosaccharide chains. Some of the carbohydrate is also
apparent at residues 22, 38 and 81 of HA1 but no carbohydrate has been
built at HA1 residue 8 or HA2 residue 154. The weak density for these
oligosaccharide chains and the other complex moieties is presumably due
to positional disorder or heterogeneity of the carbohydrate.
The conformation of the polypeptide chain around the Asn attachment sites
indicates the Asn residues are in various secondary structures. Two of
the Asn residues are in beta-strands, three in beta-turns and one in an

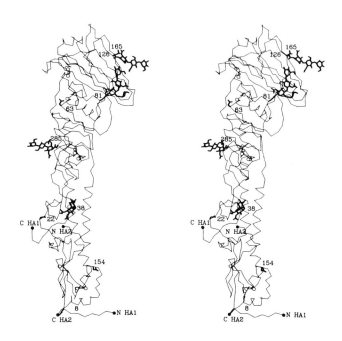

FIG. 2. Stereo drawing of the location of the carbohydrate on the influenza virus haemagglutinin monomer. The oligosaccharide moieties of the
carbohydrate presently interpreted in the electron density map together
with all the Asn sidechains of the attachment sites are shown in thick
lines. In the 1975 haemagglutinin the oligosaccharide at position 81 is
lost and two new sites at 63 and 126 are added. The alpha-carbon trace
is shown in the thinner line and the view is 90 degrees different from
Fig. 1. Picture produced by a program written by A.M. Lesk and K.D.
Hardman (1982).

alpha-helix. The most well-defined carbohydrate in the structure is that
of the NAG-NAG-MAN core attached to Asn 165. The three hexoses are essentially in an extended planar conformation with hydrogen bonds between the
sugars. The C3 hydroxyl of the NAG appears to hydrogen bond with either
the ring oxygen and/or the C6 hydroxyl of the adjacent hexose in a similar
manner to that identified in the Fc of IgG (Deisenhofer, 1981). However,
details of the remaining carbohydrate conformation await further analysis
of the refined structure presently being calculated.

4. ROLE OF THE CARBOHYDRATE

The number of glycosylation sites varies in the different strains of haemagglutinin with from five to nine attachment sites per monomer. The carbohydrate may protect the molecule against degradative enzymes. The Hong
Kong structure indicates seven potential chymotryptic and tryptic cleavage sites may be partially covered by the oligosaccharide. The carbohydrate may also play a structural role. The carbohydrate at 165 from one
subunit interacts closely with the adjacent subunit and hence may stabil-
ise the trimer. The carbohydrate also seems to be important in modulating
recognition of the viral protein by the immune system. This role is discussed further below.

5. RECEPTOR BINDING SITE

The location of the receptor binding site was proposed to be at the distal
tip of the haemagglutinin from an analysis of the invariant residues among
different strains and subtypes (Wilson *et al.*, 1981), (see Figure 1). A
number of conserved residues Tyr 98, Trp 153, His 183, Glu 190, Leu 194
and Tyr 195 form a shallow pocket on the surface of the haemagglutinin.
Recent experiments on selecting mutant viruses with different receptor
specificities have confirmed this pocket to be the receptor binding site.
If the virus is grown in the presence of inhibitors, mutants can be selected which bind preferentially to 2-3 sialic acid linkages to galactose
or to 2-6 linkages or to both. Sequence analysis of these variants indicates a single mutation results in the change in specificity (Paulson, Skehel *et al.*, unpublished). Viruses with a Leu residue at position 226 in
HA1 results in the 2-6 specificity whereas substitution to a Gln results
in the 2-3 specificity. A Met residue at this location has some specificity for both. This residue is located at the mouth of the previously

postulated receptor binding pocket. A structure determination of 2-6
sialyllactose diffused into the crystals of the Hong Kong haemagglutinin
is currently under investigation.

6. MEMBRANE FUSION

The fusion activity of the haemagglutinin requires cleavage activation of
the haemagglutinin HA0 into the two chains HA1 and HA2. The sequence of
the amino terminus of HA2 is analogous to that identified as fusion peptide in the F1 protein of Sendai Virus. This peptide in the haemagglutinin is partially buried in the trimer interface and the sidechains of
Leu 2 and Phe 3 make contacts around the three-fold axis of the trimer.
The polypeptide chain then extends in a series of bends and wraps around
the outside of the molecule to around residue 20 (Figure 1). Membrane
fusion has been shown for influenza to occur around pH 5.0-5.5 depending
on the strain (White *et al.*, 1981, Maeda and Ohnishi, 1980). A conformational change in the haemagglutinin has been detected at this pH (Skehel
et al., 1982) and this change has been associated with the appearance of
a hydrophobic sequence which causes the bromelain-released haemagglutinin
to aggregate and form rosettes. This aggregation has been identified with
a change in conformation of the hydrophobic fusion peptide at around pH 5.
The change in conformation can be modelled by moving the fusion peptide
out from the trimer interface by rotation around the torsion angles of
Gly 16 (Figure 3). The trimer becomes very susceptible to proteolytic
cleavage by trypsin after pH 5.0 treatment and is cleaved at residue 27
of HA1 and secondarily at residue 224. The globular heads of the haemagglutinin are then cleaved from the trimer and leave behind the aggregated
stem region of the trimer. A further thermolytic cleavage at residue 23
of HA2 identified the fusion peptide residues 1-20 as causing the hydrophobic aggregation (Skehel *et al.*, unpublished).
     The role of the fusion-triggered activation has not been firmly es-
tablished but a possible mechanism may be envisaged as follows. The virus
binds to target cells through the receptor binding site on the haemagglutinin, is taken into the cell by endocytosis by way of clathrin-coated
vesicles, and eventually ends up in a lysosome (Matlin *et al.*, 1981). The
lower pH in the endosome triggers the conformational change and the membranes of the host and virus come closer together as a result of the conformational change in the HA. Whether the fusion peptide inserts into the
target membrane, its own membrane or causes the haemagglutinin to aggregate

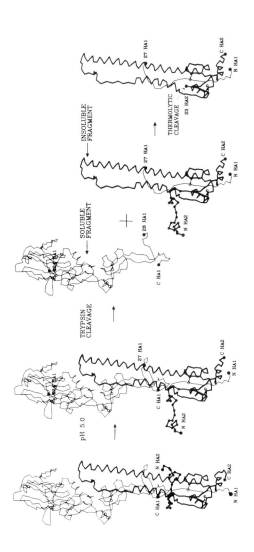

FIG. 3. Representation of the conformational change in the haemagglutinin at pH 5.0. HA1 is shown in the thinner alpha-carbon trace and HA2 in the thicker trace. Only a monomer is shown for ease of illustration although the trimer is the functional unit. The amino-terminus of HA2 has been model built to represent the change in conformation at pH 5.0 and to be accessible for aggregation of the hydrophobic fusion peptide. Trypsin cleaves the trimer after adjusting to pH 7.0 and the heads are cleaved from the trimer into a soluble fragment at residue 27 of HA1 (Skehel *et al.*, 1982). The bottom fraction aggregates and can be further cleaved by thermolysin at HA2 23 to identify the amino terminus of HA2 as being responsible for the hydrophobic aggregation. Picture by a program written by A.M. Lesk and K.D. Hardman (1982).

thereby exposing free lipid on the virus surface, has not yet been determined although the first possibility is currently favoured.

7. ANTIGENIC VARIATION

Influenza virus escapes neutralisation by circulating antibodies of the
immune system by altering the antigenicity of the haemagglutinin. Pan-
demics (antigenic shift) are associated with major changes in the haemagglutinin antigenicity when the virus acquires a substantially different
haemagglutinin by recombination or reassortment of the viral genome with
the large animal reservoir. Recurrent epidemics (antigenic drift) are
associated with the accumulation in point mutations in the haemagglutinin
gene which are reflected in changes of the amino acid sequence on the surface of the protein. An analysis of the amino acid sequence changes of
field variants from 1968-1975 and of laboratory selected variants identified four antibody-binding sites, A-D, on the surface of the haemagglutinin. At least one amino acid substitution in each site seemed to be
required for the production of a new epidemic strain during that time
(Wiley *et al.*, 1981). The subsequent sequence of the A/Bangkok/1/79 haemagglutinin (Sleight *et al.*, 1981) confirmed this prediction and also indicated a further combining site may be present in the later strains from
1975 (Figure 4) due to a change in the oligosaccharide attachment sites.
In 1975, the oligosaccharide binding site changed from 81 to 63 (Figure 2),
indicating the carbohydrate may play a role in masking or unmasking potential antibody combining sites.

The location of the antigenic sites on a different subtype (H1) has
been most extensively described for A/PR/8/34 (Caton *et al.*, 1982) by
selecting variants in the presence of monoclonal antibodies. The location
of the amino acid sequence changes which gave rise to their detection as
variants map broadly in the same areas as outlined for the Hong Kong subtype (H3) but the extent of each area varies somewhat. For example, the
size and location of two sites (B and D as designated for the Hong Kong
subtype), differ mainly due to the lack of carbohydrate at position 165 in
the PR/8/34 haemagglutinin. Thus the carbohydrate may also help modulate
the antigenicity of the haemagglutinin among the subtypes as was observed
in the variation of the Hong Kong subtype.

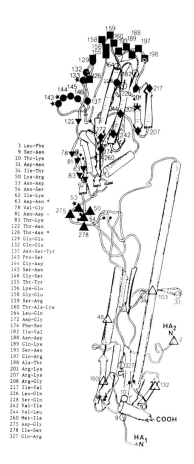

3 Leu-Phe
9 Ser-Asn
10 Thr-Lys
31 Asp-Asn
34 Ile-Thr
50 Lys-Arg
53 Asn-Asp
54 Asn-Ser
62 Ile-Lys
63 Asp-Asn *
78 Val-Gly
81 Asn-Asp -
83 Thr-Lys
122 Thr-Asn
126 Thr-Asn *
129 Gly-Glu
132 Gln-Glu
137 Asn-Ser-Tyr
143 Pro-Ser
144 Gly-Asp
145 Ser-Asn
146 Gly-Ser
155 Thr-Tyr
156 Lys-Glu
158 Gly-Glu
159 Ser-Arg
160 Thr-Ala-Lys
164 Leu-Gln
172 Asp-Gly
174 Phe-Ser
182 Ile-Val
188 Asn-Asp
189 Gln-Lys
193 Ser-Asn
197 Gln-Arg
198 Ala-Thr
201 Arg-Lys
207 Arg-Lys
208 Arg-Gly
217 Ile-Val
226 Leu-Gln
228 Ser-Gln
242 Val-Ile
244 Val-Leu
260 Met-Ile
275 Asp-Gly
278 Ile-Ser
327 Gln-Arg

FIG. 4. Schematic drawing of a monomer of the 1968 Hong Kong haemagglutinin showing the locations of amino acid substitutions from 1968-1979.
The antibody combining sites are labelled in filled symbols - site A by
a circle, site B by a square, site C by triangles pointing up, site D by
diamonds and site E by triangles pointing down. Open symbols indicate
changes which are not thought to be identified with antibody binding. Laboratory selected variants are shown by stars.

8. SYNTHETIC PEPTIDE VACCINES

Synthetic peptides have recently been shown to successfully stimulate anti-peptide antibody production when injected into animals. Peptides corresponding to sequences of the haemagglutinin have also been shown to produce antibodies which cross-react with the intact haemagglutinin and virus (Green *et al.*, 1982, Muller *et al.*, 1982). Structural investigation of the recognition of synthetic peptides of the haemagglutinin by monoclonal antibodies is in progress (Niman and Wilson, unpublished).

ACKNOWLEDGEMENTS

We would like to thank Andrew Cherenson for excellent assistance in the preparation of the illustrations. Acknowledgement is given to grant support from NIH AI-13654 (DCW) and NSF PCM-77-11398 (computing hardware).

REFERENCES

Bricogne, G.   (1976)   Acta crystallogr. A32, 832-847.

Caton, A.J., Brownlee, G.G., Yewdell, J.W. and Gerhard, W.   (1982)   Cell, 31, 417-427.

Deisenhofer, J.   (1981)   Biochemistry 20, 2361-2370.

Fang, R., Jin Jou, W., Huylebroek, D., Devos, R. and Fiers, W.   (1981)   Cell 25, 315-325.

Gething, J.J., Bye, J., Skehel, J.J. and Waterfield, M.   (1980) in The Structure and Variation of Influenza Virus (eds. Laver, W.G. and Air, G. M.) 1-10 (Elsevier, Amsterdam).

Green, N., Alexander, H., Olson, A., Alexander, S., Shinnick, T.M., Sutcliffe, J.G. and Lerner, R.A.   (1982)   Cell 28, 477-487.

Harrison, S.C.   (1968)   J. Appl. Cryst. 1, 84-89.

Hiti, A.L., Davis, A.R. and Nayak, D.P.   (1981)   Virology 111, 113-124.

Krystal, M., Elliot, R.M., Benz, E.W., Young, J.F., and Palese, P.   (1982)   Proc. Natl. Acad. Sci. USA 79, 4800-4804.

Laver, W.G., Air, G.M., Dopheide, T.M. and Ward, C.W.   (1980)   Nature 283, 454-457.

Lesk, A.M. and Hardman, K.D.   (1982)   Science 216, 539-540.

Maeda, T. and Ohnishi, S.   (1980)   FEBS Lett. 122, 283-287.

Matlin, K.S., Reggio, H., Helenius, A. and Simons, K.   (1981)   J. Cell Biol. 91, 601-613.

Muller, G.M., Shapira, M. and Arnon, R.   (1982)   Proc. Natl. Acad. Sci. USA 79, 569-573.

Porter, A.G., Barber, C., Carey, N.H., Hallewell, R.A., Threfall, G. and Emtage, J.S.   (1979)   Nature 282, 471-477.

Skehel, J.J., Bayley, P.M., Brown, E.B., Martin, S.R., Waterfield, M.D., White, J.M., Wilson, I.A. and Wiley, D.C.   (1982)   Proc. Natl. Acad. Sci. USA 79, 968-972.

Sleigh, M.J., Both, G.W., Underwood, P.A. and Bender, V.J.   (1981)   J. Virology 37, 845-853.

Verhoeyen, M., Fang, R , Min Jou, W., Devos, R., Huylebroek, D., Saman, E. and Fiers, W.   (1980)   Nature 286, 771-776.

Ward, C.W. and Dopheide, T.A.   (1981)   Biochem. J. 292, 72-75.

Ward, C.W. and Dopheide, T.A. (1980) in The Structure and Variation in Influenza Virus (eds. Laver, W.G. and Air, G.M.) 27-38 (Elsevier, Amsterdam).

White, J.M., Matlin, K. and Helenius, A. (1981) J. Cell Biol. 89, 674-679.

Wiley, D.C., Wilson, I.A. and Skehel, J.J. (1981) Nature 289, 373-378.

Wilson, I.A., Skehel, J.J. and Wiley, D.C. (1981) Nature 289, 366-373.

Winter, G., Fields, F. and Brownlee, G.G. (1981) Nature 292, 72-75.

14 Conformationally restricted analogues of neurotransmitters and drugs
Alan S. Horn

1. INTRODUCTION

X-ray crystallography provides useful and unique information for medicinal chemists-molecular pharmacologists in three basic areas i.e.

a. about the small molecule (drug),

b. about the macromolecule (enzyme, receptor) and

c. about the small molecule-macromolecule (drug - enzyme, drug-receptor) complex.

This sequence, a, b and c, is the one of increasing molecular complexity and also that of difficulty of practical effectuation. Thus it is relatively easy to obtain suitable crystals of a drug and solve its structure but it is very difficult to isolate and crystallize receptors or enzymes and also to obtain suitable drug-receptor complexes. These latter (b and c) daunting problems are dealt with by other authors in this book and will not be discussed further in this chapter. X-ray analysis yields information about the crystal and molecular structure as well as the conformation of a molecule. It is really the latter information that is of most interest to the molecular pharmacologist, assuming that the molecular structure is already known. Although it is often fairly straightforward to obtain this information its significance and interpretation can present great difficulties. These problems arise due to the molecular flexibility of most small molecules. Although single crystal analysis will provide highly accurate information about the preferred conformation in the solid-state there is no guarantee that this will correspond to the preferred conformation in solution. This problem can, of course, be approached by the application of theoretical energy calculations or N.M.R. studies. However, this is of secondary importance because what we really wish to know is something about the nature of the <u>receptor</u> preferred conformation of the bound drug. In principle, and sometimes in practice, it is possible to obtain this information from an X-ray analysis of a complex of drug plus enzyme. However, the question then again arises as to the significance of such solid-state information for the situation <u>in vivo</u>. Once

again theoretical calculations and N.M.R. can play a role but the accuracy
of this information is, in general, less than that provided by X-ray
crystallography. The main reasons for our interest in the nature of the
receptor bound conformations of small molecules are that it is hoped that
such information will lead to a better understanding of the topography of
the receptor site and the mechanism of drug action. In addition the chances of designing more potent and specific drugs are increased.

2.   CONFORMATIONALLY RESTRICTED ANALOGUES

A method of circumventing some of these problems is the synthesis of
conformationally restricted analogues (C.R.A.). This is a more accurate
definition than the often used "rigid analogue" terminology because
various authors differ with regard to what they understand by the term
"rigid". Portoghese (1970) has pointed out that this approach has two
advantages i.e. key functional groups are held in one position and the
chirality of the pharmacophoric conformation can be investigated. In
addition we shall see that this method can also lead to an increase in
pharmacological potency and selectivity, in other words to new drugs.

This approach does, however, have its drawbacks. It may, for example,
be difficult to synthesize the structural analogue. There are structural
limitations with regard to how closely one can "imitate" a possible receptor preferred conformation of a drug with a flexible side-chain or ring
system. The increased conformational restraint may increase or decrease
activity unpredictably, i.e. flexibility may in fact be required for activity. Finally it has been pointed out by Martin (1978) that similar
structures may bind in different ways to the same receptor.

Sometimes the question is raised as to the value of determining the
crystal and molecular structure of a rigid or semi-rigid molecule i.e.
it is not possible to obtain sufficient conformational information from
molecular models? The answer is that this depends on the rigidity of the
analogue and the type and accuracy of the structural information that one
requires. Thus if one needs very accurate data about interatomic distances
and torsion angles this can only be obtained with the afore mentioned
technique. In addition C.R.A. frequently have asymmetric centres and X-ray
crystallography is the best technique for determining absolute configuration.

Obviously, the new analogue should bear as close a structural resemblance as possible to the original flexible molecule and in addition its
pka and partition coefficient should be altered as little as possible.

However, in the interests of rigidification the liberties which have some-times been taken with these criteria are quite surprising. In order to produce rigid analogues of noradrenaline (NA), for example, the trans-decalin ring system has been incorporated into derivatives of this neuro-transmitter (Fig. 1)(Smissman and Borchardt, 1971). The addition of 8 extra carbon atoms will, of course, have a dramatic effect on the parti-tion coefficient and will entirely change the steric bulk of the original molecule. Not surprisingly, these analogues were less potent than NA, (Tuomisto, Tuomisto and Smissman, 1974).

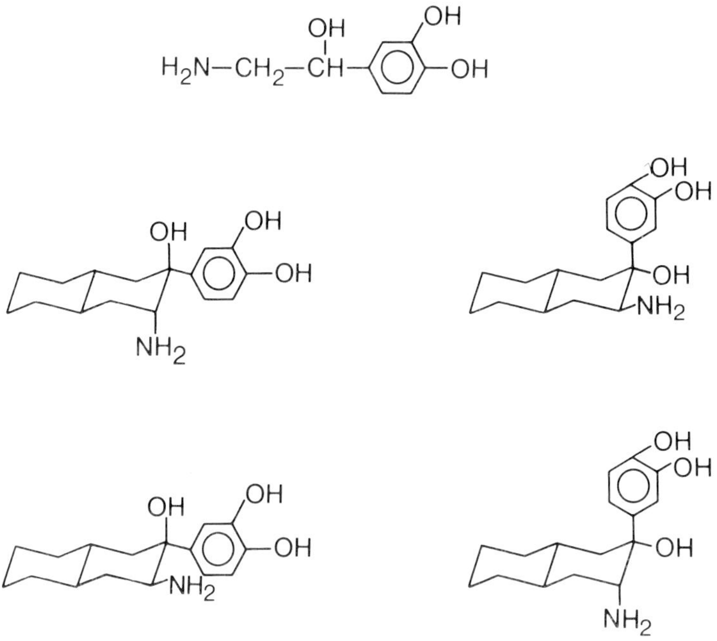

FIG. 1.   Noradrenaline and four analogues incorporating a trans-decalin moiety.

2.1. Examples of conformationally restricted analogues

This chapter is not intended to be a review of all previous work on C.R.A. (for a review of some of the older literature see Portoghese 1971). However, it is worth mentioning just three examples because they illus-trate different aspects of the basic approach.

Over the years much effort has been devoted to attempting to eluci-date the receptor preferred conformation of acetylcholine (ACh, Fig. 2)

and these studies have included both conformational and X-ray crystallographic analysis (Beers and Reich, 1970, Baker, Chothia, Pauling and
Petcher, 1971). One of the best designed C.R.A. of ACh contains the
cyclopropane ring as the constraining moiety (Fig. 3)(Chiou, Long, Cannon
and Armstrong, 1969).

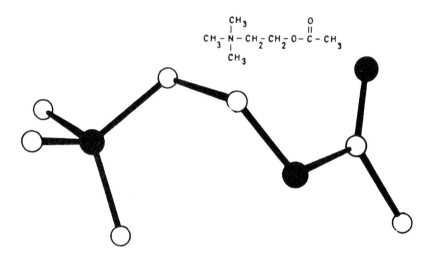

FIG. 2.   The conformation of acetylcholinebromide as found in the crystal.

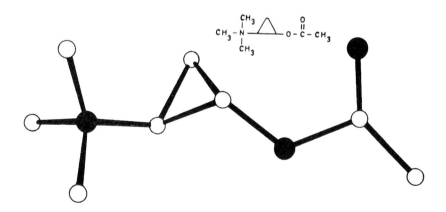

FIG. 3.   The solid-state conformation of trans-2-acetoxycyclopropyltri-
methylammonium iodide.

This is often an ideal structural element for such a purpose i.e. restricting the free rotation around a $-CH_2-CH_2-$ bond, because it is achieved by a very small structural change. The (+) enantiomer of the trans isomer i.e. trans-2-acetoxycyclopropyltrimethylammonium iodide was found to be as potent a muscarinic agent as ACh and 330 times more active than the (-) enantiomer. Both enantiomers were found to be substrates for acetylcholinesterase. The cis isomer was found to be inactive as a muscarinic agonist. An X-ray analysis of the (+) trans enantiomer added further strong support to the suggestion that the conformation of ACh relevant to the muscarinic receptor is that in which the trimethylammonium group is staggered (Chothia and Pauling, 1970; Baker et al. 1971). Pharmacological and X-ray studies with a trans decalin analogue of ACh also tended to support this general conclusion (Smissman, Nelson, LaPidus and Day, 1966, Shefter and Smissman, 1971) however, the previously noted objections to the use of this restricting moiety are again valid in this instance.

A further example of the use of the cyclopropane ring system for restricting rotational freedom is its incorporation into the amphetamine molecule (Fig. 4)(Zirkle, Kaiser, Tedeschi, Tedeschi and Burger, 1962) to yield 2-phenylcyclopropylamine i.e. (±) cis and (±) trans forms (Fig. 4). Amphetamine is an indirectly-acting sympathomimetic amine, it causes a release of catecholamines from nerve endings and inhibits their re-uptake (Horn, 1979). In addition it is a weak monoamine oxidase (M.A.O.) inhibitor (Tipton, 1979). The trans-isomer, tranylcypromine, was found to be a very potent MAO inhibitor and has been widely used as an antidepressant. (Quitkin, Rifkin and Klein, 1979). This is thus an interesting example of conformational restriction producing an enhancement of pharmacological activity and ultimately leading to a clinically useful drug. In order to gain a better insight into the active conformation of amphetamine that occurs at the catecholamine uptake site the inhibition of the transport of ^{3}H-noradrenaline (NA) and dopamine (DA) was studied in homogenates of rat brain hypothalamus and striatum, respectively (Horn and Snyder, 1972). It was found that (±) trans-2-phenylcyclopropylamine, (tranylcypromine) was more active than the (±) cis form. In addition the (-) trans enantiomer was the more potent stereoisomer. These and other results led to the suggestion that it is probably the trans extended conformation of amphetamine which is important in inhibiting catecholamine uptake. This corresponds to the conformation of amphetamine found in the crystal structure (Fig. 4)(Bergin and Carlström, 1971). Although the crystal structure of

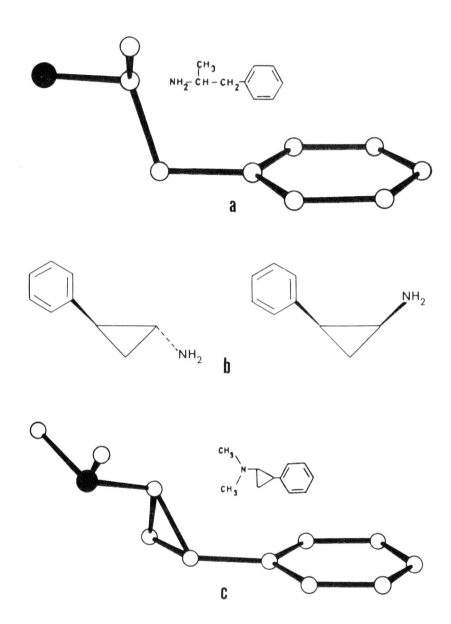

FIG. 4.a. The crystal structure of amphetamine sulphate.

    b. <u>Trans</u> and <u>cis</u> 2-phenylcyclopropylamine.

    c. The crystal structure of N,N-dimethyl-2-phenylcyclopropylamine HCl (SKF 556).

tranylcypromine has not been determined, that of the N,N-dimethyl analogue
is known (Fig. 4)(Carlström, 1975).

The prototype neuroleptic drug was chlorpromazine (Fig. 5) which is
widely used in the treatment of schizophrenia. The neuroleptics are
thought to bring about their clinical anti-schizophrenic effects by
blocking dopamine receptors in the brain (van Praag, 1979). Some years
ago we had suggested that this blockade might possibly be due to a certain
partial structural complementarity between DA and chlorpromazine (Horn and
Snyder, 1971). One of the problems with this suggestion was that it was
based on the crystal structures of DA (Bergin and Carlström, 1968) and
chlorpromazine (McDowell, 1969) both of which are flexible molecules.

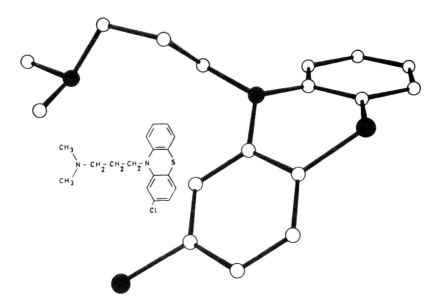

FIG. 5.   The crystal structure of chlorpromazine free base.

The synthesis of more rigid neuroleptic analogues such as butaclamol and
dexclamol (Bruderlein, Humber and Voith, 1975)(Fig. 6) was very important
because results from crystal structure analyses showed that they could
both be superimposed on a portion of the X-ray structure of the rigid
dopaminergic agonist apomorphine, thus lending some support to the original hypothesis, (Humber, Bruderlein and Voith, 1975, Bird, Bruderlein and
Humber, 1976, Giesecke, 1973).
At a more practical level butaclamol has been very useful in increasing

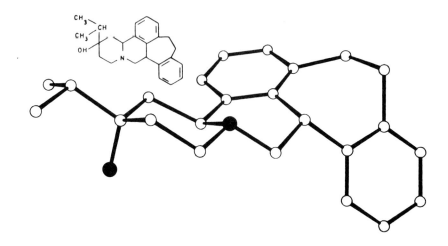

FIG. 6. The crystal structure of dexclamol. Butaclamol has a t-butyl
instead of an iso-propyl group.

our understanding of DA receptors because of its stereoselectivity of
action i.e. unlike chlorpromazine it is an asymmetric molecule that can
exist in (+) and (−) forms, the former being the active enantiomer. Thus
these isomers can be used to distinguish between specific and non-specific binding of the DA receptor antagonist ^{3}H-haloperidol in brain homogenates rich in DA receptors (Seeman, 1980).

3. THE RECEPTOR-PREFERRED CONFORMATION OF DOPAMINE

Dopamine (Fig. 7, 8a) is known to be an important neurotransmitter
in the CNS (Horn, Korf and Westerink, 1979). Disease states in which a
disturbed functioning of the dopaminergic system are known or are thought
to play a role are Parkinsonism, schizophrenia and possibly some forms of
depression (Horn et al, 1979).

The conformational analysis of the receptor preferred conformation
of DA can be divided into three-levels of complexity (Fig. 7).
1. DA can exist in a trans or two gauche forms (Fig. 7a).
2. There are two possible extremes for the trans form i.e. with the
   catechol ring perpendicular to the $-CH_2-NH_2$ bond (trans α) or coplanar
   to it (trans β)(Fig. 7b).
3. When the catechol ring is coplanar to the side chain (trans β) there

are two further possibilities depending on the orientation of the ring
i.e. the α and β rotamers (Fig. 7c).

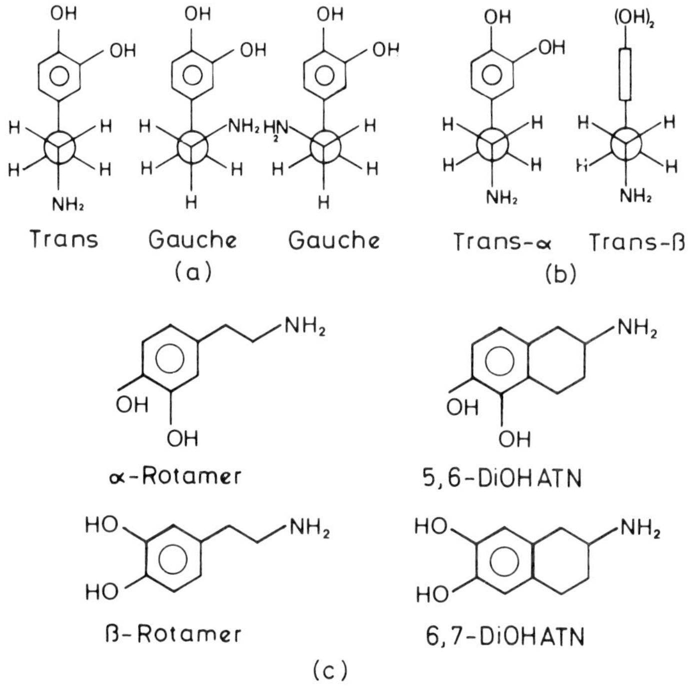

FIG. 7.   Conformational analysis of dopamine .a. and .b. are Newman
projections. c. is an analysis of the two extreme forms of the trans-β
conformer and its corresponding 2-aminotetrahydronaphthalene (ATN) analogues.

With regard to the possibilities under section 1, some theoretical studies
indicate a preference for the trans and some for the gauche form (Bustard
& Egan, 1971; Rekker, Engel and Nijs, 1972; Kier, 1973; Pullman Coubeils,
Courrière and Gervois, 1972; Pullman, Berthod and Courrière, 1974; Katz,
Heller and Jacobson, 1973; Grol and Rollema, 1977) the general conclusion
is that the energy difference between these forms is only of the order of
4-8 kJ mol^{-1}. However, all the pharmacological evidence using rigid and
semi-rigid DA analogues shows convincingly that the receptor preferred
conformation is a form of the trans species (Miller, Horn, Iversen and
Pinder, 1974; Costall, Naylor and Pinder, 1974; Komiskey, Bossart

Miller and Patil, 1978; Miller, 1978).

The second question of whether, at the receptor, the catechol ring is perpendicular (<u>trans</u> α) or coplanar (<u>trans</u> β) to the $-CH_2-NH_2$ bond is more difficult to answer. In the crystal the preferred form is <u>trans</u> α (Fig. 9a)(Bergin & Carlström, 1968). Molecular orbital studies of n.m.r. coupling constants also indicate that the preferred form is <u>trans</u> α

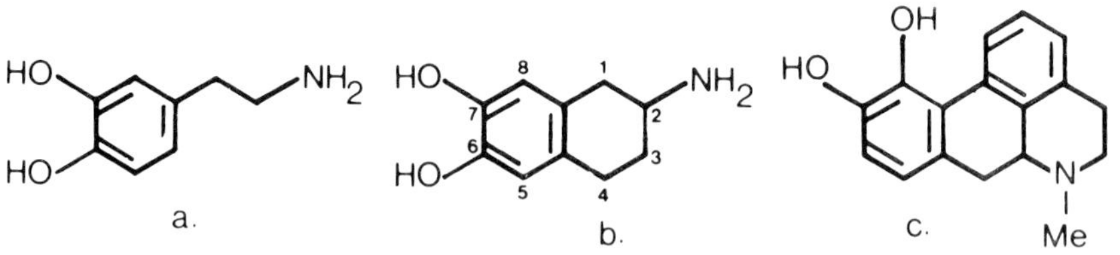

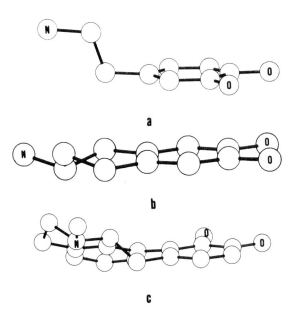

FIG. 8. a. Dopamine, b. 6,7-dihydroxy-2-aminotetrahydronaphthalene (6,7-diOH.ATN), c. Apomorphine.

FIG. 9.   Solid-state conformations of a. Dopamine HCl, b. 6,7-diOH.ATN.HBr c. Apomorphine HCl.

(Giessner-Prettre & Pullman, 1975). It is known, however, that the potential energy difference between these two forms is also quite small. The most direct evidence in favour of the hypothesis that the trans β form is the active one is the fact that such pharmacologically active analogues as 6,7-diOH.ATN (Figs. 8b and 9b) and apomorphine (Figs. 8c and 9c) are both allied to the trans β rather than the trans α form, i.e. they are fairly 'planar' molecules. Indeed, the surprising inactivity of a cyclopropane analogue of DA on peripheral (Borgman, Erhardt, Gorcyzynski and Anderson, 1978; Erhardt, Gorcyzynski and Anderson, 1979) and central dopaminergic receptors (Watling, Woodruff and Poat, 1979) has been explained by the fact that its preferred conformation may correspond more closely to the trans α than the trans β form of DA (Borgman et al. 1978). This conclusion is also supported by the X-ray analysis of the conformation of NN-dimethyl-2-phenylcyclopropylamine (Fig. 4c)(Carlström, 1975).

Similar problems of unfavourable steric bulk were also suggested as a possible explanation for the inactivity of certain 6-exo and endo-(3'4'-dihydroxyphenyl)-2-azabicyclo[2.2.2.] octanes (Law, Morgan, Masten, Borne, Arana, Kula and Baldessarini, 1982). Further evidence against the trans α conformer being the active species was provided by the results of Burn, Crooks, Heatley, Costall, Naylor and Nohria (1982) who showed that various exo-2-aminobenzonorbornenes (Fig. 10), which are rigid analogues of the trans α form, are inactive in vivo.

Thus if it is accepted that the trans coplanar form of DA (trans β) is probably the receptor-site preferred one, the final remaining uncertainty is that of the preferred orientation of the catechol ring i.e. the α or β rotamer (Fig. 7c). Fortunately a more or less direct answer can be provided to this question by studying various catechol (5,6 and 6,7 diOH) derivatives of 2-aminotetrahydronaphthalene (ATN, Fig. 7c). These semi-rigid analogues of the trans β form of DA are known to be very potent DA receptor agonists (Woodruff, Elkhawad and Pinder, 1974; Woodruff, Watling, Andrews, Poat and McDermed, 1977; Miller et al. 1974; McDermed, McKenzie and Phillips, 1975; Costall, Naylor, Cannon and Lee, 1977). In various in vitro and in vivo tests the 6,7-isomer is consistently much more active than its 5,6 positional isomer (Woodruff et al. 1977; Seeman, Titeler, Tedesco, Weinrich and Sinclair, 1978; Schorderet, McDermed and Magistretti, 1978; Sheppard, Burghardt and Long, 1978; Cannon, Costall, Laduron, Leysen and Naylor, 1978; Westerink, Dijkstra, Feenstra, Grol, Horn, Rollema and Wirix, 1979; Rick Szabo, Payne, Kovathana, Cannon and Frohman, 1979; Seiler and Markstein, 1982). Certain authors, however, are of the opinion

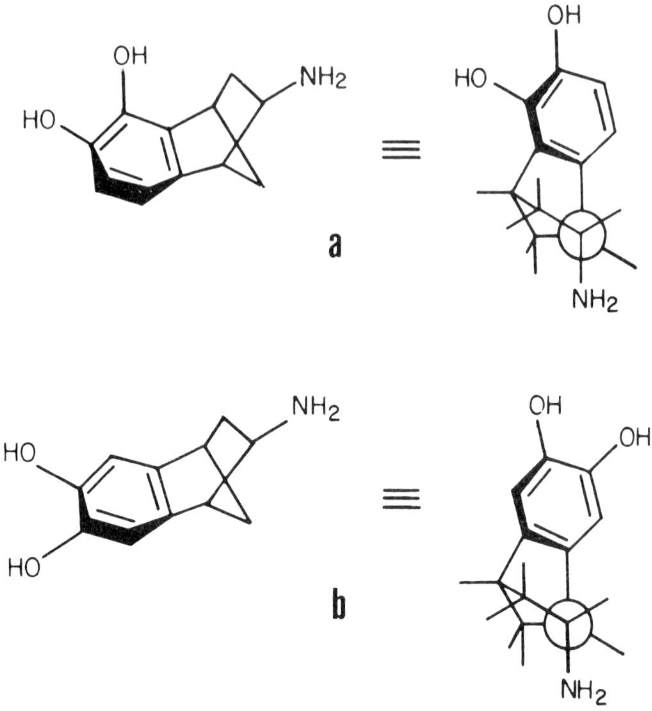

FIG. 10.   a. exo-2-amino-5,6-dihydroxybenzonorbornene, b. exo-2-amino-
6,7-dihydroxybenzonorbornene.

that in some behavioural test systems the 5,6-isomer is more potent than
the 6,7-analogue (Costall et al 1977; Cannon et al 1977). Such behavioural
experiments are always complicated by the large number of factors affecting drug action such as distribution, metabolism, uptake and actions at
multiple DA receptors (Horn, Dijkstra, Feenstra, Grol, Rollema and
Westerink, 1980). We have recently shown that some of the in vivo differences between the 6,7-ADTN and 5,6-isomers are due to the fact that the
former compound is readily metabolized by COMT (Rollema, Westerink, Mulder,
Dijkstra, Feenstra and Horn, 1980; Horn, Dijkstra, Mulder, Rollema and
Westerink, 1981). The consistent in vitro data are thus probably a much
better guide to the receptor-site preferred conformation of DA than are
the behavioural results. The third possible catechol isomer of ATN i.e.
the 7,8-diOH form, has been shown to be inactive (Costall & Naylor, personal communication). In addition, it has recently been shown that the (+)-

enantiomer of 6,7-diOHATN is about 100 times more potent than the (−) form and about 4 times more active than DA itself in stimulating the DA sensitive adenylate cyclase system (Andrews, Davis, Freeman, McDermed, Poat and Woodruff, 1978). Thus, with regard to its pharmacological activity, 6,7-diOHATN seems to be almost an ideal semi-rigid analogue of DA.

The amino group in 6,7-diOHATN can, of course, adopt an equatorial or an axial conformation. In the crystal structure, the equatorial form was found to be the preferred one. Theoretical calculations have also shown that this form is 36.4 kJ mol^{-1} more stable than the axial one (Grol & Rollema, 1977). Thus although 6,7-diOHATN is not a 'completely' rigid molecule (this is, of course, an idealized concept) its conformational mobility is sufficiently restricted to allow the suggestion that the non-bonded distances and torsion angles of the DA skeleton in it (Table 1) probably correspond closely to those of DA at its receptor. For comparison the corresponding values found in the crystal structure of DA.HCl are shown in Table 2.

Costall et al have recently again presented arguments in favour of their hypothesis that the receptor preferred conformation of DA is α rather than β (Costall, Lim, Naylor and Cannon, 1982). This is based on a series of rigid multiring analogues of DA that have been prepared over the years by Cannon and co-workers. However, these compounds are N-alkylated analogues of DA and it is known that in such analogues there is a shift of

Table 1. 6,7-diOHATN.HBr Torsion angles and interatomic distances. The values are derived from atomic coordinates obtained by X-ray analysis.

Torsion angles		Interatomic distances	
$C_8C_9C_1C_2$	165·2°	N–x	5·15 Å
$C_9C_1C_2N$	168·1°	N–Oα	7·88 Å
$C_1C_2C_3C_4$	−59·9°	N–Oβ	7·32 Å
$NC_2C_3C_4$	177·6°	N above plane of	
$C_9C_1C_2C_3$	44·7°	benzene ring	0·001 Å

Table 2. DA.HCl torsion angles and interatomic distances. The values are derived from published atomic coordinates (Bergin & Carlström 1968)

Torsion angles		Interatomic distances	
$C_6C_1C_7C_8 =$	$-99°$	N–x	$= 5·14$ Å
$C_2C_1C_7C_8 =$	$79°$	N–Oα	$7·83$ Å
$C_1C_7C_8N =$	$174°$	N–Oβ	$= 6·83$ Å
		N above plane of ring	$= 1·61$ Å

preference from the α to the β rotamer (Feenstra, Rollema, Dijkstra, Grol, Horn and Westerink 1980; Seiler and Markstein, 1982). Thus because DA itself is a primary amine the value of results obtained with secondary and tertiary amines is doubtful. Clearly an active analogue of DA such as 6,7-diOH.ATN which is a primary amine and bears a very close structural resemblance to DA is a much better guide to its receptor preferred conformation than are N-alkylated and multiring compounds. In line with the general remark by Martin (1978) about differences in the mode of binding by analogues it has been suggested (Seiler and Markstein 1982) that N-alkylated analogues may interact with accessary binding sites. For the sake of clarity, however, it should be pointed out that such compounds can be useful in attempts to obtain information about the preference for trans and gauche forms and the importance of co-planarity i.e. factors 1 and 2 in the conformational analysis.

It has also been shown that 6,7-diOH.ATN is a potent inhibitor of DA uptake into synaptosomes from the rat corpus striatum and in addition that it has itself a high affinity for the DA transport system (Horn, 1974; Davis, Roberts and Woodruff, 1978). Thus because 6,7-diOH.ATN is a more or less planar molecule it may be that the preferred conformation of amphetamine that is responsible for the inhibition of catecholamine uptake has the side chain in the same plane as the benzene ring rather than perpendicular to it as in the crystal structure.

It was the intention in this section to show that in attempting to decide what the receptor preferred conformation of a molecule is, one should ideally try and combine the information provided by various techniques i.e. X-ray crystallography, C.R.A., n.m.r. and theoretical energy calculations.

4. THE RECEPTOR PREFERRED CONFORMATION OF A DA-AUTORECEPTOR AGONIST

Most of the pharmacological results discussed so far with regard to DA and its analogues are thought to probably be due to an interaction with the postsynaptic DA receptor. However, we now know that there are also DA receptors situated on the presynaptic side of the synapse and also on the cell body and dendrites (Horn, Feenstra, Grol, Rollema, van Oene and Westerink, 1981). These sites are known collectively as autoreceptors and they are thought to control the release and/or synthesis of DA. Interest in these receptors increased when it was postulated that selective agonists for these sites might provide a useful new therapy for schizophrenia (Hjorth, Carlsson, Wikström, Lindberg, Sanchez, Hacksell, Arvidsson, Svensson and Nilson, 1981) i.e. instead of blocking postsynaptic DA receptors with the neuroleptics one could inhibit DA synthesis and release with DA autoreceptor agonists.

To date one of the most selective agonists of this type is 3-(3-hydroxyphenyl)-N-n-propylpiperidine (3-PPP, Fig. 11). However, this compound is not very potent and in order to increase potency and obtain information about the possible receptor preferred conformation of this molecule we decided to prepare more rigid analogues by adding an ethane bridge between the two rings to form the octahydrobenzo[f]quinolines (OBQ's)(Fig. 11). There are in fact two extreme rotameric forms of 3-PPP leading to the 7- and 9-OH compounds (7 and 9 OBQ's). In addition depending on the ring fusion one can have both cis(c) and trans(t) isomers of each compound. In the 7-OH series we found the trans compound to be 23 times as active as 3-PPP whilst the cis isomer was inactive (van Oene, Houwing, Dijkstra and Horn, 1983). Unfortunately, however the 7-OH trans compound had increased postsynaptic dopaminergic activity which reduced its selectivity compared to 3-PPP. Wilkström et al examined both the 7- and the 9-OH compounds and found the 9-OH-trans isomer to be the most potent of the series, however, it also had postsynaptic activity (Wilkström, Sanchez, Lindberg, Arvidsson, Hacksell, Johansson, Nilson, Hjorth and Carlsson, 1982). The tentative conclusion can be drawn, however, that the 9-OH-trans isomer may correspond to the receptor preferred conformation of 3-PPP. The fact that the

trans isomers were consistantly more potent than the cis isomers shows
that this DA receptor also has an apparent preference for planar molecules.
It is debatable as to whether the above information could have been obtained by the other methods mentioned in this chapter.

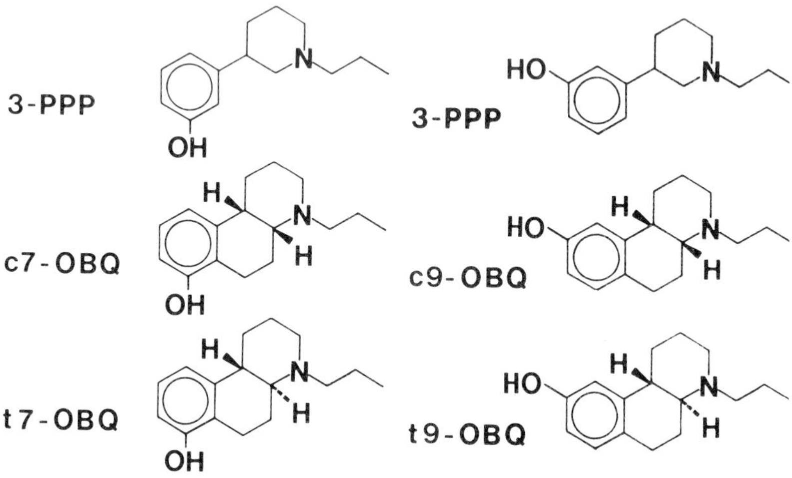

FIG. 11. Top. Two extreme rotomeric forms of 3-(3-hydroxyphenyl)-N-n-
propylpiperidine (3-PPP); Middle cis 7 and 9-hydroxy-4-n-propyl-octa-
hydrobenzo[f] quinoline (OBQ); Bottom trans 7 and 9-hydroxy OBQ.

5. POTENT AND SELECTIVE MAO INHIBITORS

As previously mentioned the enzyme monoamine oxidase (MAO) is responsible for degrading biogenic amines (Horn et al 1979). There are thought
to be two forms of this enzyme i.e. MAO A and MAO B, which differ in their
substrate specificity (Johnston 1968). In the past inhibitors of this
enzyme have been used in the treatment of depression (Quitkin et al, 1979).
However, this was discontinued when adverse reactions were found after the
concomitant ingestion of tyramine containing foods (Blackwell, Marley,
Price and Taylor, 1967).

An important advance in this area was made when it was shown that
the compound deprenyl (Fig. 12) is a selective MAO B inhibitor which does
not cause tyramine related side effects and which can be successfully
employed together with L-DOPA for the treatment of Parkinson's disease
(Knoll 1978). Because of our interest in the 2-aminotetrahydronaphthalenes
we decided to prepare a cyclic analogue (NO425) of deprenyl in the hope

of increasing its activity (Fig. 12).

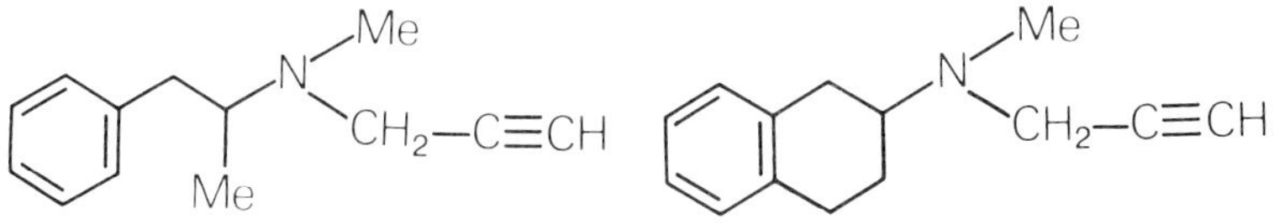

DEPRENYL   NO425

FIG. 12

In various _in vivo_ and _in vitro_ tests NO425 was found to be at least ten times as potent as deprenyl as an inhibitor of MAO. Initial experiments also indicate that it possesses the same spectrum of activity. It would therefore appear that the conformation of the deprenyl fragment in NO425 may be an optimal one for interaction with the active site of MAO. A further advantage of NO425 is that it is easy to synthesize and after toxicological evaluation could become freely available for clinical trials. One of the practical problems with deprenyl has been that supplies of it from Hungary have been very limited.

CONCLUSION

An attempt has been made to show that the synthesis of conformationally restricted analogues of neurotransmitters and drugs can provide a valuable insight into the possible receptor, as opposed to the solid-state, solution or vacuum, preferred conformation of a molecule. In addition to this information one may sometimes be able to enhance potency which in favourable cases may result in a new and useful drug. However, great care must be taken in the choice of restricting units.

In spite of recent impressive advances in the application of Q.S.A.R. and computer graphics, drug design still has a large element of luck attached to it. However, the technique discussed in this chapter can in certain instances increase our chances of a favourable outcome in the continual search for new and more selective drugs.

ACKNOWLEDGEMENT

Figs. 2, 3, 4a, 4c. 5 and 6 were based on similar illustrations by Tollenaere, Moereels and Raymaekers in "Atlas of the Three-Dimensional Structure of Drugs", Elsevier, 1979.

REFERENCES

Andrews, C.D., Davis, A., Freeman, H.S., McDermed, J.D., Poat, J.A., Woodruff, G.N. (1978) Br. J. Pharmacol. 64, 433P.

Baker, R.W., Chothia, C.H., Pauling, P. and Petcher, T. (1971) Nature 230, 439-445.

Beers, W.H. and Reich, E. (1970) Nature 228, 917-922.

Bergin, R. and Carlström, D. (1968) Acta Cryst. B24, 1506-1510.

Bergin, R. and Carlström, D. (1971) Acta Cryst. B27, 2146-2152.

Bird, P., Bruderlein, F.T. and Humber, L.G. (1976) Can. J. Chem. 54, 2715-2711.

Blackwell, B., Marley, E., Price, J. and Taylor, D. (1967) Br. J. Psychiat. 113, 349-365.

Borgman, R.J., Erhardt, P.W., Gorcyzynski, R.J. and Anderson, W.G. (1978) J. Pharm. Pharmacol. 30, 193-194.

Bruderlein, F.T., Humber, L.G. and Voith, K. (1975) J. Med. Chem. 18, 185-188.

Burn, P., Crooks, P.A., Heatley, F., Costall, B., Naylor, R.J. and Nohria, V. (1982) J. Med. Chem. 25, 363-368.

Bustard, T.M. and Egan, R.S. (1971) Tetrahedron 27, 4457-4469.

Cannon, J.G., Costall, B., Laduron, P.M., Leysen, J.E. and Naylor, R.J. (1978) Biochem. Pharmacol. 27, 1417-1420.

Cannon, J.G., Lee, T., Goldman, H.D., Costall, B. and Naylor, R.J. (1977) J. Med. Chem. 20, 1111-1116.

Carlström, D. (1975) Acta Cryst. Sect. B, B31, 2185-2191.

Chiou, C.Y., Long, J.P., Cannon, J.G. and Armstrong, P.D. (1969) J. Pharmacol. Exp. Ther. 166, 243-248.

Chothia, C.H. and Pauling, P.J. (1970) Nature 226, 541-542.

Costall, B., Naylor, R. and Pinder, R. (1974) J. Pharm. Pharmacol. 26, 753-762.

Costall, B., Lim, S.K., Naylor, R.J. and Cannon, J.G. (1982) J. Pharm. Pharmacol. 34, 246-254.

Costall, B., Naylor, R., Cannon, J.G. and Lee, T. (1977) Eur. J. Pharmacol. 41, 307-319.

Davis, A., Roberts, P.J. and Woodruff, G.N. (1978) Br. J. Pharmacol. 63, 183-190.

Erhardt, P.W., Gorczynski, R.J. and Anderson, W.G. (1979) J. Med. Chem. 22, 907-911.

Feenstra, M.G.P., Rollema, H., Dijkstra, D., Grol, C.J., Horn, A.S. and Westerink, B.H.C. (1980) Arch. Pharmacol. 313, 213-219.

Giesecke, J. (1973) Acta Cryst. B29, 1785-1791.

Giessner-Prettre, C. and Pullman, B. (1975) J. Mag. Resonance 18, 564-568.

Grol, C.J. and Rollema, H. (1977) J. Pharm. Pharmacol. 29, 153-156.

Hjorth, S., Carlsson, A., Wikström, H., Lindberg, P., Sanchez, D., Hacksell, U., Arvidsson, L.E., Svensson, U. and Nilsson, J.L.G. (1981) Life Sci. 28, 1225-1238.

Horn, A.S. (1974) J. Pharm. Pharmacol. 26, 735-737.

Horn, A.S. (1979) in The Neurobiology of Dopamine, (eds. A.S. Horn, J. Korf and B.H.C. Westerink) pp 217-235. Academic Press, London.

Horn, A.S. and Snyder, S.H. (1971) Proc. Nat. Acad. Sci. U.S.A. 68, 2325-2328.

Horn, A.S. and Snyder, S.H. (1972) J. Pharmacol. Exp. Ther. 180, 523-530.

Horn, A.S., Korf, J. and Westerink, B.H.C. (1979) eds, The Neurobiology of Dopamine, Academic Press, London.

Horn, A.S., Dijkstra, D., Mulder, T.B.A., Rollema, H. and Westerink, B.H.C. (1981) Eur. J. Med. Chem. 16, 469-472.

Horn, A.S., Dijkstra, D., Feenstra, M.G.P., Grol, C.J., Rollema, H. and Westerink, B.H.C. (1980) Eur. J. Med. Chem. 15, 387-392.

Horn, A.S., Feenstra, M.G.P., Grol, C.J., Rollema, H., van Oene, J.C. and Westerink, B.H.C. (1981) Pharm. Weekblad Sci. Ed. 3, 1021-1041.

Humber, L.G., Bruderlein, F.T. and Voith, K. (1975) Mol. Pharmacol. 11, 833-840.

Johnston, J.P. (1968) Biochem. Pharmacol. 17, 1285-1297.

Katz, R., Heller, S.R. and Jacobson, A.E. (1973) Mol. Pharmacol. 9, 486-494.

Kier, L.B. (1973) J. Theor. Biol. 40, 211-217.

Knoll, J. (1978) J. Neural. Transmn. 43, 177-198.

Komiskey, H.L., Bossart, J.F., Miller, D.D. and Patil, P.N. (1978) Proc. Nat. Acad. Sci. U.S.A. 75, 2641-2643.

Law, S.J., Morgan, J.M., Masten, L.W., Borne, R.F., Arana, G.W., Kula, N.S. and Baldessarini, R.J. (1982) J. Med. Chem. 25, 213-216.

Martin, Y.C. (1978) in Quantitative Drug Design pp 364-365, Marcel Dekker New York.

McDermed, J.D., McKenzie, G.M. and Phillips, A.P. (1975) J. Med. Chem. 18, 362-367.

McDowell, J.J.H. (1969) Acta Cryst. B25, 2175-2181.

Miller, D.D. (1978) Fed. Proc. Fed. Am. Soc. Exp. Biol. 37, 2392-2395.

Miller, R.J., Horn, A.S., Iversen, L.L. and Pinder, R. (1974) Nature, 250, 238-241.

Portoghese, P.S. (1970) Annu. Rev. Pharmacol. 10, 51-76.

Pullman, B., Berthod, H. and Courrière, Ph. (1974) Int. J. Quant. Chem. 1, 93-108.

Pullman, B., Coubeils, J.L., Courrière, Ph. and Gervois, J.P. (1972) J. Med. Chem. 15, 17-23.

Quitkin, F., Rifkin, A. and Klein, D.F. (1979) Arch. Gen. Psychiatry, 36 749-760.

Rekker, R.F., Engel, D.J.C. and Nijs, G.S. (1972) J. Pharm. Pharmacol. 24, 589-591.

Rick, J., Szabo, M., Payne, P., Kovathana, N., Cannon, J.G. and Frohman, L.A. (1979) Endocrinology 104, 1234-1242.

Rollema, H., Westerink, B., Mulder, T., Dijkstra, D., Feenstra, M.G.P., Horn, A.S. (1980) Eur. J. Pharmacol. 64, 313-323.

Schorderet, M., McDermed, J. and Magistretti, P. (1978) J. Physiol. (Paris) 74, 509-513.

Seeman, P. (1980) Pharmacol. Rev. 32, 229-313.

Seeman, P., Titeler, M., Tedesco, J., Weinreich, P. and Sinclair, D. (1978) in Dopamine (Advances in Biochemical Pharmacology) vol. 19 (eds. P. Roberts, G.N. Woodruff and L.L. Iversen).

Seiler, M.P. and Markstein, R. (1982) Molec. Pharmacol. 22, 281-289.

Shefter, E. and Smissman, E.E. (1971) J. Pharm. Sci. 60, 1364-1367.

Sheppard, H., Burghardt, C.R. and Long, J.P. (1978) Res. Commun. Chem. Pathol. Pharmacol. 19, 213-224.

Smissman, E.E. and Borchardt, R.T. (1971) J. Med. Chem. 14, 377-382.

Smissman, E.E., Nelson, W.L., La Pidus, J.B. and Day, J.L. (1966) J. Med. Chem. 9, 458-465.

Tipton, K.F. (1979) in The Neurobiology of Dopamine (eds. A.S. Horn, J. Korf and B.H.C. Westerink) pp. 145-156 Academic Press, London.

Tuomisto, I., Tuomisto, J. and Smissman, E.E. (1974) Eur. J. Pharmacol. 25, 351-361.

Van Oene, J.C., Houwing, H.A., Dijkstra, D. and Horn, A.S. (1983) Eur. J. Pharmacol. 87, 491-495.

Van Praag, H.M. (1979) in The Neurobiology of Dopamine (eds. A.S. Horn, J. Korf and B.H.C. Westerink) pp. 655-677, Academic Press, London.

Watling, K.J., Woodruff, G.N. and Poat, J.A. (1979) Eur. J. Pharmacol. 56, 45-49.

Westerink, B.H.C., Dijkstra, D., Feenstra, M.G.P., Grol, C.J., Horn, A.S., Rollema, H.and Wirix, E. (1979) Eur. J. Pharmacol. 61, 7-15.

Wikström, H., Sanchez, D., Lindberg, P., Arvisson, L.E., Hacksell, U.,

Johansson, A., Nilsson, J.L.G., Hjorth, S. and Carlsson, A. (1982) J. Med. Chem. 25, 925-931.

Woodruff, G.N., Elkhawad, A.O. and Pinder, R.M. (1974) Eur. J. Pharmacol. 25, 80-86.

Woodruff, G.N., Watling, C.C., Andrews, C.D., Poat, J.A. and McDermed, J.D. (1977) J. Pharm. Pharmacol. 29, 422-427.

Zirkle, C.L., Kaiser, C., Tedeschi, D.H., Tedeschi, R.E. and Burger, A. (1962) J. Med. Chem. 5, 1265-1284.

15 Antiepileptic drugs: stereochemical basis of activity

Norman Camerman and Arthur Camerman

The receptor concept describes pharmacological action in terms of a drug binding to a receptor, probably a protein or protein complex, followed either by a conformational change in the receptor which triggers a specific response if the drug is an agonist, or by inhibition of a conformational event and blockade of a response if the agent is an antagonist. In either case an understanding of drug-receptor binding is highly desirable for efficient design of new pharmacological agents and for elucidating drug mechanisms. Drug-receptor complexes are three-dimensional entities and their interactions must be characterized in three-dimensional vectorial terms in order to understand drug action. Ideally this could be done by determining the structures of drug-receptor complexes and identifying the parts of the drug molecule that take part in interacting with the receptor; however, receptors for most drug molecules are not known, let alone isolatable. In the absence of this information how can systematic determination be made of the three-dimensional features of a drug which are responsible for a particular pharmacological effect?

The approach we have taken is the following: We have sought out compounds which differ from each other chemically, but which have in their spectra of pharmacological actions one clinically or laboratory demonstrated activity in common; we have then determined the structures of several of these compounds and sought three-dimensional stereochemical features in common to them which could be the basis of their one common pharmacological action. This approach can be quite general and is applicable to drugs for a wide range of pathologies; the system we first applied it to, and which we have studied in most detail is drugs with activity against grand mal epilepsy.

Epilepsy is a collective term used to describe "a group of chronic convulsive disorders having in common the occurrence of brief episodes (seizures) associated with loss or disturbance of consciousness, usually but not always with characteristic body movements (convulsions) and sometimes autonomic hyperactivity, and always correlated with abnormal and excessive EEG discharges" (Toman, 1969). There are a number of systems of classification of epilepsies, based on various clinical criteria; one of the most common of the epilepsies characterized by generalized seizures is grand mal, which is manifested by major convulsions, usually a sequence of tonic spasm of all body musculature followed by synchronous clonic jerking, and prolonged post-seizure depression of all central functions. Anticonvulsant drugs which have activity against grand mal epilepsy appear, by and large, not to prevent abnormal firing of neurons of seizure foci, but rather to prevent the spread of the abnormal activity to normal neurons. Therefore drugs which may be potentially useful as anti-grand mal agents can be screened by determining their ability to abolish the generalized spread of electrical activity (characterized by tonic muscle extensions) in brains of animals undergoing electroshock or chemically induced seizures. In our studies of anticonvulsant structures we have thus been able to select different families of drugs which have been shown by experiments such as these to possess (among any other pharmacological properties) activity against grand mal epilepsy.

1. PHENYTOIN AND DIAZEPAM

Our initial stereochemical investigations involved the drugs phenytoin (PHT) (I) and diazepam (DZP) (II). PHT has been effective and widely used in the treatment of epilepsy for over 40 years. The drug has no effect upon seizure foci, but likely acts upon normal neurons to prevent their firing by the abnormal foci. PHT is widely regarded as the drug of choice for most forms of epilepsy and is the drug against which most other anticonvulsants are compared. Diazepam and other 1,4-benzodiazepine derivatives were first successfully employed in medicine as tranquillizing agents, and later shown to possess specific anticonvulsant activity against a number of epilepsies. The benzodiazepines derive their antiepileptic action from their ability to limit spread of abnormal electrical firing in the brain and to elevate seizure thresholds. Anticonvulsant activity is the only pharmacological property PHT and diazepam appear to share.

We determined the crystal and molecular structres of PHT and diazepam; perspective drawings of the molecular conformations are shown in Figure 1.

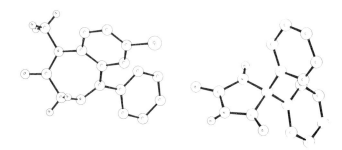

Fig. 1. Perspective drawings of the molecular
conformations of diazepam (left) and phenytoin.

Although these two clinically useful antiepileptic agents are chemically very different, comparison of their three-dimensional molecular conformations revealed that certain stereochemical features are common to both drugs (Camerman and Camerman, 1970, 1971a, 1972a). Specifically, both compounds contain two bulky hydrophobic groups (two phenyl rings in phenytoin and a phenyl and chlorophenyl ring in diazepam) with similar mutual orientations in space. Furthermore, when the phenytoin and diazepam molecules are brought into coincidence so that these hydrophobic groups most nearly superpose, two electron-donating functions in each (the two carbonyl oxygen atoms in phenytoin and the carbonyl oxygen and trigonal nitrogen atoms in diazepam) also occupy similar positions and orientations in three-dimensional space. These stereochemical similarities are also evident in the space-filling models of PHT and diazepam illustrated in Figure 2. It can also be seen in Figures 1 and 2 that if the chlorine atom (or another atom of comparable size) were not substituted where it is on the chlorophenyl ring of diazepam, stereochemical similarity of that part of the molecule to PHT would be reduced. (Alternatively, absence of the chlorine atom could lead to metabolic oxidation at this position, changing the hydrophobic character of the group). This correlates well with the fact that diazepam minus the chlorine atom demonstrates greatly reduced anti-epileptic activity.

2. PROCYCLIDINE AND TRIHEXPHENIDYL

The next compounds we studied crystallographically were procyclidine (III) and trihexyphenidyl (IV). These drugs exhibited significant anticonvulsant activity in laboratory trials, and procyclidine has demonstrated by itself (Millichap et al, 1968) and in combination with phenobarbital (Majkowski and Lysakowska, 1977), clinical usefulness against generalized and partial seizures. The two compounds bear little chemical resemblance to phenytoin or diazepam and share no other pharmacological properties with them.

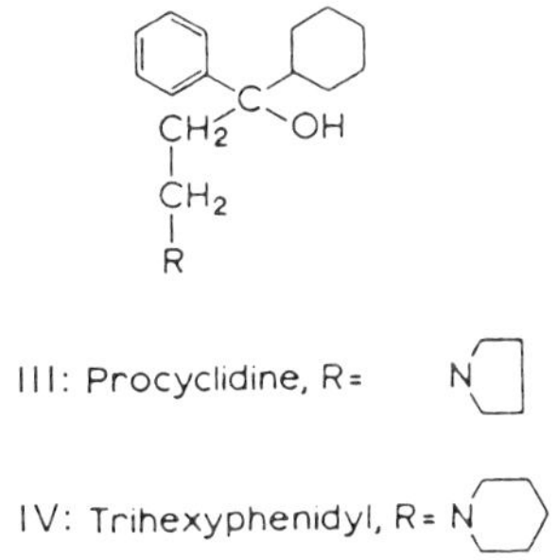

III: Procyclidine, R=

IV: Trihexyphenidyl, R=

The crystal structure analyses of procyclidine and trihexyphenidyl (Camerman and Camerman 1971b, 1972b) revealed that the
stereochemical features we found common to phenytoin and diazepam
are also present to a certain extent in the three dimensional
conformations of these molecules: (1) the two bulky hydrophobic
groups, phenyl and cyclohexyl, are oriented in a manner similar to
that of the aromatic rings in phenytoin and diazepam, (2) two
electron-donating or hydrogen-bondable atoms, the hydroxyl oxygen
and heterocyclic nitrogen, occupy positions in space similar to
those occupied by the electron-donating atoms of phenytoin and
diazepam.

Space-filling molecular models of phenytoin, diazepam,
procyclidine, and trihexyphenidyl, constructed from the crystal
structure results, are shown in Figure 2. Examination of the scale
models shows that the nitrogen in trihexyphenidyl is considerably
more buried within the molecule and thus less accessible to a
receptor than is the case with procyclidine. Since an electron donor
in this position is one of the stereochemical features common to the
anticonvulsants we have studied and is probably essential for
receptor binding, this steric inhibition of the nitrogen in
trihexyphenidyl could well account for the greater anticonvulsant
potency of procyclidine. (In addition to the shielding, O....N
separation is 2.76 Å in trihexyphenidyl, vs 3.4-4.6 Å between
electron donors in the other three anticonvulsants studied).

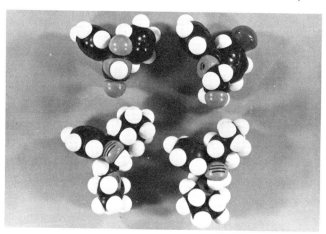

Fig. 2   Molecular models of the crystal structure
         conformations of phenytoin (top left), diazepam (top
         right), procyclidine (bottom left) and trihexyphenidyl
         (bottom right).

3. PHENACEMIDE AND ETHYLPHENACEMIDE

In compounds such as hydantoins, benzodiazepines, and barbiturates, the electron-donating functional groups that we suggest correlate with anticonvulsant activity are located on cyclic structures (five, six, or seven-membered rings) so that their mutual orientation and separation are fixed. A more stringent test of the necessity for these stereochemical features for anticonvulsant action would come from structural investigations on antiepileptic agents in which these functional groups are not part of rigid structures, so that a wider range of conformational variability is possible. Accordingly, we have determined (Camerman and Camerman, 1977) the crystal and molecular structures of phenacemide (V) and ethylphenacemide (VI), two ring-opened analogs of hydantoins. They are potent anticonvulsants even though they contain straight-chain acetylurea groupings rather than closed ring heterocycles.

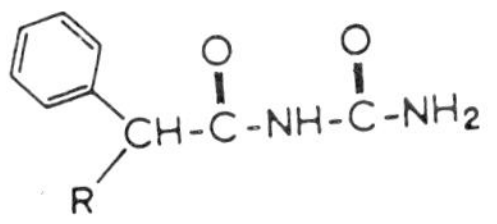

V: Phenacemide, R = H
VI: Ethylphenacemide, R = C_2H_5

Although phenacemide and ethylphenacemide are closely similar chemically, crystallographically they differ considerably. They crystallize in different space groups and the molecular packing patterns, intermolecular forces, and molecular environments in the two crystal structures are therefore quite different.

The molecular structure determinations (Figure 3) show that the three-dimensional conformations of phenacemide and ethylphenacemide are almost identical, differing only in the rotation of the phenyl ring about the C-C bond. As can be seen in the figure, the acetylurea portion of both molecules folds into a six-atom ring, with the open end joined by an intramolecular hydrogen bond between a hydrogen atom on the terminal amino group and one of the ketonic oxygens. The distance between the oxygen and hydrogen atoms is 2.20 Å and the N-H$\cdots$O angle is 103°. This 'open-ended ring' is approximately planar, with the maximum deviation of any atom from the plane being 0.06 Å. The two ketonic oxygens are oriented in opposite directions with respect to the C–N bonds. A consequence of this conformation is that both phenacemide and ethylphenacemide bear a striking stereochemical resemblance to phenytoin and the other anticonvulsants whose structures have been studied. Figure 4 shows space-filling scale models of phenytoin and ethylphenacemide in the conformations observed in the crystal structures. The models clearly illustrate that if the molecules are superposed so that the hydrophobic groups of each most closely coincide, the ketonic groups of each drug also closely match.

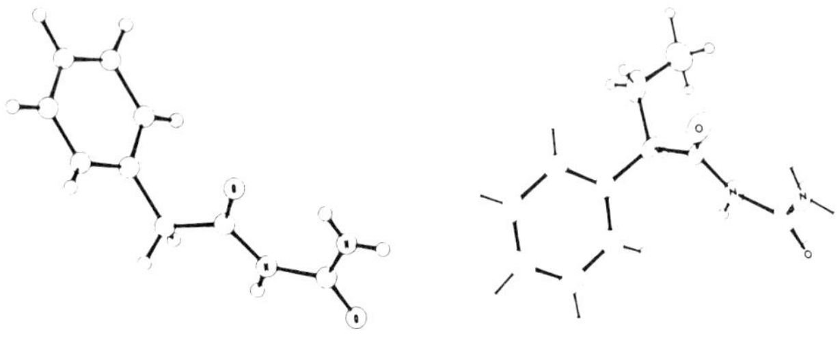

Fig. 3   Perspective drawings of the three-dimensional
         conformations of ethylphenacemide (right) and
         phenacemide (left).

It is clear that phenylhydantoins and phenylacetylureas are extremely similar in conformation despite the presence of the heterocyclic ring in the former and the straight-chain configuration in the latter class of compounds. Therefore, these results provide powerful support for the concept that it is the stereochemical features that we have noted to be common to the different drugs that confer to them anticonvulsant action against grand mal epilepsy and that this activity likely results from interaction with the same class of receptors.

A quantitative measure of the stereochemical resemblances between the antiepileptic drugs whose structures have been elucidated is given in Table 1, which lists the distances between hydrophobic groups and electron-donor atoms. The numbers in the table are scalar distances; it should be kept in mind that because relative orientations of the functional parts of the molecules as well as relative positions determine activity, vector distances are the important quantities for comparative purposes.

Fig. 4.   Molecular models of the crystal structure confor-
          mations of phenytoin (left) and ethylphenacemide
          (right).

TABLE 1. *Distances (Å) between centroids of the aromatic rings and electron-donor atoms in phenytoin (PHT), diazepam (DZP), procyclidine (PRO), phenacemide (PCA), ethylphenacemide (EPA), and phenobarbital (PB)*[a]

	PHT	DZP	PRO	PCA	EPA	PB
Ring–atom *1*	5.51	5.35	5.36	6.39	5.85	5.63
Ring–atom *2*	4.23	4.06	3.63	3.75	4.02	4.19, 4.26
Atom *1*–atom *2*	4.56	3.35	3.55	4.13	4.12	4.51, 4.53

[a] In DZP the ring is the chlorophenyl group and atom *1* is the oxygen, in all other compounds the ring is a phenyl group; in PRO atom *1* is the nitrogen. In PB there are two equivalent "atom *2*" oxygens.

4.   PHENOBARBITAL

Phenobarbital was the first effective organic anticonvulsant agent. It has been used in the treatment of epilepsy since 1912 and is still a very potent and widely used drug. Like the other barbiturates, phenobarbital is a powerful sedative, but unlike almost all other barbiturates, it has a selective anticonvulsant action useful especially in the treatment of grand mal epilepsy. The anti-convulsant activity is unrelated to sedation since non-sedative doses are often effective and amphetamines can counteract the sedation space of these atoms seems to be an important stereochemical feature common to the chemically different antiepileptic drugs, it is easy to appreciate why the 1,3-benzodiazepines do not display anti-epileptic properties. Quantitatively, the carbonyl oxygen and the nitrogen atom are separated by only 2.24 Å; evidence from the molecular structure determination of sulthiame (Camerman and Camerman, 1975) suggests that this value is below the lower limit needed for anticonvulsant activity. Phenylbutansultam (sulthiame minus the sulfonamide moiety) resembles the antiepileptic drugs in the positioning of the phenyl ring with respect to two electon-donor oxygen atoms. However, the oxygen-oxygen separation is only 2.45 Å, and phenylbutansultam exhibits only very weak anticonvulsant properties.

Distances between aromatic rings and electron donor atoms in the 1-3-benzodiazepine, the 2,4-benzodiazepine, diazepam, and phenytoin are listed in Table 2.

In the 2,4-benzodiazepines, the 1-nitrogen and 2-carbonyl group of diazepam have been interchanged; the carbonyl is now situated at the 1-position of the heterocyclic ring. The consequence of this shift is again to alter the distance between the two electron-donor atoms from what it is in diazepam, but in this case the distances between the important features are similar to those found in the other antiepileptic drugs and, in fact, the separation of the two electron-donor atoms is even closer to the comparable distance in phenytoin than is the value found in diazepam. It is not surprising, therefore, in view of its stereochemistry, that the 2,4-benzodiazepine analog of diazepam displays significant anti-convulsant properties.

TABLE 2. Distances (Å) between the centroids of the aromatic rings and the oxygen and nitrogen atoms in diazepam (DZP) and 1,3 and 2,4-benzodiazepines and comparable distances in phenytoin (PHT)[a]

	DZP	1,3	2,4	PHT
Cl Ph – Phenyl	5.06	5.13	5.10	4.84
Cl Ph – Carbonyl O	5.35	5.44	4.08	5.51
Cl Ph – Trigonal N	4.06	4.63	4.08	4.23
Phenyl – Carbonyl O	6.68	5.16	6.27	5.68
Phenyl – Trigonal N	3.64	3.88	3.61	3.97
O – N	3.35	2.24	4.15	4.56

[a] In PHT the distances listed are between the two phenyl rings and the two carbonyl oxygen atoms (Cl Ph: chlorophenyl).

6. SULTHIAME

Sulthiame (IX) is one of a group of sulfonamide compounds, including acetazolamide, which have been shown to possess anticonvulsant activity. This activity has been closely correlated with these compounds' ability to inhibit carbonic anhydrase in the brain (Tanimukai et al, 1965), with a possible mechanism being accumulation of carbon dioxide which inhibits nerve conduction (Nishimura et al., 1963). These compounds then presumably act quite differently from the other antiepileptic drugs we have examined, which do not inhibit carbonic anhydrase.

$$\text{(structure) } N-\!\!\bigcirc\!\!-SO_2NH_2$$

IX

When sulthiame is given to patients already receiving PHT it elevates serum levels of PHT (Hansen et al, 1968). It has been suggested that this may be due to the displacement by sulthiame of PHT from storage depots in the erythrocytes (Olesen and Jensen, 1969).

We determined the crystal structure of sulthiame in order to compare the conformation of this molecule to those of other anticonvulsant drugs. The results are shown in Figure 6 which shows a scale model of the sulthiame molecule constructed from the atomic positions compared with a similar model of phenytoin. It is apparent from the two models that the three-dimensional structure of sulthiame, as a whole, is not very similar to those of the other anticonvulsant drugs previously described. These results are wholly

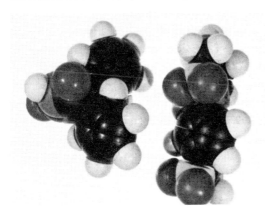

Fig. 6.
Molecular models
of phenytoin (left)
and sulthiame.

(d)   Acting on our suggestion "that the compounds might exert antiepileptic effects through interactions with the same unspecified receptor in the brain", Matthews and Connor (1976) compared PHT and DZP with regard to their effects on post-tetanic potentiation of evoked potentials within the hippocampus of anaesthetized rats. They used doses within the range which protects rats against the tonic phase of maximal electroshock seizures, i.e. anticonvulsant doses, and found qualitatively similar patterns of action for both drugs in the various hippocampal regions tested.   Thus in their test situations in vivo, phenytoin and diazepam cause very similar end-effects on complex systems, and show similar selective actions which are site and dose dependent.

(e)   The discovery of specific benzodiazepine (BZ) receptors in the brain has resulted in a number of important receptor-binding experiments which have provided further corroboration of our conclusions.   Paul et al, (1979) found an excellent correlation between benzodiazepine receptor occupancy by diazepam and protection against chemically induced seizures, thus strongly implicating these receptors to be integral to the anticonvulsant mechanism of action of benzodiazepines. Subsequently (Mimaki et al, 1980) Yamamura's laboratory, acting on our suggestion that the similarities in molecular stereochemistries of DZP and PHT implied a common class of receptors for the anticonvulsant action of these drugs, studied the effect of phenytoin on benzodiazepine receptors in rat cerebral cortex and cerebellum.   They verified our prediction by reporting that phenytoin acts at the same receptor sites as the benzodiazepines, and that chronic phenytoin administration produces a dose-dependent decrease in the density of benzodiazepine receptors.

(f)   Again acting on our suggestion that these compounds may act at the same CNS sites, Gallager et al, (1980) performed a series of experiments to investigate possible interactions between PHT and benzodiazepine systems in brain using electrophysiological and binding techniques. Their results "indicate that one site of action of PHT in brain is the BZ binding site", and that "this interaction is relevant in the intact animal". Further, they conclude that "since the effects of PHT on BZ binding are dose-related and significant only for doses with anticonvulsant potency, these findings may also suggest a correlation between BZ binding effects and the anticonvulsant properties of PHT".

(g) Very recently Bowling and DeLorenzo (1982) reported the identification and characterization of micromolar affinity benzodiazepine receptors in rat brain. The receptors exhibit saturable, stereospecific binding, and the binding order of benzodiazepines correlates with their laboratory demonstrated anti-convulsant properties; thus the criteria for true receptors are satisfied. The authors also report that PHT binds to the same receptors; this provides additional strong evidence for a commmon mechanism of anticonvulsant action for BZ's and PHT, as we had earlier postulated from our structural results.

8. SUMMARY

Previous widespread research on anticonvulsant drugs led to numerous theories which ascribed their central nervous system activity to a variety of simple physiochemical properties, but none of these theories is satisfactory. Our crystallographic investigations have resulted in the discovery and the identification of stereochemical features common to the three-dimensional conformations of a number of chemically diverse drugs, features that may be the basis of their one mutual pharmacological action, anti-convulsant activity. We have found that the drugs all possess two hydrophobic regions of structure, and when their conformations are compared so that these regions are maximally superposed, each compound has two electron-donor groups situated in similar orientations and positions with respect to each other and to the hydrophobic groups.

Having made these identifications, we can now suggest stereochemical guidelines for the design of new potential antiepileptic drugs: compounds that retain those molecular features deemed responsible for anticonvulsant activity but have alterations in other molecular parameters may prove to be more effective and safer drugs than those currently used in the treatment of epilepsy.

It is perhaps necessary that we end this chapter with further points to be kept in mind about the applications of stereochemical principles to drug design. When considering stereochemical alterations to parts of molecules, one must consider as well what

effects those changes will have on properties other than receptor
fitting which are also important for drug action, i.e. properties
such as solubility, transport, metabolic fate, etc. For example, it
is recognized that to be useful in the treatment of grand mal
epilepsy a potential anticonvulsant must possess a phenyl ring as one
of its hydrophobic groups; some drugs, phenytoin foremost among
them, contain two aromatic rings, whereas others, including
phenobarbital and ethylphenacemide, have only one. The question
arises, in light of our stereochemical findings, why are not all
diphenyl compounds more effective anticonvulsants than their
respective phenyl, alkyl analogs? The answer is probably that
whether or not a particular type of compound is a better
antiepileptic drug when substituted with two phenyl rings rather
than one depends largely on solubility factors. Aromatic
substituents decrease aqueous solubility and therefore decrease
absorption of a drug from the gastrointestinal tract. Since
hydantoins are generally more soluble than the corresponding
acetylureas, diphenylhydantoin (phenytoin) retains sufficient water
solubility to be effectively absorbed on oral administration;
diphenylacetylurea, in contrast, is not a useful antiepileptic drug,
despite its appropriate stereochemistry, because it is very poorly
absorbed into the bloodstream (Swinyard and Toman, 1950).
Diphenylbarbituric acid has historically been regarded as a weak
anticonvulsant, but recent retesting has shown that inefficient mouse
gastrointestinal absorption of this compound leads to misleading
results. When administered intraperitoneally to mice (Raines et al,
1973) and by gavage to rats (Raines et al, 1975), diphenylbarbituric
acid is a highly effective anticonvulsant in both the maximal
electroshock and pentylenetetrazol laboratory tests and appears to
have considerable potential as a possible antiepileptic drug.

A particular strength of the stereochemical conclusions we have
drawn from crystallographic analyses is that we could have predicted
the presence and absence, respectively, of anticonvulsant properties
in the 2,4- and 1,3-benzodiazepines. Distances between electron-
donor atoms in the two structures obviously suggest a priori that
the 2,4-benzodiazepine stereochemistry, in terms of the
anticonvulsant activity-determining parameters, matches the
structures of the traditional antiepileptic compounds more closely

than do the 1,3-benzodiazepines. Thus the correlations between specific conformational features and specific pharmacological activity we have made from the structures of phenytoin, diazepam, and the other drugs can clearly be useful in predicting, in a qualitative sense, anticonvulsant potential in other compounds. Design of new molecules to be tested for use in generalized epilepsies should take these correlations into account: all of the necessary stereochemical features of the other successful drugs should be retained and those features that do not appear to be essential for activity can be changed. When appropriate considerations for solubility and polarizability are included, we are confident that this approach can lead to development of new drugs for the control of epilepsy.

ACKNOWLEDGEMENT   Support for these studies was from the USPHS through NIH grant NS 09839. A.C. is a Klingenstein Senior Fellow in the neurosciences.

REFERENCES

BIDEAU, J.P., MARLY, L. and HOUSTY, J. (1969) Compt. Rend. Acad. Sci. 269, 549-551.

BOWLING, A.C. and DeLORENZO, R.J. (1982) Science 216, 1247-1250.

CAMERMAN, A. and CAMERMAN, N. (1970) Science 168, 1457-1458.

CAMERMAN, A. and CAMERMAN, N. (1971a) Acta Crystallogr. 27, 2205-2211

CAMERMAN, N. and CAMERMAN, A. (1971b) Mol. Pharmacol. 7, 406-412.

CAMERMAN, A. and CAMERMAN, N. (1972a) J. Am. Chem. Soc. 94, 268-272.

CAMERMAN, N. and CAMERMAN, A. (1972b) J. Am. Chem. Soc. 94, 8553-8556

CAMERMAN, A. and CAMERMAN, N. (1975) Can. J. Chem. 53, 2194-2198.

CAMERMAN, A. and CAMERMAN, N. (1977) Proc. Natl. Acad. Sci USA 74, 1264-1266.

GALLAGER, D.W., MALLORGA, P. and TALLMAN, J.F. (1980) Brain Res. 189, 209-220.

GOLIK, U. (1975) J. Heterocyclic Chem. 12, 903-908.

HANSEN, J.M., KRISTENSEN, M. and SKOVSTED, L. (1968) Epilepsia 9, 17-22.

JULIEN, R.M. and HALPERN, L.M. (1971) Life Sci. 10, 575-582.

JULIEN, R.M. (1972) Neuropharmacology 11, 683–691.

KIM, S.-H. and RICH, A. (1968) Proc. Natl. Acad. Sci. USA 60, 402–408

MAJKOWSKI, J. and LYSAKOWSKA, K. (1977) 13th Congress of the
    International League Against Epilepsy, Amsterdam, Holland
    (Abstract).

MATTHEWS, W.D. and CONNOR, J.D. (1976) Neuropharmacology 15, 181–186.

MILLICHAP, J.G., PITCHFORD, G.L. and MILLICHAP, M.G. (1968)
    Proc. Soc. Exp. Biol. Med. 127, 1187–1190.

MIMAKI, T., DESHMUKH, P.P. and YAMAMURA, H.I. (1980) Acta Neurol.
    Scand. Suppl. 79. 62, 11.

NISHIMURA, T., TANIMUKAI, H. and NISHINUMA, H. (1963) J. Neurochem.
    10, 257–263.

OLESEN, O.V. and JENSEN, O.N. (1969) Dan. Med. Bull. 16, 154–158.

PAUL, S.M., SYAPIN, P.J., PAUGH, B.A., MONCADA, V. and SKOLNICK, P.
    (1979) Nature 281, 688–689.

RAABE, W. and GUNMIT, R.J. (1977) Epilepsia 18, 117–120.

RAINES, A., NINER, J.M. and PACE, D.G. (1973) J. Pharmacol. Exp.
    Ther. 186, 315–322.

RAINES, A., BAUMEL, I., GALLAGHER, B.B. and NINER, J.M. (1975)
    Epilepsia 16, 575–581.

SCHUSSLER, G.C. (1971) J. Pharmacol. Exp. Ther. 178, 204–209.

SWINYARD, E.A. and TOMAN, J.E.P. (1950) J. Pharmacol. Exp. Ther. 100,
    151–157.

TANIMUKAI, H., INUI, M., HARIGUCHI, S. and KANEKO, Z. (1965) Biochem.
    Pharmacol. 14, 961–970.

THOMAS, W.A. (1974) (Private Communication).

TOMAN, J.E.P. (1969) in The Pharmacological Basis of Therapeutics,
    L.S. Goodman and A. Gilman, Eds., Macmillan Co., Chapter 13.

WILLIAMS, P.P. (1973) Acta Crystallogr. B29, 1572–1579.

16 Structure–activity relationships in a series of 5-phenyl-1,4-benzodiazepines

Thomas A. Hamor and Ian L. Martin

1. INTRODUCTION

Since the early 1960's with the introduction of first chlordiazepoxide (1.1) [marketed as Librium (Roche)] and then diazepam (2.1) [marketed as Valium (Roche)], the 1,4-benzodia-
zepines have become the most frequently prescribed of all
psychoactive drugs, being used as anxiolytics, sedative/
hypnotics anticonvulsants and muscle relaxants (Sternbach
1978, 1979; Randall, Schallek, Sternbach and Ning, 1974).
The 1,3-dihydro-5-phenyl-2H-1,4-benzodiazepin-2-one system
(2) of diazepam has proved to be clinically the most useful.
In addition to diazepam, camazepam (2.2), clonazepam (2.3),
clorazepate potassium (2.4), desmethyldiazepam (2.5), flunitrazepam (2.6), flurazepam (2.7), lorazepam (2.8), lormet-
azepam (2.9), nimetazepam (2.10), nitrazepam (2.11), oxazepam
(2.12), pinazepam (2.13), prazepam (2.14), temazepam (2.15)
and the closely related bromazepam (3), medazepam (4) and the
5-cyclohexene derivative tetrazepam are also marketed for
clinical use. Other 1,4-benzodiazepines which have found
clinical use include the triazolobenzodiazepines, estazolam
(5.1) and triazolam (5.2), which have an additional ring
fused to the "a" face of the benzodiazepine system, and oxa-
zolam, cloxazolam and ketazolam which are "d" face fused
(Sternbach, 1978).

Following the success of diazepam, by the mid-1960's a
sufficient number of 1,4-benzodiazepines had been prepared
and tested for biological activity in laboratory animals for
it to become possible to draw up certain empirical structure-
activity relationships of a qualitative nature (Childress and
Gluckman, 1964; Sternbach, Randall, Banziger and Lehr, 1968).
It was obvious that the compounds of type (2) were generally
superior to those in which the ring system had been altered,
by, for example, reducing the N4-C5 double bond, as in (6.1)-

(6.3) or replacing the carbonyl function at C2 as in chlordiazepoxide (1.1) and medazepam (4), or replacing the 5-phenyl
ring by another substituent. Further, within the class of
1,3-dihydro-5-phenyl-1,4-benzodiazepin-2-ones of type (2) it
was possible to delineate the effect on activity of various
substituents at various positions in the basic framework.
Electron-withdrawing groups at C7 (e.g., halogen, NO_2, CF_3)
are of paramount importance for high activity, 7-dechlorodia-
zepam (2.16) being relatively inactive. Electron-releasing
groups at C7 (e.g., CH_3, OCH_3), or any substituent at C6, C8
or C9 lead to a marked decrease in activity. Methylation at
N1 increases activity but a larger group tends to decrease
activity, as in (2.17). A hydroxyl group at C3 generally
increases activity somewhat, but other substituents have
generally not proved beneficial. In the 5-phenyl ring an
ortho-fluoro- or ortho-chloro-substituent leads to increased
activity, but a meta- or para-halo-substituent causes a sharp
drop in activity, as in (2.18) (Sternbach, Sancilio and
Blount, 1974).

The triazolo-benzodiazepines of type (5) first prepared
somewhat later (1970) (Meguro and Kuwada, 1970; Hester, Rudzik
and Kamdar, 1971) are in some respects superior to the benzodiazepines of type (2). However, here also an electron-withdrawing substituent at C8 [corresponding to C7 of the type
(2) benzodiazepines] is required for high activity and an
ortho-chloro-substituent in the phenyl ring leads to an additional increase in activity.

At about the same time structural studies by NMR spectroscopy and X-ray crystallography began to appear in the literature (Linscheid and Lehn, 1967; Karle and Karle, 1967). By
the end of 1982, the crystal structures of at least 13 of the
clinically used benzodiazepines had been determined, as had
those of a number of analogues covering a wide range of biological activity. Many of these and related compounds have
also been studied by NMR spectroscopy. Recently molecular
orbital calculations have been applied in the derivation of
electronic parameters such as π-electron charge densities,
dipole moment and molecular orbital energies. The object of
these studies has generally been to establish relationships
between structural features and activity in order to provide
a theoretical basis for the empirical relationships already

established. This would allow prediction of compounds with
biological properties superior to those currently available,
and also lead to a better understanding of their mode of action.

2.   BIOLOGICAL ACTIVITY

The mechanism of action of the benzodiazepines is not fully
understood though it is clear from electrophysiological
evidence that they produce their effects by facilitation of
GABA mediated transmissions in the central nervous system
(Haefely, 1978). A considerable advance was made in our understanding when specific high affinity binding sites for these
compounds were discovered (Squires and Braestrup, 1977; Möhler
and Okada, 1977), the properties of which indicated that these
were distinct from the GABA binding sites but part of the
receptor through which this neurotransmitter modifies neuronal
excitability (Study and Barker, 1981).

Although the most frequent use of the benzodiazepines is
in the treatment of anxiety, the most easily quantifiable
aspect of their biological spectrum is anticonvulsant activity,
which can be measured by the dose of benzodiazepine required
to give protection against convulsions induced by pentylene-
tetrazole (anti-pent test). An alternative measure of biological activity is based on measurements of affinity for the
benzodiazepine receptor. Quantitative measurements of affinity are obtained from the ability of the compound to displace
tritium-labelled diazepam bound to the receptor in CNS membrane preparations. The affinity constants are highly correlated with the results of standard pharmacological tests, but
appear to give a more objective measure of activity than biological tests, where the pharmacokinetics and formation of
biologically active metabolites may complicate interpretation
of results.

3.   STRUCTURAL FEATURES
. X-ray crystallographic data
A crystal structure analysis of the benzodiazepine-receptor
complex would obviously be of vital interest; no such analysis
appears to have been carried out. Information regarding the

differing biological activity by the methods which have been outlined above and which we consider now in greater detail.

Full crystal structure data are currently (February, 1983) available for a large number of 1,4-benzodiazepines, including sixteen compounds of type (2), chlordiazepoxide (1.1) and the analogue (1.2), bromazepam (3), medazepam (4), estazolam (5.1) and the three compounds (6.1)-(6.3). References to the X-ray analyses and selected geometrical parameters are listed in Table 1; computer drawings of the highly active molecules (2.1), (2.3), (2.8) and (5.1) together with the relatively inactive (2.16)-(2.18) and (1.2) are shown in Figures 1-8. It must however be remembered that the results apply to the molecule in the solid state. Crystal packing forces may affect the geometry of flexible portions of the molecule in a manner different from solvent interactions which would be operative in vivo. Nevertheless X-ray crystallographic results generally agree well with spectroscopic results obtained from study of solutions. In a number of cases where the structure contains two or more independent molecules, which are exposed to different packing forces (multiple entries in Table 1), only the orientation of the 5-phenyl ring differs significantly, but not generally by large amounts (see Table 1). The X-ray results can therefore be accepted with some confidence as representing something close to the optimum geometry of the free molecule.

Considering first the benzodiazepines of type (2), to a first approximation the molecular framework of all the compounds is very similar. Although the seven-membered heterocyclic ring contains only one formal double bond and one shared aromatic bond, it adopts a cycloheptatriene-like boat conformation. The third "double" bond is the amide bond, N1-C2. This bond is shortened by electron delocalisation between the nitrogen lone pair and the carbonyl oxygen atom to a length of about half-way between the C-N pure single and double bond values. The disposition of bonds at the nitrogen atom is near planar and its overall geometry closely resembles that of a normal double bond.

The molecules of this type can be described in terms of four planes; the fused benzene (benzo) moiety together with the linked atoms N1 and C5, the central plane of the "boat"

of the boat, atoms C2, C3, N4, and the 5-phenyl ring. The
5-phenyl ring is oriented at angles within the range $54-86°$
to that of the benzo plane, constituting the major geometrical
differences between these molecules. The larger dihedral
angles $(>75°)$ occur in compounds with a halogen substituent
at the 2'-position of the phenyl ring (θ_1 in Table 1). The
"stern" plane of the boat-shaped ring, formed by atoms N1,
C5, C10 and C11 coincides approximately with the benzo plane.
The stern and bow angles of the "boat" are listed under θ_2
and θ_3. Stern angles are all in the range $32-40°$ and bow
angles in the range $58-64°$.

In an ideal cycloheptatriene boat, with mirror (C_s) symm-
try, the ring torsion angles about N1-C2, N4-C5 and C10-C11
are zero and the other four torsion angles taken in pairs are
equal in magnitude but of opposite sign. The "deviation parameter", Δ,

$$[1/5(T1^2 + T4^2 + T6^2 + (T2 + T3)^2 + (T5 + T7)^2]^{\frac{1}{2}}$$

where T1-T7 are the ring torsion angles about N1-C2, C2-C3,
C3-N4, N4-C5, C5-C11, C11-C10, C10-N1 respectively is a
measure of distortion; in the ideal case $\Delta = 0$ [cf. asymmetry
parameter of Duax et al., (1976)]. The larger distortions
$(>6°)$ all occur in compounds with a methyl substituent at N1,
although the N1-Me substituted compounds (2.15), (2.16) and
(2.22) are only slightly distorted (see Table 1). (2.14) and
(2.17) where N1 carries the bulky cyclopropylmethyl and methyl-
acetamido groups, have only moderately distorted rings, as has
(2.23) where C3 carries a large substituent. The torsion
angles about N1-C2 generally parallel the Δ values.

A halo-substituent at the 2'-position of the 5-phenyl
ring generally increases the biological activity, and affects
the geometry of the molecule by increasing θ_1 and torsion
angle C11-C5-C1'-C2' and decreasing torsion angle N4-C5-
C1'-C2' (see Table 1). Any electronic interaction between
the 5-phenyl ring and the benzodiazepine framework would affect
the length of the C5-C1' bond. The mean length of this bond
for unsubstituted rings is 1.49 Å, the same as for the C2'-
halo-substituted ones. The X-ray analysis of (2.22) is of

relatively low accuracy and the length of 1.56 Å for this
compound was not included in the averaging. There is also
no correlation between the length of C5-Cl' and the C11-C5-
Cl'-C2' or N4-C5-Cl'-C2' torsion angles. It corresponds in
length to a normal single bond between trigonally hybridised
carbon atoms, and there is no evidence of electron delocalisation across it.

Bromazepam (3) differs from the type (2) benzodiazepines
only in having a pyridinyl substituent at C5 rather than phenyl,
and geometrical parameters are similar.

The seven-membered ring of medazepam (4) which does not
have a keto function at C2, is essentially a cycloheptadiene-
like pseudo-boat. The torsion angles about N1-C2 are 26.5°
and 29.3° in the two independent molecules in the crystal.
The length of the N1-C2 bond, 1.46 Å, in both cases, corresponds to a C-N single bond.

The seven-membered ring of chlordiazepoxide (1.1) and the
C3-hydroxy compound (1.2) adopts a cycloheptatriene boat-like
conformation. The N1-C2 bond is a formal double bond in these
compounds, but is somewhat longer (1.30-1.31 Å) indicating
some electron delocalisation.

The 5-phenyl-1,4-benzodiazepine framework of estazolam
(5.1) is quite similar to that of the 1,4-benzodiazepinones
of type (2) so that the additional ring fused to the "a" face
of the benzodiazepine system does not have any great effect
on the geometry (see Table 1).

In the 5-phenyl-1,3,4,5-tetrahydro-2H-1,4-benzodiazepin-
2-ones (6.1)-(6.3) the N4-C5 bond is a pure single bond and
the disposition of bonds is pyramidal at N4 and tetrahedral
at C5, and the overall geometry of the seven-membered ring is
that of a cycloheptadiene pseudo-boat. The substituents at
N4 and C5 are oriented pseudo-equatorial. Benzodiazepines of this type exhibit only low activity.

3.2. Spectroscopic data

In contrast to X-ray crystallographic data the results pertain
to the molecule in solution, an environment closer to the
actual conditions under which it interacts with its receptor
in vivo. However, the spectroscopic data are generally harder

to interpret and the results less definitive.

A number of NMR spectroscopic studies by several groups of workers have shown that the seven-membered heterocyclic ring of benzodiazepines of type (2), for example, (2.1), (2.3) (2.5)-(2.7), (2.10), (2.11) and (2.13) undergoes inversion between two energetically equivalent forms (Linscheid and Lehn, 1967; Bley, Nuhn and Benndorf, 1968; Sadee, 1969; Sarrazin, Bourdeaux-Pontier, Briand and Vincent, 1975; Raban et al., 1975; Romeo et al., 1979; Haran and Tuchagues, 1980; Kovar, Linden and Breitmaier, 1981). At 37°C the inversion is rapid on the NMR time scale for the N1-H compounds and slow for those in which N1 carries a methyl or larger group. The free energies of activation for the inversion are ca. 42-52 kJ mol^{-1} and 65-75 kJ mol^{-1} respectively. A 2'-chloro or fluoro-substituent reduces the barrier to ring inversion slightly. The geometry of the invertomers is considered to be intermediate between a cycloheptatriene boat and a cycloheptadiene pseudo-boat. C3-OH and C3-Me substituted compounds, for example (2.12) and (2.15), are considered to exist in only one conformation at 37°C with the substituent quasi-equatorial and H quasi-axial; this is the energetically favoured form (Sadee, Schwandt and Beyer, 1973; Sunjic et al., 1979), and is the conformation found in the solid state (see Figure 3). Medazepam (4) and its desmethyl derivative both exhibit conformational equilibrium involving rapid inversion of two pseudo-boat forms (Romeo et al., 1981). It has been suggested that high biological activity is associated with a relatively high barrier to ring inversion (Romeo et al., 1979). The 5-phenyl ring is considered to undergo rapid rotation about the C5-C1' bond (Haran and Tuchagues, 1980). In a recent study lanthanide-induced shifts were used for the signal assignments of the ^{13}C spectra of (2.1), (2.5), (2.7) and (2.14), and to characterise the orientation of the 5-phenyl ring (Paul, Sapper, Lohmann and Kalinowski, 1982). The agreement with the solid state conformations of (2.1) and (2.14) is reasonably good.

Spectroscopic methods have also been applied to the study of benzodiazepines bound to human serum albumin (Müller and Wollert, 1973a, 1973b, 1974; Sjöholm and Sjödin, 1972; Sjödin,

Roosdorp and Sjöholm, 1976; Alebic-Kolbah et al., 1979; Konowal et al., 1979; Sarrazin, Sari, Bourdeaux-Pontier and Briand, 1979) and nucleobases (Paul, Sapper and Lohmann, 1980).

4. STRUCTURE-ACTIVITY RELATIONSHIPS

Camerman and Camerman (1974) have found that there are certain common stereochemical features in a number of chemically different anticonvulsant drugs including diazepam, phenytoin and phenobarbital, and have proposed a common mechanism of anticonvulsant activity (Camerman and Camerman, 1981). This is, however, not universally accepted (Olsen, 1981). Phenobarbital exhibits less than one thousandth of the affinity of diazepam for the benzodiazepine receptor (Braestrup and Squires, 1978), although phenytoin is reported to act as a competitive inhibitor (Camerman and Camerman, 1981).

In the first attempts to establish relationships between structure and activity within the class of 1,4-benzodiazepines on a systematic basis, three independent studies appeared during 1976-77. In the first of these, Sarrazin, Bourdeaux-Pontier and Briand (1976) calculated certain electronic parameters by the CNDO/2 molecular orbital method for a series of fourteen compounds of type (2). Some correlation ($r = 0.67$) was observed between biological activity and the energy of the lowest empty molecular orbital, E_{LEMO}, but little correlation ($r = 0.47$) with the energy of the highest occupied orbital, E_{HOMO}. These energy terms can be equated to electron affinity and ionisation potential, respectively. The net electronic charge on N4, however, showed a strong ($r = 0.97$), but non-linear correlation with activity. This was interpreted as indicating that a hydrogen bond, N4...H-receptor, is critical in the drug-receptor interaction. Lipophilicity was found not to be a factor in activity.

Blair and Webb (1977) carried out CNDO/2 calculations on 59 benzodiazepines. These were mainly of type (2) with various substituents at C6, C7, C8, C9, C2', C3' and C4', but included also some 5-pyridinyl derivatives of type (3). Correlations were observed between certain measures of CNS activity (inclined screen test, footshock test, anti-pent test) and the net charge on the C2 carbonyl oxygen atom (r = 0.75, 0.85, 0.73) and the total molecular dipole moment (r = 0.75, 0.87, 0.82). Lipophilicity was again not correlated with activity.

In both these studies the molecular orbital calculations were based essentially on the atomic coordinates obtained from the crystal structure analysis of diazepam (2.1), on the assumption that the geometry of the molecular framework is not affected by substituents - hardly a valid assumption (see Table 1). Sarrazin et al. (1976) did attempt to allow for the geometric and postulated electronic effects of a C2'-halo-substituent. The latter effect, decreased electron delocalisation across C5-Cl' and hence increased charge density at N4 with increasing rotation of the 5-phenyl ring out of the N4-C5-Cl' plane which was assumed in the analysis for the C2'-substituted compounds, is however, not supported by the X-ray crystallographic data.

In the third of these early studies, Gilli, Borea, Bertolasi and Sacerdoti (1977) concluded that there was no correlation between molecular geometry and anti-pent activity on the basis of a comparison of the crystal structure of diazepam, lorazepam, nitrazepam, oxazepam and the relatively inactive (2.16) and (2.18); however they found from CNDO/2 calculations on seven benzodiazepines that anti-pent activity is correlated (inversely) with E_{HOMO}, E_{LEMO} and $E_{LEMO}-E_{HOMO}$.

In contrast, in a later study, Lucek, Garland and Dairman (1979) found from CNDO/2 calculations on eighteen benzodiazepines that there was no significant correlation between anti-pentylenetetrazole activity and the charge on any atom, E_{HOMO} or E_{LEMO}. A highly significant correlation was observed, however, between activity and the electron density in the p_y orbital at the aromatic carbon atom adjacent to the amide nitrogen (C10 in our numbering system).

The present study differs from these in three respects; firstly, as a measure of biological activity the affinity of the specific benzodiazepine receptor for the individual compounds has been used. This we believe, provides a more

$r = 0.92$, $r^2 = 0.84$, $F = 19.12$, $p < 0.001$

$$\hat{V1} = 37.22 \ (\pm 9.53) \ V9 + 0.72 \ (\pm 0.17) \ V21 - 0.03$$
$$(\pm 0.01) \ V33 - 8.23 \ (\pm 2.27) \tag{9}$$
$r = 0.91$, $r^2 = 0.83$, $F = 18.18$, $p < 0.001$

Calculations were then carried out considering only the
variables appearing in equations 1-9 and taking electronic
parameters (V8, V9, V17, V21, V24) and geometric parameters
(V27, V31, V33) separately. This led to the best three-
variable regression which has been obtained; it involves
only electronic parameters.

$$\hat{V1} = 32.89 \ (\pm 7.87) \ V8 + 32.69 \ (\pm 8.25) \ V9 +$$
$$0.71 \ (\pm 0.14) \ V21 + 1.12 \ (\pm 1.56) \tag{10}$$
$r = 0.94$, $r^2 = 0.89$, $F = 29.91$, $p < 0.001$

Equations 1-10 indicate that some nine parameters may be
important in explaining biological activity; V8 (charge on
C1', qC1'), V17 (charge on C3', qC3'), V24 (HOMO energy),
V27 (sternangle of the boat-shaped seven-membered ring, θ_2),
V31 (deviation of N4 from the plane of the fused benzene
ring, dN4), V33 (deviation of C2' from this plane, dC2'),
($E_{HOMO} + E_{LEMO}$), V21 (dipole moment component in the direction parallel to the C10-C11 bond, μ_x) and V9 (charge on C10,
qC10). According to equation 10, qC1', qC10 and μ_x together
explain 89% of the variability in the data. Equation 8 again
involves qC1' (V8) and μ_x(V21), together with E_{HOMO} (V24).
Equation 9 includes qC10 (V9), μ_x (V21) and also a geometric
parameter dC2' (V33). The dipole moment component μ_x thus
occurs in all three equations and qC1' and qC10 each occur
in two equations. Equations 8 and 9 account for 84 and 83%
of the variability respectively. Strong cross-correlations,
however, occur between qC1' (V8) and qC3' (V17), and between
dC2' (V33) and dC1' (V32) and dC3' (V34). The apparent
importance of the charge on C1' is interesting in relation
to the importance of the presence of a 5-phenyl substituent
for biological activity and the appearance of a dipole moment
component is gratifying as it is thought that this is norm-

ally involved in aligning the drug and the receptor in the initial interaction.

Plots of experimental affinities v. values calculated by equations 8-10 are shown in Figure 9.

Affinity and structural data are available also for four other benzodiazepines, chlordiazepoxide (1.1), flunitrazepam (2.6), bromazepam (3) and compound (6.3). None of these was included in the regression analysis, as structural data for the first three and affinity data for the fourth compound became available only very recently. They have, however, been used to test the regression equations. Data are listed in Table 4. Affinity values for the compounds, calculated by equations 8-10 are listed in Table 5, together with the experimental affinities.

Bromazepam unlike all the other compounds is not a 5-phenyl derivative. The presence of the heteroatom at the 2'-position of the 5-pyridinyl substituent has a dramatic effect on qCl' (V8) and calculation of affinity by means of equations 8 and 10 leads to values of 9.2 and 8.3 compared with the experimental 4.3 (Braestrup and Squires, 1978). Equation 9 which does not involved qCl' leads to an affinity value of 4.2. Equation 9 also gives the best agreement for flunitrazepam and for the C4'-chloro compound (6.3), but leads to an unsatisfactorily low affinity of 0.6 for chlordiazepoxide (experimental value 3.0). Equations 8 and 10 also underestimate affinity for chlordiazepoxide. Equation 9 gives the best overall agreement for these four compounds although if bromazepam is excluded, equation 10 is more satisfactory.

It is however noteworthy that the agreement between observed and calculated values for the four test compounds is not nearly as good as for the fifteen compounds used to set up the regression equations. This emphasises the limitations of the present analysis.

In order not to prejudge the importance of any particular parameter, the analysis had to be carried out on a data set which is very large (59 variables) relative to the number of compounds available (15). The characterisation of each variable is therefore limited and the analysis can in no way be considered rigorous. Nevertheless, the parameters whose

KONOVAL, A., SNATZKE, G., ALEBIC-KOLBAH, T., KAJFEZ, F., RENDIC, S. and SUNJIC, V. (1979). Biochem. Pharmacol. 28, 3109.

LINSCHEID, P. and LEHN, J.M. (1967). Bull. Soo. Chim. France 3, 992.

LUCEK, R.W., GARLAND, W.A. and DAIRMAN, W. (1979). Fed. Proc. 38, 541.

MEGURO, K. and KUWADA, Y. (1970). Tetrahedron Lett. p.4039.

MÖHLER, H. and OKADA, J. (1977). Science 198, 849.

MOTHERWELL, W.D.S. (1978). PLUTO 78. Program for plotting molecular and crystal structures. University of Cambridge, England.

MÜLLER, W. and WOLLERT, V. (1973a). Naunyn Schmiedebergs Arch. Pharmacol. 278, 301.

MÜLLER, W. and WOLLERT, V. (1973b). Naunyn Schmiedebergs Arch. Pharmacol. 280, 229.

MÜLLER, W. and WOLLERT, V. (1974). Naunyn Schmiedebergs Arch. Pharmacol. 283, 67.

NIE, N.H., HULL, C.H., JENKINS, J.G., STEINBRENNER, K. and BENT, D.H. (1975). SPSS Statistical Package for the Social Sciences (Second Edition). McGraw-Hill, New York.

OLSEN, R.W. (1981). J. Neurochem. 37, 1.

PAUL, H.H., SAPPER, H. and LOHMANN, W. (1980). Biochem. Pharmacol. 29, 137.

PAUL, H.H., SAPPER, H., LOHMANN, W. and KALINOWSKI, H.O. (1982). Org. Magn. Reson. 19, 49.

RABAN, M., CARLSON, E.H., SZMUSZKOVICZ, J., SLOMP, G. CHIDESTER, C.G. and DUCHAMP, D.J. (1975). Tetrahedron Lett. p.139.

RANDALL, L.O., SCHALLEK, W., STERNBACH, L.H. and NING, R.Y. (1974). In: Psychopharmacological Agents (ed. M. Gordon) Vol.3, p.175. Academic Press, New York.

ROMEO, G., AVERSA, M.C., GIANNETTO, P., FICARRA, P. and VIGORITA, M.G. (1981). Org. Magn. Reson. 15, 33.

ROMEO, G., AVERSA, M.C., GIANNETTO, P., VIGORITA, M.G. and FICARRA, P. (1979). Org. Magn. Reson. 12, 593.

SADEE, W. (1969). Arch. Pharm. 302,769.

SADEE, W., SCHWANDT, H.J. and BEYER, K.H. (1973). Arch. Pharm. 306, 751.

SARRAZIN, M., BOURDEAUX-PONTIER, M. and BRIAND, C. (1976). Ann. Phys. Biol. Med. 9, 211.

SARRAZIN, M., BOURDEAUX-PONTIER, M., BRIAND, C. and VINCENT, E.J. (1975). Org. Magn. Reson. 7, 89.

SARRAZIN, M., SARI, J.C., BOURDEAUX-PONTIER, M. and BRIAND, C. (1979). Mol. Pharmacol. 15, 71.

SJÖDIN, T., ROOSDORP, N. and SJÖHOLM, I. (1976). Biochem. Pharmacol. 25, 2131.

SJÖHOLM, I. and SJÖDIN, T. (1972). Biochem. Pharmacol. 21, 3041.

SQUIRES, R.F. and BRAESTRUP, C. (1977). Nature 266, 732.

STERNBACH, L.H. (1978). Progr. Drug Res. 22, 229.

STERNBACH, L.H. (1979). J. Med. Chem. 22, 1.

STERNBACH, L.H., RANDALL, L.O., BANZIGER, R. and LEHR, H. (1968). In: Drugs Affecting the Central Nervous System (ed. A. Burger) Vol.2, p.237. Marcel Dekker, New York.

STERNBACH, L.H., SANCILIO, F.D. and BLOUNT, J.F. (1974). J. Med. Chem. 17, 374.

STUDY, R.E. and BARKER, J.L. (1981). Proc. Natl. Acad. Sci., U.S.A. 78, 7180.

SUNJIC, V., LISINI, A., SEGA, A., KOVAC, T., KAJFEZ, F. and RUSCIC, B. (1979). J. Heterocycl. Chem. 16, 757.

ACKNOWLEDGEMENTS

We thank Hoffmann - La Roche and Richter-Gideon Company for samples, and one of us (TAH) is indebted to the Medical Research Council for the award of a Project Grant.

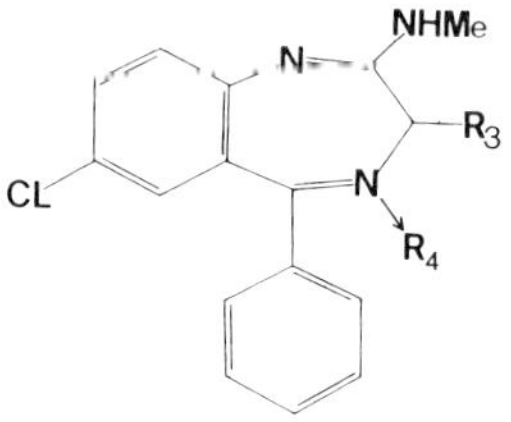

	R_3	R_4
(1.1)	H	O
(1.2)	OH	–

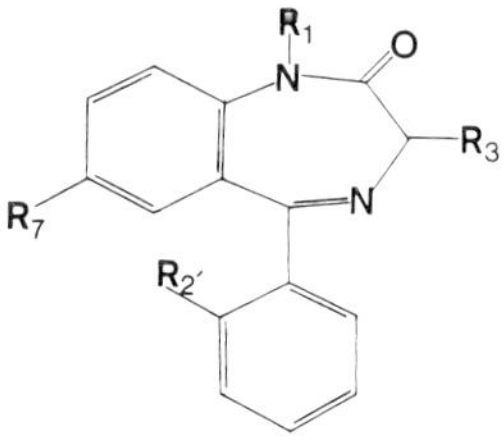

(2)

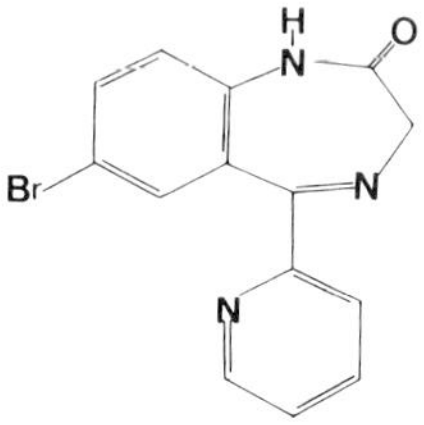

(3)

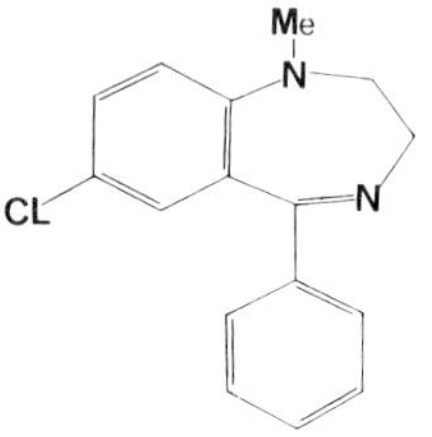

(4)

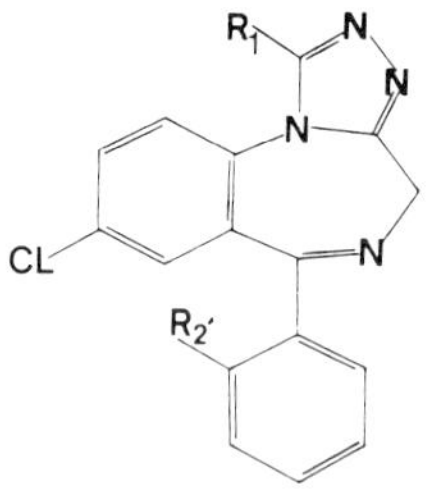

	R_1	$R_{2'}$
(5.1)	H	H
(5.2)	Me	Cl

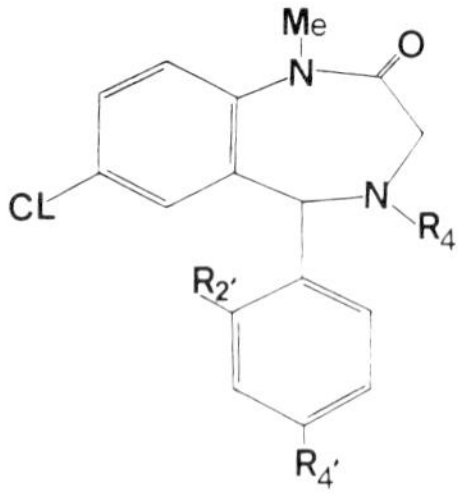

	R_4	$R_{2'}$	$R_{4'}$
(6.1)	H	H	H
(6.2)	Me	F	H
(6.3)	Me	Cl	Cl

		R_1	R_3	R_7	$R_{2'}$
Diazepam	(2.1)	Me	H	Cl	H
Camazepam	(2.2)	Me	$OCONMe_2$	Cl	H
Clonazepam	(2.3)	H	H	NO_2	Cl
Clorazepate	(2.4)	H	COOK	Cl	H
Desmethyldiazepam	(2.5)	H	H	Cl	H
Flunitrazepam	(2.6)	Me	H	NO_2	F
Flurazepam	(2.7)	$CH_2CH_2NEt_2$	H	Cl	F
Lorazepam	(2.8)	H	OH	Cl	Cl
Lormetazepam	(2.9)	Me	OH	Cl	Cl
Nimetazepam	(2.10)	Me	H	NO_2	H
Nitrazepam	(2.11)	H	H	NO_2	H
Oxazepam	(2.12)	H	OH	Cl	H
Pinazepam	(2.13)	$CH_2C \equiv CH$	H	Cl	H
Prazepam	(2.14)	$CH_2 \triangleleft$	H	Cl	H
Temazepam	(2.15)	Me	OH	Cl	H
	(2.16)	Me	H	H	H
	(2.17)	$CH_2CONHMe$	H	Cl	H
	(2.18)	Me	H	Cl	4'-F
	(2.19)	Me	H	Cl	Cl
	(2.20)	H	H	Cl	Cl
	(2.21)	H	H	Br	Cl
	(2.22)	Me	H	Br	H
	(2.23)	H	NHCOCONHMe	Cl	H

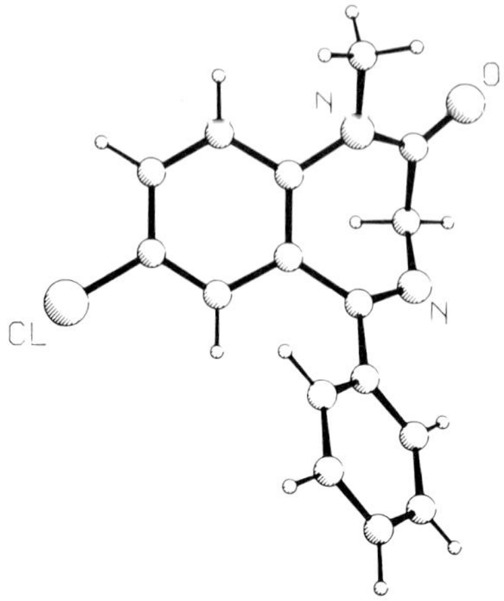

Figure 1.
Compound (2.1)
(diazepam)

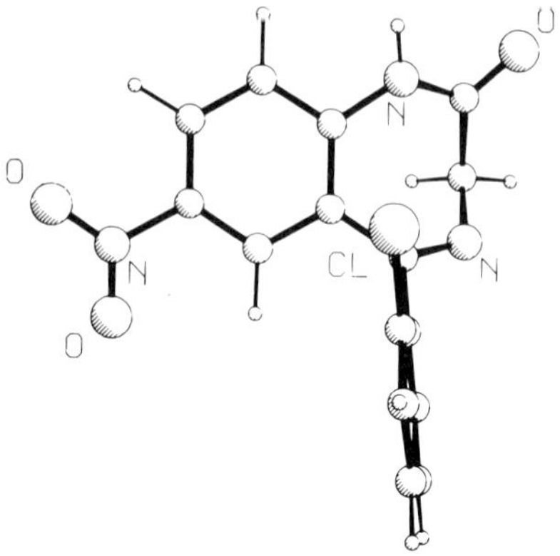

Figure 2.
Compound (2.3)
(clonazepam)

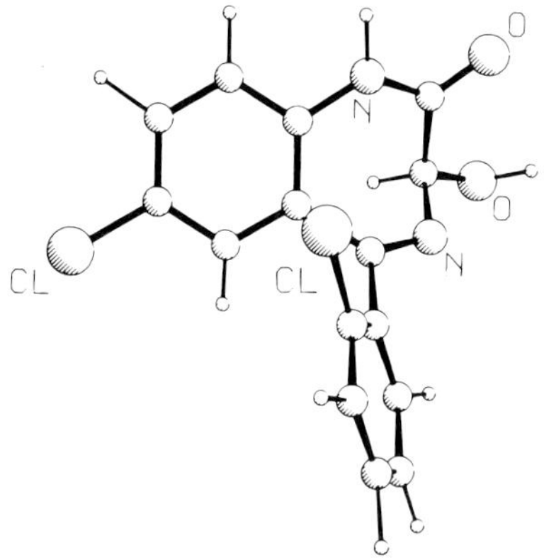

Figure 3.
Compound (2.8)
(lorazepam)

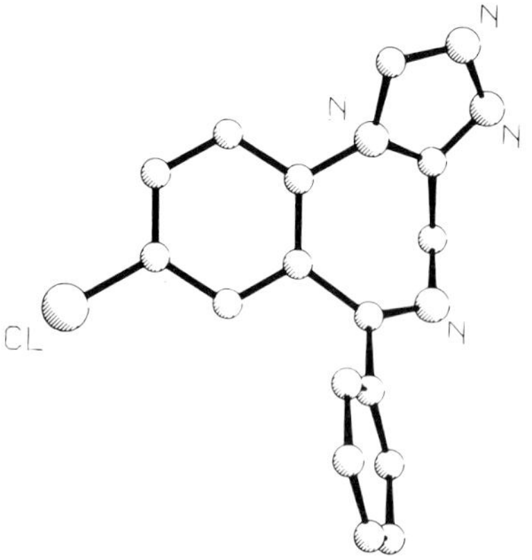

Figure 4.
Compound (5.1)
(estazolam)

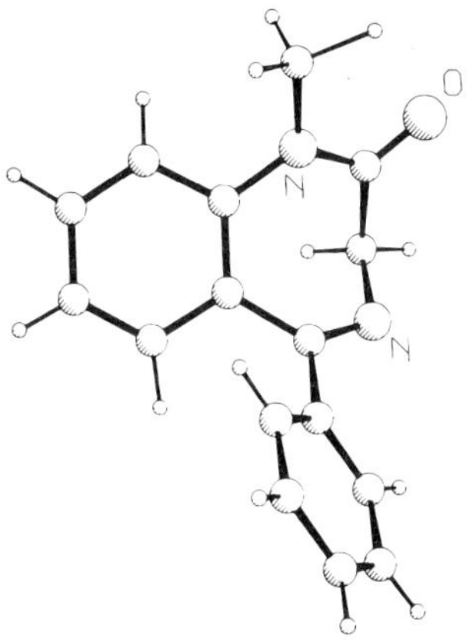

Figure 5.
Compound (2.16)

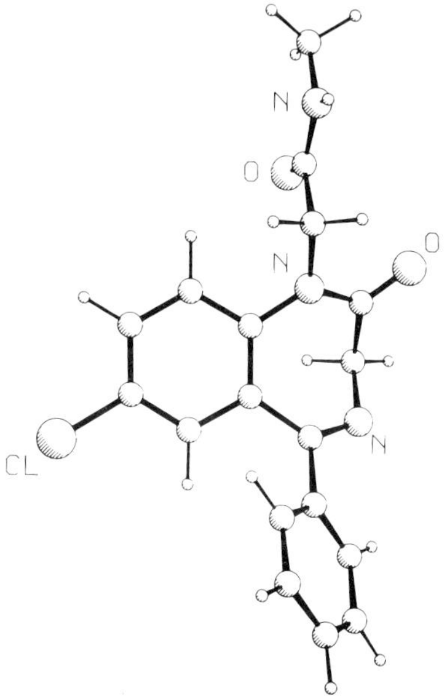

Figure 6.
Compound (2.17)

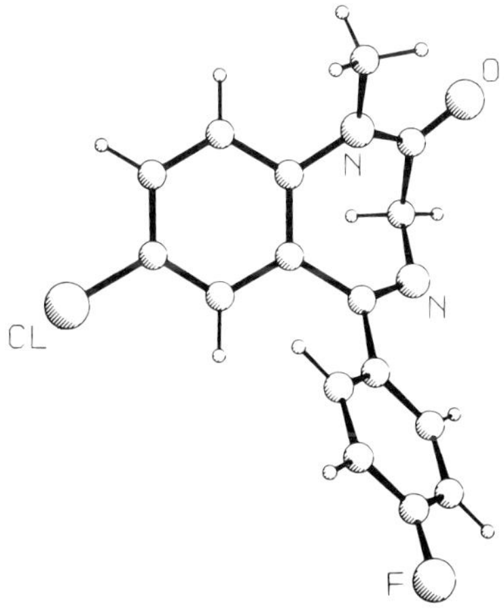

Figure 7.
Compound (2.18)

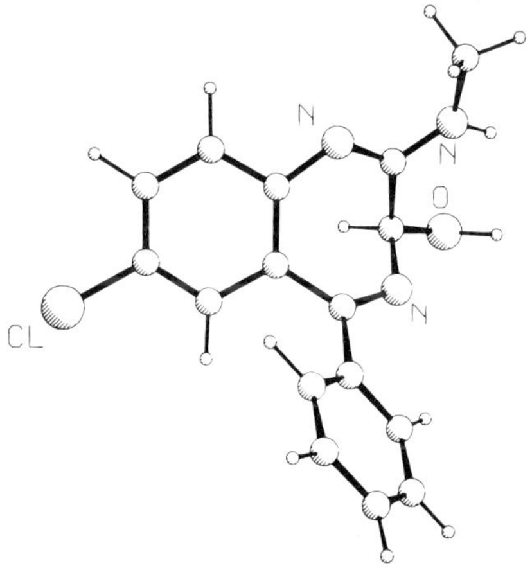

Figure 8.
Compound (1.2)

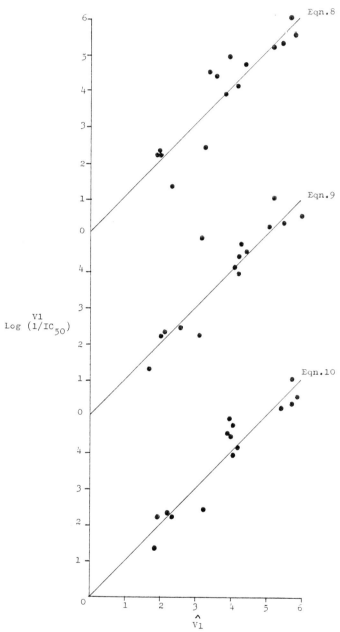

Figure 9. Correlations between measured affinity expressed as $\log[1/IC_{50}(mM)]$ and those calculated by regression equations 8-10 $(\hat{V}1)$.

TABLE 1

Selected geometric parameters of 1,4-benzodiazepines

θ_1 is the angle between the planes of the 5-phenyl ring and the fused benzene moiety.

θ_2 and θ_3 are the stern and bow angles of the boat shaped seven-membered ring.

Δ is deviation parameter (see text).

T(N1-C2) is the endocyclic torsion angle about N1-C2.

L(N1-C2) and L(C5-C1') are the lengths (Å) of the N1-C2 and C5-C1' bonds.

Under T(C5-C1') are listed the torsion angles Cl1-C5-C1'-C2' and N4-C5-C1'-C2' [N2' in bromazepam (3)].

Estimated standard deviations are generally <0.01 Å for lengths and <1° for angles; in (2.20) and (2.22) e.s.d.'s are greater by factors of ca. 2 and 4, respectively.

	θ_1	θ_2	θ_3	Δ	T(N1-C2)	L(N1-C2)	L(C5-C1')	T(C5-C1')	Reference
(1.1)	67	34	58	2.1	6.2	1.30	1.48	45,136	17
	61	33	56	3.7	5.6	1.30	1.48	40,139	
	70	37	57	5.7	6.6	1.30	1.47	40,138	
	63	33	55	7.7	9.8	1.30	1.48	45,138	
(1.2)	61	35	59	2.2	2.2	1.30	1.49	29,149	18
	65	32	61	5.8	5.8	1.31	1.50	42,137	
(2.1)	55	38	58	9.3	13.4	1.37	1.49	25,157	19
(2.3)	84	34	58	4.4	7.2	1.36	1.50	66,116	20
	78	33	59	2.4	4.0	1.36	1.51	61,123	
(2.6)	75	37	61	3.9	5.8	1.37	1.48	48,134	21
(2.8)	73	35	62	3.1	3.2	1.35	1.48	52,131	22
	81	33	58	2.9	4.0	1.31	1.49	58,121	
(2.11)	62	32	60	1.3	0.9	1.36	1.49	35,144	23
(2.12)	54	33	63	2.3	0.2	1.36	1.49	30,152	24
	66	33	63	1.6	2.8	1.36	1.49	42,138	
(2.14)	71	37	62	4.6	3.3	1.37	1.49	44,138	25
	67	40	60	5.3	8.8	1.37	1.49	29,151	
(2.15)	66	34	63	2.8	0.5	1.38	1.49	40,142	26
	67	36	64	2.1	4.0	1.37	1.50	37,144	
	59	40	63	3.5	5.6	1.37	1.49	24,160	
(2.16)	66	37	61	3.8	7.2	1.36	1.49	35,145	4
(2.17)	62	36	61	4.8	3.4	1.37	1.50	32,151	27
(2.18)	62	39	59	9.0	11.8	1.36	1.49	24,156	4
	68	39	60	7.0	10.1	1.36	1.49	35,146	
(2.19)	86	39	58	8.2	14.8	1.37	1.50	63,120	28
(2.20)	80	33	59	1.7	1.0	1.34	1.48	63,122	28
(2.21)	75	34	60	3.7	2.8	1.38	1.49	54,131	29
(2.22)	74	36	58	2.3	2.6	1.31	1.56	38,138	30
(2.23)	69	32	59	4.4	6.3	1.35	1.51	48,134	31
(3)	60	31	60	2.1	1.0	1.35	1.50	40,142	32
(4)	63	44	57	15.1	26.5	1.46	1.49	26,153	33
	55	46	56	17.1	29.3	1.46	1.49	14,164	
(5.1)	75	31	54	3.2	3.6	1.34	1.53	48,130	34
(6.1)	88	54	48	29.3	3.1	1.37	1.52	87,152	35
(6.2)	82	51	54	18.2	8.0	1.36	1.52	82,151	36
(6.3)	88	55	55	25.4	0.3	1.38	1.51	87,151	8

[4] Sternbach et al., 1974; [8] Karle and Karle, 1967; [17] Bertolasi et al., 1982; [18] Chananont et al., 1980a; [19] Camerman and Camerman, 1972; [20] Chananont et al., 1979; [21] Butcher, Hamor and Martin, to be published; [22] Bandoli and Clemente, 1976; [23] Gilli et al., 1977; [24] Gilli et al., 1978a; [25] Brachtel and Jansen, 1981; [26] Galdecki and Glowka, 1980; [27] Chananont et al., 1980b; [28] Chananont et al., 1981; [29] Karapetyan et al., 1979; [30] Dunphy and Lynton, 1971; [31] Fryer et al., 1977; [32] Butcher, Hamor and Martin, to be published; [33] Gilli et al., 1978b; [34] Kamiya et al., 1973; [35] Czugler et al., 1977; [36] Chananont et al., 1980c.

TABLE 2

List of variables used in regression analysis

V1, Biological activity
V2, Atomic charge N1
V3, Atomic charge C2
V4, Atomic charge O2
V5, Atomic charge C3
V6, Atomic charge N4
V7, Atomic charge C5
V8, Atomic charge C1'
V9, Atomic charge C10
V10, Atomic charge C11
V11, Atomic charge p_y orbital of C10
V12, Atomic charge C6
V13, Atomic charge C7
V14, Atomic charge C8
V15, Atomic charge C9
V16, Atomic charge C2'
V17, Atomic charge C3'
V18, Atomic charge C4'
V19, Atomic charge C5'
V20, Atomic charge C6'
V21, Dipole moment component parallel C10-C11 (μ_x)
V22, Dipole moment component parallel C10-C8 (μ_y)
V23, Dipole moment component perpendiculat to μ_x
     and μ_y (μ_z)
V24, Mol orbital energy HOMO
V25, Mol orbital energy LEMO
V26, θ_1 (see Table 1)
V27, θ_2 (see Table 1)
V28, θ_3 (see Table 1)
V29, Deviation C2 from plane of benzo ring
V30, Deviation C3   "   "   "   "   "
V31, Deviation N4   "   "   "   "   "
V32, Deviation C1'   "   "   "   "   "
V33, Deviation C2'   "   "   "   "   "
V34, Deviation C3'   "   "   "   "   "
V35, Deviation C4'   "   "   "   "   "
V36, Deviation C5'   "   "   "   "   "
V37, Deviation C6'   "   "   "   "   "
V38, Deviation O2   "   "   "   "   "
V39, Bond length N1-C10
V40, Bond length N1-C2
V41, Bond length C5-C11
V42, Bond length C5-C1'
V43, Bond length C2-C3
V44, Bond length C3-N4
V45, Bond length N4-C5
V46, Bond length C11-C6
V47, Bond length C6-C7
V48, Bond length C7-C8
V49, Bond length C8-C9
V50, Bond length C9-C10
V51, Bond length C10-C11
V52, Bond length C1'-C2'
V53, Bond length C2'-C3'
V54, Bond length C3'-C4'
V55, Bond length C4'-C5'
V56, Bond length C5'-C6'
V57, Bond length C6'-C1'
V58, Bond length C2-O2
V59, Distance between centres of "benzo" and phenyl rings
V60, Torsion angle C10-N1-C2-C3

TABLE 3

Correlation Matrix

	V1	V6	V7	V8	V9	V17	V19	V21	V24	V25	V27	V30	V31	V32	V33	V34	V35	V45
V1	1																	
V6	0.62	1																
V7	-0.60	-0.92	1															
V8	0.75	0.76	-0.69	1														
V9	0.52	0.51	/	/	1													
V17	0.73	0.55	/	0.90	/	1												
V19	0.63	0.56	-0.55	0.66	/	0.81	1											
V21	0.52	/	/	/	/	/	/	1										
V24	-0.64	/	/	/	-0.75	/	/	/	1									
V25	-0.61	/	/	/	-0.73	/	/	/	0.77	1								
V27	-0.66	-0.50	0.57	-0.60	/	/	/	/	/	0.65	1							
V30	-0.59	/	0.51	-0.52	-0.62	/	/	/	/	0.61	0.83	1						
V31	-0.69	-0.50	0.51	-0.64	-0.54	/	/	/	/	0.55	0.89	0.90	1					
V32	-0.60	-0.61	0.59	-0.68	/	-0.50	/	/	/	/	0.71	0.67	0.88	1				
V33	-0.68	-0.62	0.61	-0.76	/	-0.73	-0.68	/	/	/	0.57	0.57	0.73	0.87	1			
V34	-0.62	-0.68	0.67	-0.78	/	-0.64	-0.52	/	/	/	0.67	0.60	0.80	0.96	0.95	1		
V35	-0.51	-0.65	0.64	-0.70	/	/	/	/	/	/	0.72	0.63	0.83	0.97	0.79	0.94	1	
V45	-0.56	-0.61	0.57	-0.75	/	/	/	/	/	/	0.83	0.68	0.85	0.82	0.58	0.77	0.88	1

/ is < 0.5

17 Structure–activity relationships in opioids of the N-alkoxyalkyl-normetazocine series

Herbert Merz

1. INTRODUCTION

A universal, unpleasant experience of mankind is pain. In certain cases, it is possible to remove pain by causal treatment. Generally, however, therapy is limited to symptomatic pain relief by analgesic drugs. Depending on the nature of the pain concerned, various chemically differentiated groups of analgesics with different modes of action are used. The group formed by morphine and its congeners ('narcotic analgesics', also known as 'strong analgesics', 'central analgesics' or 'opiates') provides the most powerful analgesics capable of relieving severe pain.

Morphine is one of the major ingredients of opium whose effects are supposed to have already been known to the ancient Sumerians more than 5000 years ago. The recent history of morphine begins with the isolation of the pure alkaloid from opium by Sertürner (1805). During the subsequent years of therapeutic use of morphine, its desirable analgesic effect proved to be accompanied by servere side-effects. In particular, its addiction liability comprising the phenomena of tolerance and physical and psychological dependence, created the need for a better analgesic lacking these side-effects.

The search for such a 'better analgesic' (Eddy and May 1973) started towards the end of the nineteenth century with the development of heroin. Soon after its introduction into clinical medicine in 1898, however, it became obvious that heroin was an even more dangerous drug than morphine. During the first half of our century, strong efforts were made to achieve the goal of a non-addicting, strong analgesic. This endeavour resulted in the syntheses of innumerable new com-

pounds and the discovery of novel structures with morphine-
like properties but a significant dissociation of strong analgesic activity and abuse liability was not accomplished.

In the early 1950's, exciting results obtained in clinical studies with the narcotic antagonist nalorphine marked
a new era in the struggle for a better analgesic. Nalorphine,
which counteracts virtually all the effects of morphine in
animals, was demonstrated to be a potent analgesic (Lasagna
and Beecher 1954) without abuse potential (Isbell 1956) in
humans. Only the psychotomimetic side-effects of nalorphine
prevented its medical use as the long-sought, non-addicting,
strong analgesic. These striking findings, however, stimulated syntheses and pharmacological evaluation of new narcotic
agonist-antagonists designed as strong analgesics devoid of
abuse potentials and psychotomimetic properties.

Just at that time, derivatives of a new partial structure of morphine, known as 6,7-benzomorphan, were synthesized
(May and Murphy 1955). During the following years, May and
his co-workers, the pioneers in benzomorphan chemistry, developed new, elegant syntheses paving the way to those novel
relatives of morphine and explored structure-activity relationships (Eddy and May 1966). Then, other groups of medicinal
chemists became interested in the 6,7-benzomorphans. Thus,
Freter and Zeile (1959) initiated research in the 6,7-benzo-
morphan series at Boehringer Ingelheim and filed a patent application claiming substituted 6,7-benzomorphans with mixed
agonist-antagonist properties as advantageous analgesics.
Identical and related compounds were prepared independently
and almost at the same time in the laboratories of Smith, Kline & French (Gordon, Lafferty, and Tedeschi 1961) and Sterling
-Winthrop (Archer, Albertson, Pierson, Bird, Keats, Telford,
and Papadopoulos 1962). Continuation of this work at Sterling-
Winthrop led to promising substances (Archer, Albertson, Harris, and Pierson 1964) some of which underwent clinical trials. One of the latter which later became known as pentazocine, was introduced into the market in 1967 under the pro-
prietary name Talwine (for a review see Brogden, Speight, and
Avery 1973). The decision of the WHO Expert Committee that
the abuse liability of pentazocine was low, and that narcotics

control was not necessary, was reinforced by two subsequent
reports of the same authority and is still valid today.

The success of pentazocine, a landmark in the search for
a better analgesic, further increased the interest in 6,7-
benzomorphans. In a review on numerous congeners of morphine
(Villarreal 1971), the correlation between analgesic activity
(mouse hot-plate test) and physical dependence capacity (substitution test in monkeys) generally seen in various groups of
such compounds, was shown to be lacking or even negative in the
6,7-benzomorphan series. Thus, the 6,7-benzomorphan structure
was a most promising starting point for the design of strong
analgesics with low, insignificant abuse potentials.

The great interest in the 6,7-benzomorphans was poten-
tiated when Martin and his associates provided strong evidence
of the existence of multiple opioid receptors (Martin, Eades,
Tompson, Huppler, and Gilbert 1976; Gilbert and Martin 1976).
They suggested that the essentially differentiated action profiles of morphine and the 6,7-benzomorphan derivatives ketazocine and SKF 10047, observed in the chronic spinal dog, might
arise from selective interactions of these opioid prototypes
with different receptors which they named μ, κ, and σ receptors, respectively. The subsequent, rapidly increasing research on multiple opioid receptors revealed that selective
ligands of these receptors were most likely to be found in the
6,7-benzomorphan series. In particular, κ agonists and antagonists are predominantly, although not exclusively, encountered
in the 6,7-benzomorphan derivatives (Merz 1982).

In conclusion, the 6,7-benzomorphans are one of the most
interesting groups of opioids for the medicinal chemist. It
would be impossible here to give a comprehensive review on
structure-activity relationships in the 6,7-benzomorphan series. Various aspects of this topic have been reviewed elsewhere (see, e.g., Eddy and May, 1966; Janssen and van der Eycken 1968; Archer and Michne 1976; Casy 1978). The following
discussion of structure-activity relationships concerns a group
of N-alkoxyalkyl substituted 6,7-benzomorphans, compounds with
unusual, differentiated action profiles which have been synthesized in our laboratories during the past decade.

2. CHEMISTRY

It is not intended to give a comprehensive review on 6,7-ben-
zomorphan chemistry (for such a review see Palmer and Strauss
1977) but rather to present chemical fundamentals needed for
the discussion of structure-activity relationships.

As mentioned above and shown in Fig. 2.1., the 6,7-ben-
zomorphan ring-system (2) represents a partial structure of

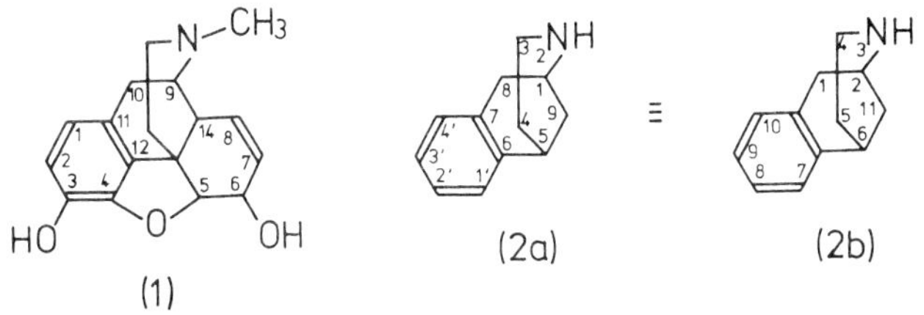

FIG. 2.1.   6,7-Benzomorphan (2),
            a partial structure of morphine (1).

morphine (1). According to the rules of the Ring Index, this
structure is documented in Chemical Abstracts as 1,2,3,4,5,6-
hexahydro-2,6-methano-3-benzazocine with the numbering shown
in (2b). However, the traditional and simpler 6,7-benzomor-
phan nomenclature with the numbering shown in (2a) is still
widely used and will be adhered to in the following discussion.

The unsubstituted parent compound (2) of numerous known
derivatives was first synthesized relatively late in the history of the 6,7-benzomorphans (Kanematsu, Parfitt, Jacobson,
Ager, and May 1968). Previous work had already shown that closer approximations to the structure of morphine considerably
increased analgesic activity (Eddy and May 1966). Prominent
representatives of those potent analgesics are the metazocines
(3), shown in Fig. 2.2., which share with morphine not only
the 6,7-benzomorphan ring-system but also the N-methyl and
the phenolic hydroxy groups in corresponding positions, and
whose 5,9-dimethyl substitution simulates the corresponding
ring of morphine.

It has been concluded from chemical evidence (Eddy and
May 1966) and shown in physicochemical studies (NMR-spectro-

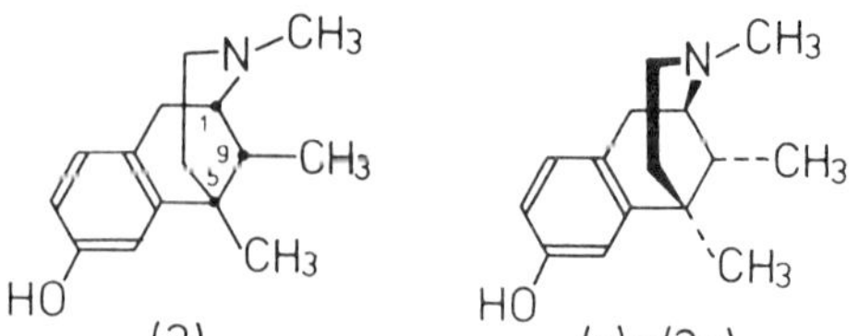

FIG. 2.2.   General formula of the metazocines (3)
            and stereoformula of (-)-α-metazocine (-)-(3a).

scopy, ORD-CD measurements, and particularly X-ray analyses)
that the three rings forming the 6,7-benzomorphan system are
spatially arranged in the same manner as the corresponding
rings of morphine (for a review see Palmer and Strauss 1977).
More precisely, it is (-)-α-metazocine, shown in the stereo-
formula (-)-(3a), that corresponds to natural (-)-morphine.
        Stereoformulas of 6,7-benzomorphans deduced from X-ray
analyses have been presented by Tollenaere, Moereels, and Raymaekers (1979). A complete list of the 6,7-benzomorphans which
have so far been submitted to crystal structure determinations
by X-ray studies is given in Table 2.1. For four of these compounds (cyclazocine, gemazocine, NCBME, and NME), the absolute
configurations were proven by coordinate inversion according
to Ibers and Hamilton (1964). The results of these X-ray studies show that in all the compounds listed in Table 2.1. the
6,7-benzomorphan skeletons are fairly rigid and have virtually
identical conformations corresponding to the T-shaped 6,7-ben-
zomorphan segment of morphine derived from crystal structure
determinations of the latter (Mackay and Crowfoot Hodgkin
1955): The aromatic benzene ring is strictly planar, the alicyclic ring adopts a sofa form, and the piperidine ring has a
slightly distorted chair conformation. That means, the aromatic benzene ring and the adjacent alicyclic ring are arranged
roughly in a plane while the cis (1,3-diaxial) fused piperidine ring protrudes perpendicularly beyond this plane as shown
in formula (-)-(3a). With respect to the piperidine ring, the
substituents at C-5 and at the nitrogen are in equatorial positions. These structural features of the eleven 6,7-benzo-

TABLE 2.1.

6,7-Benzomorphans which have been submitted to
crystal structure determinations by X-ray analyses

6,7-Benzomorphan derivative	References
2-Cyclopropylmethyl-α-normetazocine (cyclazocine)	Karle, Gilardi, Fratini, and Karle (1969).
2-Allyl-α-normetazocine (SKF 10047)	Fedeli, Giacomello, Cerrini, and Vaciago (1970).
2-Cyclopropylmethyl-9,9-dimethyl-5-ethyl-2'-hydroxy-6,7-benzomorphan (gemazocine)	Gelders, De Ranter, and Schenk (1979).
2-Cyclobutylmethyl-9,9-dimethyl-5-ethyl-2'-hydroxy-6,7-benzomorphan (NCBME) and 5-Ethyl-2'-hydroxy-2,9,9-trimethyl-6,7-benzomorphan (NME)	Gelders and De Ranter (1979).
(-)-9,9-Dimethyl-5-ethyl-2'-hydroxy-2-isobutyl-6,7-benzomorphan	Gelders and De Ranter (1980a).
9,9-Dimethyl-5-ethyl-2'-hydroxy-2-phenethyl-6,7-benzomorphan (dimephen) and 2-Phenethyl-α-normetazocine (phenazocine)	Gelders and De Ranter (1980b).
(-)-2-(4-Methoxybutyl)-α-normetazocine (MRZ 2436)	Peeters, De Ranter, and Blaton (1982b).
(-)-(1R,5R,9R,2"R)- and (-)-(1R,5R,9R,2"S)- N-Tetrahydroturfuryl-normetazocine (Mr 1526 and Mr 2034, respectively)	Peeters, De Ranter, and Blaton (1982a).

morphans can be easily deduced from their torsion angles which
are compared to those of morphine (Gylbert 1973) in Table 2.2.

TABLE 2.2.
Torsion angles (°) of 6,7-benzomorphans and morphine

Ring	Torsion angles	
Aromatic benzene ring (A)	6,7-benzomorphans	morphine
C(7) - C(6) - C(1') - C(2')	-1.5 - 3.1	-2.1
C(6) - C(1') - C(2') - C(3')	-2.6 - 3.3	-3.7
C(1') - C(2') - C(3') - C(4')	-3.9 - 3.2	11.1
C(2') - C(3') - C(4') - C(7)	-2.2 - 2.3	-11.0
C(3') - C(4') - C(7) - C(6)	-2.5 - 3.1	-1.0
C(4') - C(7) - C(6) - C(1')	-3.2 - 3.0	5.4
Hydroaromatic ring (C)		
C(8) - C(1) - C(9) - C(5)	-66.1 - -60.4	-57.8
C(1') - C(9) - C(5) - C(6)	54.9 - 60.2	58.3
C(9) - C(5) - C(6) - C(7)	-32.8 - -24.4	-35.6
C(5) - C(6) - C(7) - C(8)	-3.0 - 10.7	2.8
C(6) - C(7) - C(8) - C(1)	-14.9 - 3.2	3.2
C(7) - C(8) - C(1) - C(9)	30.1 - 45.2	25.9
Piperidine ring (E)		
C(9) - C(1) - N(2) - C(3)	-64.0 - -55.0	-65.3
C(1) - N(2) - C(3) - C(4)	48.5 - 57.1	57.6
N(2) - C(3) - C(4) - C(5)	-56.3 - -49.9	-52.7
C(3) - C(4) - C(5) - C(9)	54.0 - 58.5	57.7
C(4) - C(5) - C(9) - C(1)	-61.7 - -56.3	-64.0
C(5) - C(9) - C(1) - N(2)	61.3 - 64.4	66.3

According to well-known rules of stereochemistry, the
maximum number of stereoisomers (n) is given by the equation
$n = 2^x$, where x is the number of centres of chirality present
in the molecule. In the metazocines (3) showing 3 centres of
chirality located at the carbons C-1, C-5, and C-9, the calculated maximum number of 8 stereoisomers is restricted to 4
because of the cis (1,3-diaxial) iminothano bridge between

C-1 and C-5 which prevents independent configurational changes
of these bridgehead carbons.

The four stereoisomeric metazocines are shown in Fig.
2.3. In the α-forms (3a), the two methyl groups in 5- and 9-
position are cis-orientated with respect to the alicyclic ring
whereas, in the β-forms (3b), these methyl groups are trans-
orientated. These diastereomeric forms exist as racemic substances (3a) and (3b) or the corresponding (-)- and (+)- enantiomers.

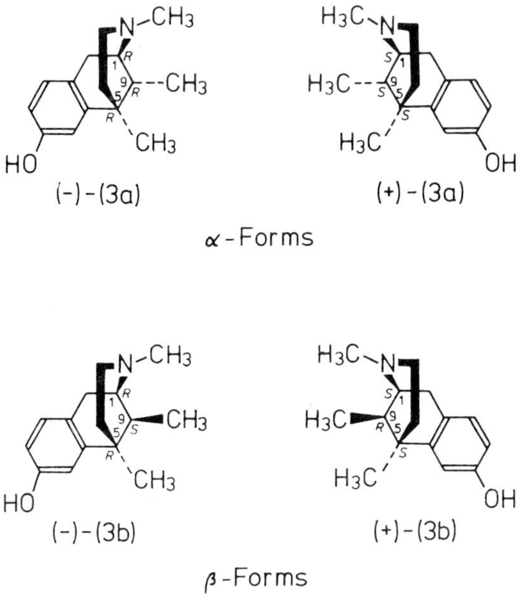

α-Forms

β-Forms

FIG. 2.3.   The four stereoisomeric metazocines

The levo α-form (-)-(3a) and its levo β-diastereomer
(-)-(3b) have been reported to possess the absolute configurations 1R,5R,9R and 1R,5R,9S, respectively (Casy and Parulkar
1969). Accordingly, the dextro counterparts (+)-(3a) and (+)-
(3b) have 1S,5S,9S and 1S,5S,9R configurations, respectively.
This configurational assignment has been confirmed by X-ray
studies of (-)-cyclazocine, the N-cyclopropylmethyl analogue
of (-)-α-metazocine (Karle, Gilardi, Fratini, and Karle 1969).

The introduction of an additional centre of chirality
into the metazocine molecules doubles the number of possible
stereoisomers. Only a few examples of such compounds are known,
e.g., a sub-group of N-alkoxyalkyl-nometazocines shown by the
general formula (5) in Fig. 2.4., which will be discussed in
detail in the next section. The additional centre of chirality

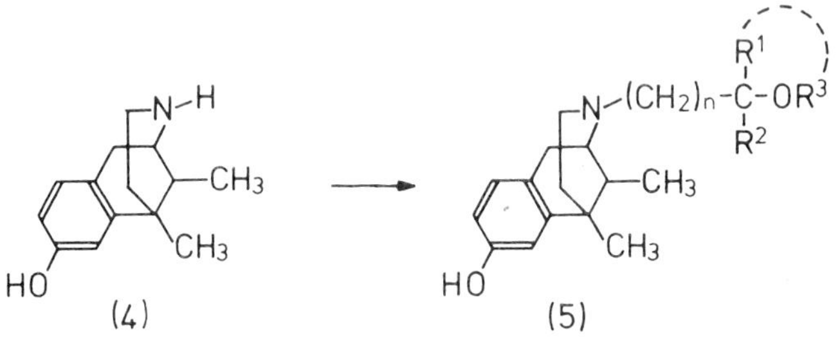

FIG. 2.4.   Syntheses of N-alkoxyalkyl-normetazocines (5)
            starting from the corresponding normetazocines (4)

is located in the N-alkoxyalkyl substituent and gives rise to
the existence of eight stereoisomers. These were synthesized
by alkylation of the corresponding stereoisomeric normetazo-
cines (4) (Tullar, Harris, Perry, Pierson, Soria, Wetterau,
and Albertson 1966) with the appropriate alkoxyalkyl bromides
according to conventional chemical methods (Merz, Stockhaus,
and Wick 1975; Merz and Stockhaus, 1979).

3.   STRUCTURE-ACTIVITY RELATIONSHIPS

The designation 'N-alkoxyalkyl-normetazocines' is used here in
a rather wide sense comprising N-furylalkyl- and N-tetrahydro-
furylalkyl-normetazocines where the oxygen is a constituent
part of an aromatic or saturated ring, respectively, and N-
alkoxyalkyl-normetazocines in the proper sense where the oxygen is a constituent part of an open-chain ether. These sub-
structures will be discussed separately.

3.1.   N-Furylalkyl-normetazocines

Exchange of the N-methyl groups of narcotic analgesics for N-

allyl, -propargyl, -cyclopropylmethyl, and related groups is
the classical strategy in the design of narcotic antagonists
(Harris 1974). We wondered whether there might be other, un-
conventional N-substituents capable of confering antagonist
properties upon opioid structures. The furfuryl group (= 2-
furylmethyl group, (7)) seemed to be a promising candidate as
shown in Fig. 3.1.1. Of course, the π electrons of the double

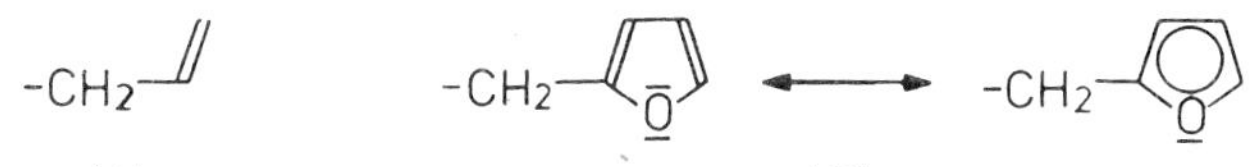

(6)   (7)

FIG. 3.1.1. Allyl character of the furfuryl group

bonds and the lone pair of the oxygen are delocalized in the
furan ring to form an aromatic system. On the other hand, however, furan is known to exhibit residual olefinic properties.
Thus, the furfuryl group may be considered to possess allyl
character, and N-furfuryl analogues of narcotic analgesics
might show antagonist properties.

This hypothesis prompted us to synthesize N-furfuryl-α-
normetazocine (8b) which, indeed, proved to be an opioid antagonist about half as potent as its N-allyl-analogue ((8a),
Gordon, Lafferty, and Tedeschi 1961) as shown in Table 3.1.1.
In this table, N-furfuryl-α-normetazocine (8b) is also compared to other N-arylmethyl-α-normetazocines (8c) - (8e) with
graduated allyl character of the N-substituents which may be
estimated from the aromatic resonance energies of the respective aryl rings. For these compounds, the resonance energies
in question gradually increase from 16 kcal/mol for the weakly
aromatic furan to 36 kcal/mol for the highly aromatic benzene
(Wheland 1955). In contrast, the allyl character of the corresponding N-arylmethyl substituents decreases in this order,
and so does antagonist potency (relative potencies, mouse tail-
clip test, s.c. application). The N-furfuryl derivative (8b)
is about half as potent as the N-allyl analogue (8a) or about
equipotent to nalorphine. These findings (Merz, Langbein,
Stockhaus, Walther, and Wick 1974) furnished strong evidence

of a causal correlation between the allyl character of the N-substituent and the antagonist potency in this group of N-arylmethyl-α-normetazocines.

TABLE 3.1.1.

Opioid antagonist properties of N-arylmethyl-normetazocines

(8)	(8a)	(8b)	(8c)	(8d)	(8e)
	$-CH_2-$ allyl	$-CH_2-$ furyl	$-CH_2-$ pyrrolyl	$-CH_2-$ thienyl	$-CH_2-$ phenyl
Aryl resonance energy		16	21	28	36
			kcal/mol →		
Allyl character		←			
Morphine antagonism	1.0	0.5	0.1	0.05	0.005
			Relative potency ←		

Of all the N-arylmethyl-α-normetazocines shown in Table 3.1.1., only the N-furfuryl compound proved also to possess analgesic activity which, however, was very low. We wondered whether modification of the N-furfuryl group might afford agonist-antagonists with higher analgesic potencies. Thus, we prepared α-normetazocine derivatives with various types of N-furylmethyl substituents. Most of them were devoid of opioid properties but several compounds exhibited markedly differentiated action profiles (Merz et al. 1974). Four interesting prototypes are shown in Table 3.1.2.

Mr 1256 (9b) is an antagonist about equipotent to the classical morphine antagonist nalorphine. However, it is completely devoid of analgesic activity and thus represents the pure antagonist type. Mr 1029 (9c) and Mr 1268 (9d) are mixed agonist-antagonists, the former with predominant antagonist,

TABLE 3.1.2.

Agonist and/or antagonist properties

of N-furylmethyl-α-normetazocines

	(9)	(9a)	(9b)	(9c)	(9d)	(9e)
Substance		PENTAZOCINE	Mr 1256	Mr 1029	Mr 1268	Mr 1353
Antagonist action		+	+ + +	+ + +	+	-
Agonist action		+ +	-	+	+ +	+ + +

the latter with predominant agonist activity. The action profile of Mr 1268 was shown (Stockhaus, Merz, and Wick 1976) to
be quite similar to that of the agonist-antagonist analgesic
pentazocine (9a). Mr 1353 (9e), finally, is a pure agonist
about equipotent to morphine. While the analgesic activities
of agonist-antagonists can only be detected in sensitive anti-
nociceptive assays like the writhing test (Taber 1974), Mr
1353, like morphine, is also active in the more rigorous hot-
plate and tail-clip assays. Most surprisingly, however, Mr
1353 fails to exert typical morphine-like side-effects. Thus,
it does not elicit the Straub tail phenomenon in mice nor does
it show morphine-like physical dependence capacity in monkeys
(Villarreal and Seevers 1972).

    This unexpected dissociation of high analgesic activity
and morphine-like physical dependence capacity in a pure opioid agonist was also observed with another compound (Villarreal and Seevers 1974) which later became known as ketazocine,
the prototype of a κ receptor agonist. It has been shown in
isolated organs (Hutchinson, Kosterlitz, Leslie, Waterfield,
Terenius 1975) and binding studies (Kosterlitz and Leslie 1978)
that Mr 1353 is a κ agonist like ketazocine. Selectivity to-

wards κ receptors was also reported for the pure antagonist Mr 1256 (9b), its (-)-enantiomer Mr 1452 (Smith 1978), and particularly for the more potent α-noretazocine homologue Mr 2266 of the latter compound (Kosterlitz, Paterson, and Robson 1981).

Results of modifications of the α-normetazocine moiety of the prototypes (9b) - (9e) (e.g., alteration of the 2'-hydroxy function, the 5- and/or 9-methyl groups, or the configuration; Merz et al. 1974) confirmed well-established structure-activity relationships in the 6,7-benzomorphan series (Eddy and May 1966). In any case, however, such structural variations did not essentially affect the respective action profiles which are thus determined primarily by the nature of the N-furylmethyl group. Moreover, these N-furylmethyl groups proved to confer similar action profiles on certain other opioid structures (e.g., 3-hydroxymorphinan, normorphine or noroxymorphone). However, the differentiation of the action profiles is less pronounced than in the normetazocine derivatives.

Conversion of N-furfuryl-α-normetazocine (9c) into its N-furylethyl homologue (10) abolishes the allyl character of the N-substituent. Accordingly, the latter (Gordon and Lafferty 1960) should lack antagonist properties. Pharmacological examinations confirmed this expectation and revealed that

(10) is a strong analgesic about 25 times more potent than morphine (writhing test) and elicits the Straub tail phenomenon. These findings support the suggestion that it is the allyl character of the N-substituent that induces the antagonist action of N-furfuryl-α-normetazocine (9c).

(9c): n = 1   (Mr 1029)

(10): n = 2

3.2. N-Tetrahydrofurfuryl-normetazocines

Another structural modification designed to prove the suggested causal correlation between the allyl character of the N-stituent and antagonist properties of the N-furylmethyl-normetazocines was the replacement of the quasi-allylic N-furfu-

ryl group by the non-allylic N-tetrahydrofurfuryl group. By
this structural change, an additional center of chirality
is introduced into the molecule giving rise to the existence
of twice as many stereoisomeric N-tetrahydrofurfuryl-normet-
azocines as there are N-furfuryl analogues. We synthesized all
the existing four racemic and eight optically active stereoisomers and studied their opioid properties (Merz et al. 1975).

According to general experience, the interactions of
opioid receptors with their ligands are highly stereoselective. As a rule, receptor affinity and resulting opioid effects reside largely in one of two enantiomers, predominantly
in the (-)-forms. In the 6,7-benzomorphan series, the effective enantiomers are known to be the (-)-(1R,5R) forms while
the (+)-(1S,5S) forms are essentially inactive. Accordingly,
the four (+)-N-tetrahydrofurfuryl-normetazocines proved to be
devoid of opioid activity and the four racemic substances to
be half as potent as the corresponding (-)-(1R,5R) forms. Consequently, the discussion can be limited to the four (-)-
(1R,5R)-N-tetrahydrofurfuryl-normetazocines which are shown
in Table 3.2.1.

The compounds listed in Table 3.2.1. are derived either
from (-)-α-normetazocine (1R,5R,9R configuration) or (-)-ß-
normetazocine (1R,5R,9S configuration). The N-furfuryl derivatives (-)-(9c) and (-)-(11) are α-/ß-diastereomers differing
only in the configuration of the 9-methyl group. The corresponding N-tetrahydrofurfuryl derivatives (-)-(12) - (-)-(15)
possess an additional center of chirality in the carbon 2" of
the N-substituent which may have either R or S configuration,
thus giving rise to the existence of 2"R and 2"S diastereomers
in the α- as well as in the ß-series.

The reference compounds with N-furfuryl substituents are
rather potent antagonists with marginal ((-)-(9c)) or lacking
((-)-(11)) analgesic activity. In contrast, the N-tetrahydrofurfuryl analogues are pure agonists, the 2"S compounds (-)-
(13) and (-)-(15) being considerably more potent than the corresponding 2"R diastereomers (-)-(12) and (-)-(14). These results provide further evidence of a correlation between the
allyl character of the N-substituent and antagonist activity
in the N-furfuryl-normetazocines.

TABLE 3.2.1.

Relative analgesic potencies
of the (-)-(1R,5R)-N-tetrahydrofurfuryl-normetazocines

	SUBSTANCE	REL. POTENCY Writhing Test, s.c. (Morphine = 1)
(-)-(12)	Mr 1526	0.2
(-)-(13)	Mr 2034	50
(-)-(14)	Mr 2093	5
(-)-(15)	Mr 2092	25

The (-)-N-tetrahydrofurfuryl-normetazocines, particular- ly the by far more potent 2"S forms Mr 2034 and Mr 2092 and the corresponding racemic substances Mr 2033 and Mr 2184, exhibit interesting action profiles. In spite of their high analgesic activities which can be demonstrated not only in the writhing test but also in the hot-plate and tail-clip assays, these compounds do not elicit the Straub tail phenomenon nor do they possess morphine-like physical dependence capacities in monkeys (Swain and Seevers 1976; Aceto, Harris, Dewey, and Balster 1976). Thus, their pharmacological properties are qui- te similar to those of ketazocine and Mr 1353 (9e) discussed above (page 12). Indeed, studies in isolated organs and binding assays with Mr 2034 and Mr 2092 (Hutchinson et al. 1975; Kosterlitz and Leslie 1978) and Mr 2033 and Mr 2184 (Woods, Smith, Medzihradsky, and Swain 1979) revealed that these compounds interact selectively with κ receptors.

Extended studies have been performed with Mr 2033 which proved its κ agonist properties in the chronic spinal dog (Gilbert, Martin, and Jessee 1977), showed distinctly no morphine-like primary physical dependence capacity in monkeys (Swain and Seevers 1976), and was not self-injected by monkeys (Woods et al. 1979). Moreover, Mr 2033 has been shown to possess an unusually favourable therapeutic ratio (LD_{50}/ED_{100}, s.c., hot-plate, mice) of 26.800 compared to that of morphine (417) (Merz et al. 1975). Further advantages of Mr 2033 over morphine have been reported in a review on the pharmacology of the former substance (Stockhaus, Ensinger, Gaida, Jennewein and Merz 1982). These compiled findings suggest that Mr 2033 might be a most promising candidate for the development as a strong analgesic with minimal side-effects and a very low, insignificant abuse potential. Clinical trials are in progress.

Obviously, it is the N-tetrahydrofurfuryl group of these compounds that induces their unique action profiles. Thus, it was tempting to modify this crucial N-substituent systematically and to see what changes in the opioid properties would result. The well-investigated (-)-(1R,5R,9R,2"S)-N-tetrahydrofurfuryl-normetazocine (Mr 2034, (-)-(13)) was chosen for such modifications which yielded striking changes in the action profile, analgesic potency, and stereoselectivity (Merz and Stockhaus 1979). The observed activities are defined in Table 3.2.2. Of course, the profiles A, B, and C correspond to those of μ agonists (morphine), κ agonists (ketazocine), and nalorphine, respectively. However, conclusive evidence of corresponding receptor selectivities has only been furnished so far for the reference compound Mr 2034.

The effects of lengthening of the methylene bridge between the nitrogen and carbon 2" of the tetrahydrofuran ring or introduction of a methyl group into position 2" are shown in Table 3.2.3. With these structural modifications, the tetrahydrofuran ring and its center of chirality in C-2" remain intact. Consequently, the resulting homologues exist in diastereomeric 2"S and 2"R forms like the reference compounds.

TABLE 3.2.2.

Action profiles of

(-)-(1R,5R,9R)-N-tetrahydrofurylalkyl-normetazocines

Action profile	Analgesic activity			Straub tail	Antagonist activity, TCT
	WT	HPT	TCT		
Morphine-like agonist (A)	+	+	+	+	-
Non-morphine-like agonist (B)	+	+	+	-	-
Agonist-antagonist (C)	+	-	-	-	+

WT: Writhing test; HPT: Hot-plate test; TCT: Tail-clip test
Straub tail: present (+) or absent (-) in the analgesic
dose range in the PQ writhing test (mice).

Reinvestigation of these reference compounds have shown
that the 2"S form, the κ agonist Mr 2034 ((-)-(16), identical
with (-)-(13)), is about 300 times more potent than its 2"R
diastereomer (-)-(16b), identical with (-)-(12), which proved to be a weak agonist-antagonist roughly comparable with
pentazocine.

Lengthening of the above mentioned methylene bridge of
(-)-(16a) to an ethylene bridge leading to the homologue (-)-
(16c) is accompanied by a profound drop of analgesic potency
by nearly two orders of magnitude. Moreover, the pronounced
stereoselectivity seen in the reference pair is essentially
abolished in the homologous pair (-)-(16c) and (-)-(16d). Interestingly enough, the action profiles of the reference compounds (B and C, respectively) are shifted to that of morphine-like compounds (A). Going on to a propylene bridge between
the nitrogen and the tetrahydrofuran ring affording the compounds (-)-(16e) and (-)-(16f) causes a further decrease of
the analgesic potencies and a complete loss of stereoselectivity. The action profiles appear to be no longer morphine-like.

TABLE 3.2.3.

Structure-activity relationships in

(-)-(1R,5R,9R)-N-tetrahydrofurylalkyl-normetazocines

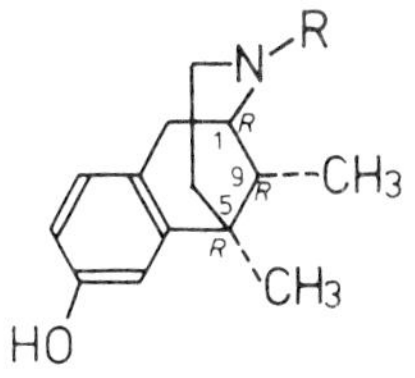

-R	C-2''		REL. POTENCY Writing Test, s.c. (Morphine = 1)	ACTION PROFILE
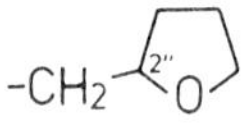 -CH₂ (Mr 2034)	S	(-)-(16a)	31.3	B
-CH₂ (Mr 1526)	R	(-)-(16b)	0.1	C
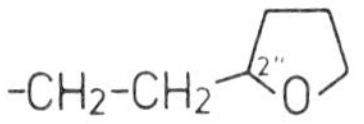 -CH₂-CH₂	S	(-)-(16c)	0.4	A
-CH₂-CH₂	R	(-)-(16d)	1.2	A
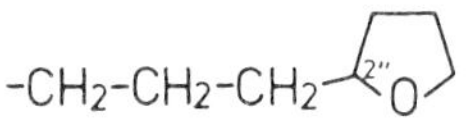 -CH₂-CH₂-CH₂	S	(-)-(16e)	0.1	B
-CH₂-CH₂-CH₂	R	(-)-(16f)	0.1	B
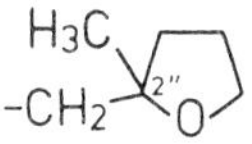 -CH₂ (H₃C)	S	(-)-(16g)	9.2	C
-CH₂ (H₃C)	R	(-)-(16h)	0.1	C

(-)-(16)

(-)-(16a) ≡ (-)-(13)

Introduction of a methyl group into position 2'' of the reference compound (-)-(16a) leading to (-)-(16g) decreases analgesic potency to about 1/3. The superiority of this 2''S compound to its 2''R diastereomer (-)-(16h) is not as high as observed with the reference pair but is still pronounced. Most surprisingly, the N-(2''-methyl-tetrahydrofurfuryl) compounds exhibit agonist-antagonist action profiles, the former showing appreciable analgesic potency. The antagonist potencies which are not given in Table 3.2.3. are about 1/2 and 1/6 of that of nalorphine for the 2''S form (-)-(16g) and its 2''R diastereomer (-)-(16h), respectively. The appearance of considerable antagonist activity in the above compounds was quite unexpected and cannot be attributed to a quasi-allylic N-substituent.

riation is accompanied by a fivefold increase in analgesic
potency and a shift of the action profile to that of morphine-
like compounds. Enlargement of the alkoxy function, stepwise
from methoxy to ethoxy and propoxy, results in a considerable
loss of analgesic potency, particularly in the first step to
the also morphine-like compound (-)-(17g). The even less potent homologue (-)-(17h), however, is not morphine-like.

Stepwise lengthening of the ethylene chain between the
nitrogen and the methoxy function of (-)-(17f) to a propylene
and a butylene chain affording the compounds (-)-(17i) and (-)-
(17j), drastically reduces analgesic potency which drops to
about 1/100 in the first step and decreases further in the second one. In spite of their relatively weak analgesic actions,
(-)-(17i) and (-)-(17j) are morphine-like compounds.

3.4. Summary of structure-activity relationships
      and derived drug-receptor interactions

In summary, the following structure-activity relationships
are observed in the N-tetrahydrofurylalkyl- and N-(open-chain
alkoxyalkyl)-normetazocines (Fig. 3.4.1.):

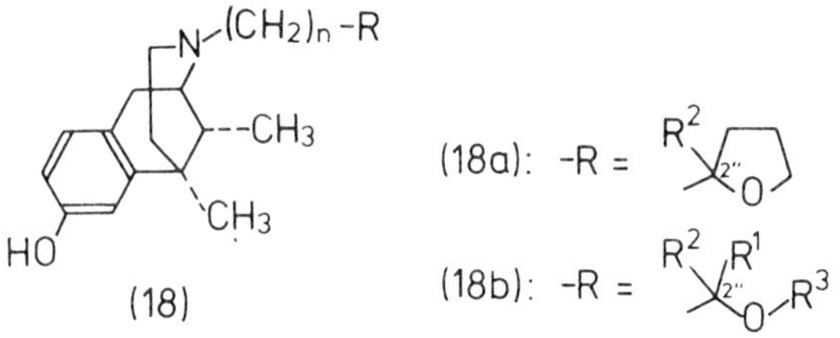

FIG. 3.4.1. Modified structural parameters of the
             N-substituents of N-tetrahydrofurylalkyl-
             and N-(open-chain alkoxyalkyl)-normetazocines

Maximum analgesic activity (independent on the action
profile) is obtained if nitrogen and oxygen are linked by a
two-carbon chain ((18), n = 1 >> 2 > 3) and if the alkoxy residue is small ((18b), $-OR^3$ = $-OCH_3$ >> $-OC_2H_5$ > $-OC_3H_7$) or
constituent part of a tetrahydrofuran ring ((18a)). Compounds
with these structural features for maximum analgesic activity

are non-morphine-like if the two-carbon chain in question is branched in β position to the nitrogen, either by one or two methyl groups ((18b), $-R^1$ = $-CH_3$, $-R^2$ = -H or $-CH_3$) or by incorporation into a tetrahydrofuran ring ((18a)). If diastereomers arise from such branchings of C-2", the 2"S forms are by far superior to the 2"R counterparts which may even be inactive ((18), 2"S >> 2"R for n = 1). The tetrahydrofurfuryl group and particularly its 2"-methyl derivative ((18a, $-R^2$ = -H or $-CH_3$) is a prerequisite for antagonist properties which are not encountered in the open-chain alkoxyalkyl analogues.

Such structure-activity relationships emphasizing the great importance of stereochemical parameters reflect drug-receptor interactions of high sterical demand. According to the classical concept of Beckett and Casy (1954), a three-point association of the "analgesic receptor" and its "opiate" ligand is essential for morphine-like analgesia. However, additional receptor sites may exist which are able to accomodate and bind appropriate functions of certain opioid structures, thus enhancing the analgesic effect (Casy 1973). Interestingly enough, the surprisingly high analgesic potencies of certain N-tetrahydrofurylalkyl- and N-alkoxyalkyl-norpethidines have been attributed to hydrogen bonding between the ether oxygen of the N-substitutent and such an additional binding site of the receptor (Blair and Stephenson 1960). Of course, the structure-activity relationships are different in the above norpethidine derivatives and the discussed normetazocine analogues with regard to the N-O distance providing maximum analgesic effect and the fact that non-morphine-like action profiles are seen only in the latter. However, such differences in potency and action profile in the norpethidine and normetazocine series are also seen in other parallel modifications of the N-substituent and can be explained by major differences in the molecular rigidity of the two series (Portoghese 1965; 1966). These findings and the discussed structure-activity relationships suggest the following drug-receptor interactions for the normetazocine derivatives (18).

The N-O distance provided by an ethylene chain (n = 1) is most favorable for hydrogen bonding between the ether oxygen and the hypothetical additional binding site of the recep-

tor and is thus a prerequisite for maximum analgesic potency.
Compounds showing this crucial feature may greatly differ,
however, in their action profiles. Morphine-like activity was
demonstrated to be correlated with unbranched, and non morphine-like activities with branched carbons C-2". According
to sophisticated opioid receptor models (for recent reviews
see Archer and Michne 1976; Casy 1978), different receptor
conformations may account for differentiated opioid action
profiles. We suggest that only an unbranched C-2" might allow
unhindered access and binding of the ether oxygen to the hypothetical additional binding site of the receptor while steric hindrance by branching of C-2" would allow such binding
only with conformational changes of the receptor. The pronounced stereoselectivity as to the configuration of C-2" for
n = 1 supports this concept of drug-receptor interaction.

3.5.   Structural features of κ agonists

Recently, research on multiple opioid receptors has been focused on κ receptors and their ligands. Only some twenty of
those ligands have become known so far, the great majority of
them being agonists. It should be noted that these compounds
are selective rather than specific κ agonists or antagonists.
Until 1982, when novel structures with κ selectivity were discovered (Römer, Büscher, Hill, Maurer, Petcher, Zeugner, Benson, Finner, Milkowski, and Thies 1982; Scmuszkovicz and Von
Voigtlander 1982), all the fifteen known κ ligands were 2'-hy-
droxy-6,7-benzomorphan derivatives (for a review see Merz
1982). Four very prominent representatives of κ agonists of
the 6,7-benzomorphan series are shown in Fig. 3.5.1.

In looking for common structural features of these κ agonists one will recognize that they show an oxygen as a constituent part of either the 8-oxo group or hydroxy and ether
functions in the N-substituent. In discrete conformations of
these N-substituents (as indicated in Fig. 3.5.1.), the oxygen
in question occupies almost the same place as the oxygen fixed
in the 8-oxo group in the case of ketazocine. It is suggested
that this position is crucial for the interaction of the respective oxygen with an adjacent auxiliary receptor-site, resulting in the supposed change of receptor conformation and,

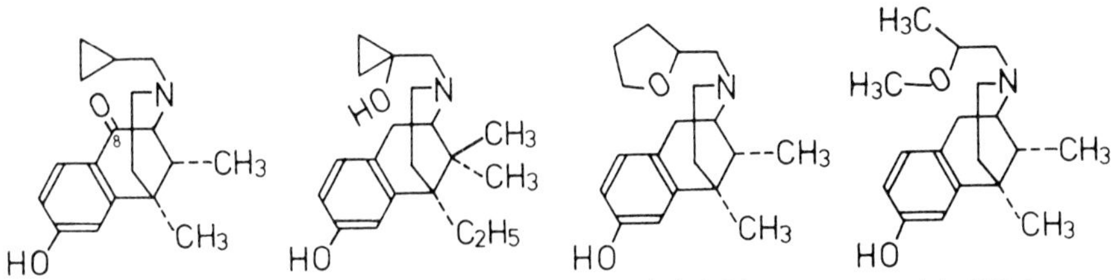

FIG. 3.5.1.   Representatives of κ agonists of the 2'-hydroxy-
              6,7-benzomorphan series: ketazocine (19), brem-
              azocine (20), Mr 2034 ((-)-(13)), and MRZ 2549
              ((-)-(17c)).

in consequence, to κ agonist action (Merz and Stockhaus 1979).
Quite recently, new evidence has been provided for the hypothesis that μ, δ, and κ receptors are different conformations
of a single allosteric opiate receptor complex which can assume various ligand selectivity patterns ((Quirion, Bowen, and
Pert 1982).

     It can be demonstrated by means of molecular models that
the conformations shown in Fig. 3.5.1. with the concerned oxygen in the crucial position, can be attained without strain.
Moreover, recent crystallographic studies support the suggested drug-receptor interaction. According to X-ray analyses of
ketazocine, its ring system does not show any outstandingly
different features in comparison to other crystallographically
investigated 6,7-benzomorphans (Verlinde and De Ranter 1983).
It should be noted, however, that the oxygen of the 8-oxo
group lies about 7° out of the plane of the benzene ring. It
can be deduced from X-ray studies of the 2"S compound Mr 2034
and its 2"R diastereomer Mr 1526 (Peeters, De Ranter, and Blaton 1982a) that the conformations of their 6,7-benzomorphan
ring systems do not show any significant deviations from those
of the other 6,7-benzomorphans with known crystal structures
(see Table 2.1., page 6). Again, in accordance with these
other 6,7-benzomorphans, the N-substituents of Mr 2034 and Mr
1526 are in equatorial positions. The conformations of these
N-substituents in the crystal structures do not exhibit significant differences that could account for the strikingly

different pharmacological properties of the two diastereomers.
Conformational energy calculations by the same authors have
shown, however, that the crucial conformation of the respective N-tetrahydrofurfuryl substituent is energetically favourable in the 2"S compound Mr 2034 but unfavourable in its 2"R
diastereomer Mr 1526. These results are in good agreement with
the observation that the former compound is a very potent κ
agonist while the latter is a rather weak agonist-antagonist.
Similar results have also been obtained with the highly potent
κ agonist MRZ 2549 (2"S form) and its completely inactive 2"R
diastereomer MRZ 2547 (Peeters, De Ranter, and Blaton 1983).
ORTEP computer plots of Mr 2034 and Mr 1526 with the crucial
conformations of the N-tetrahydrofurfuryl substituents (Verlinde and De Ranter 1983) are shown in Fig. 3.5.2. Apart

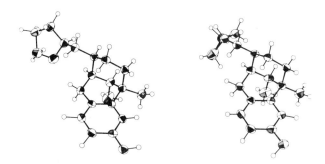

FIG. 3.5.2.   ORTEP computer plots of the diastereomeric com-
              pounds Mr 2034 and Mr 1526 with the crucial con-
              formations of the N-substituents

from the discussed differences in the conformational energies
of Mr 2034 and Mr 1526 in the distinct conformations shown
above, other structural parameters might account for the different pharmacological properties of these compounds. Of course,
the oxygen in question is located in almost the same position
in both compounds. It is obvious, however, that their tetrahydrofuran rings are arranged in different planes. That means
that the lobes of the lone electron pairs of the ring oxygen
point in different directions in Mr 2034 and Mr 1526, a fact
which must be of major importance for the hypothetical hydrogen bonding with the receptor by means of these lobes.

It is obvious that the suggested binding of the oxygen in question to the hypothetical receptor site located in the very neighbourhood of position 8 of the 6,7-benzomorphan system must be highly sensitive with regard to enlargement of the N-O distance and branching and configuration of C-2". Thus, the suggested mode of drug-receptor interaction of κ agonists of the 6,7-benzomorphan series is in good accordance with the observed structure-activity relationships. Of course, there are differences in the hydrogen bonding capacity of the oxo, hydroxy, and ether oxygen. These differences might be responsible for the well-known differentiation of κ agonist action profiles (Schulz and Wüster 1981).

4. ACKNOWLEDGEMENT

This work is dedicated to my honoured mentor, Professor Theodor Wieland, on the occasion of his 70th birthday in gratitude for his enthusiastic introduction into chemical research.

I am most grateful to Professor Camiel De Ranter for fruitful discussions of the X-ray data.

5. REFERENCES

Aceto, M.D., Harris, L.S., Dewey, W.L., and Balster, R.L. (1976). In Addendum to the proceedings of the 38th annual scientific meeting of the Committee on Problems of Drug Dependence, pp 774-76. National Academy of Sciences, Washington, D.C.

Aceto, M.D., Harris, L.S., May, E.L., Grove, F.T., Jones, R.J., and Tucker, S.M. (1982). In Problems of Drug Dependence (ed. L.S. Harris), in press. U.S. Government Printing Office, Washington, D.C.

Archer, S. and Michne, W.F. (1976). Prog. Drug Res. 20, 45.

Archer, S., Albertson, N.F., Harris, L.S., Pierson, A.K., and Bird, J.G. (1964). J. Med. Chem. 7, 123.

Archer, S., Albertson, N.F., Pierson, A.K., Bird, J.G., Keats, A.S., Telford, J., and Papadopoulos, C.N. (1962). Science 137, 541.

Beckett, A.H., and Casy, A.F. (1954). J. Pharm. Pharmacol. 6, 986.

Blair, A.M.J.N. and Stephenson, R.P. (1960). Br. J. Pharmacol. 15, 247.

Brogden, R.N., Speight, T.M., and Avery, G.S. (1973). Drugs 5, 6.

Casy, A.F. (1973). Mod. Pharmacol. 217-78.

Casy, A.F. (1978). Prog. Drug. Res. 22, 149-227.

Casy, A.F. and Parulkar, A.P. (1969). J. Med. Chem. 12, 178.

Eddy, N.B. and May, E.L. (1966). In Synthetic analgesics, Part IIB, 6,7-benzomorphans (ed. D.H.R. Barton and W. v. Doering), Pergamon Press, New York, N.Y.

Eddy, N.B. and May, E.L. (1973). Science 181 (407).

Fedeli, W., Giacomello, G., Cerrini, S., and Vaciago, A. (1970). J. Chem. Soc. (B), 1190.

Freter, K. and Zeile, K. (1959). German Pat. 1,420,015 (to C.H. Boehringer Sohn, Ingelheim). See C.A. 75, 129686f.

Gelders, Y.G. and De Ranter, C.J. (1980a). Acta Cryst. B36, 744.

Gelders, Y.G. and De Ranter, C.J. (1980b). Acta Cryst. B36, 1141.

Gelders, Y.G., De Ranter, C.J., and Overbeek, A.R. (1979). Acta Cryst. B35, 1111.

Gelders, Y.G., De Ranter, C.J., and Schenk, H. (1979). Acta Cryst. B35, 699.

Gilbert, P.E. and Martin, W.R. (1976). J. Pharmacol. Exp. Ther. 198, 66.

Gilbert, P.E., Martin, W.R., and Jessee, C.A. (1977). Reported to the Committee on Problems of Drug Dependence at the 39th annual scientific meeting, Cambridge, Mass.

Gordon., M. and Lafferty, J.J. (1960). US Patent 2.924.603, (to Smith, Kline & French Laboratories). See C.A. 54, P 1855b.

Gordon, M., Lafferty, J.J., and Tedeschi, D.H. (1961). Nature (London) 92, 1089.

Gylbert, L. (1973). Acta Cryst. B29, 1630.

Harris, L.S. (1974). Adv. Biochem. Psychopharmacol. 8, 13.

Hutchinson, M., Kosterlitz, H.W., Leslie, F.M., Waterfield, A.A., and Terenius, L. (1975). Br. J. Pharmacol. 55, 541.

Ibers, J.A. and Hamilton, W.C. (1964). Acta Cryst. 17, 781.

Isbell, H. (1956). Fed. Proc. 15, 442.

Janssen, P.A.J. and van der Eycken (1968). In Drugs affecting the central nervous system (ed. A. Burger), Vol. 2, pp 25-66. Marcel Dekker, Inc., New York, N.Y.

Kanematsu, K., Parfitt, R.T., Jacobson, A.E., Ager, J.H., and May, E.L. (1968). J. Am. Chem. Soc. 90, 1064.

Karle, I.L., Gilardi, R.D., Fratini, R.V., and Karle, J. (1969). Acta Cryst. B25, 1469.

Kosterlitz, H.W. and Leslie, F.M. (1978). Br. J. Pharmacol. 55, 541.

Kosterlitz, H.W., Paterson, S.J., and Robson, L.E. (1981). Br. J. Pharmacol. 73, 939.

Lasagna, L. and Beecher, H.K. (1954). J. Pharmacol. Exp. Ther 112, 356.

Mackay, M. and Crowfoot Hodgkin, D. (1955). J. Chem. Soc. 3261.

Martin, W.R., Eades, C.G., Thompson, J.A., Huppler, R.E., and Gilbert, P.E. (1976). J. Pharmacol. Exp. Ther. 197, 517.

May, E.L. and Murphy, J.G. (1955). J. Org. Chem. 20, 257.

Merz, H. (1982). In Kappa receptors and their ligands, proceedings of the symposium on this topic (Toulouse, 1982), in press.

Merz, H. and Stockhaus, K. (1979). J. Med. Chem. 22, 1475.

Merz, H., Stockhaus, K., and Wick, H. (1975). J. Med. Chem. 18, 996.

Merz, H., Stockhaus, K., and Wick, H. (1977). J. Med. Chem. 20, 844.

Merz, H., Langbein, A., Stockhaus, K., Walther, G., and Wick, H. (1974). Adv. Biochem. Psychopharmacol. 8, 91.

Palmer, D.C. and Strauss, M.J. (1977). Chem. Rev. 77, 2.

Peeters, O.M., De Ranter, C.J., and Blaton, N.M. (1982a). Acta Cryst. B38 3055.

Peeters, O.M., De Ranter, C.J., and Blaton, N.M. (1982b). Acta Cryst. B38, 3168.

Peeters, O.M., De Ranter, C.J., and Blaton, N.M. (1983). Personal communication, Professor De Ranter.

Portoghese, P.S. (1965). J. Med. Chem. 8, 609.

Portoghese, P.S. (1966). J. Pharm. Sci. 55, 865.

Quirion, R., Bowen, W.D., and Pert, C.B. (1982). In Advances in endogenous and exogenous opioids (ed. H. Takagi and E.J. Simon), pp 63-65. Elsevier/North-Holland Biomedical Press, Amsterdam, New York, Oxford.

Römer, D., Büscher, H.H., Hill, R.C., Maurer, R., Petcher, T.J., Zeugner, H., Benson, W., Finner, E., Milkowski, W., and Thies, P.W. (1982). Nature (London) 298, 759.

Schulz, R. and Wüster, M. (1981). Eur. J. Pharmacol. 76, 61.

Sertürner, F.W.A. (1805). J. Pharmazie 13, 134.

Smith, C.B. (1978). In Characteristics and functions of opioids (ed J.M. Van Ree and L. Terenius), pp 237-38. Elsevier/ North-Holland Biomedical Press, Amsterdam, New York, Oxford.

Szmuszkovicz, J. and Von Voigtlander, P.F. (1982), J. Med. Chem. 25, 1125.

Stockhaus, K., Merz, H., and Wick, H. (1976). Reported to the Committee on Problems of Drug Dependence at the 38th annual scientific meeting, Richmond, Va.

Stockhaus, K., Ensinger, H.A., Gaida, W., Jennewein, H.M., and Merz, H. (1982). Reported to the Committee on Problems of Drug Dependence at the 44th annual scientific meeting, Toronto.

Swain, H.H. and Seevers, M.H. (1974). In Addendum to the proceedings of the 36th annual scientific meeting of the Committee on Problems of Drug Dependence, p 1171. National Academy of Sciences, Washington, D.C.

Swain, H.H. and Seevers, M.H. (1976a). In Addendum to the proceedings of the 38th annual scientific meeting of the Committee on Problems of Drug Dependence, pp 775-76. National Academy
of Sciences, Washington, D.C.

Swain, H.H. and Seevers, M.H. (1976b). In Addendum to the proceedings of the 38th annual scientific meeting of the Committee on Problems of Drug Dependence, pp 784-85. National Academy
of Sciences, Washington, D.C.

Taber, R.I. (1974). Adv. Biochem. Psychopharmacol. 8, 191.

Tollenaere, J.P., Moereels, H., and Raymaekers, L.A. (1979). In
Atlas of the three-dimensional structure of drugs, pp 177-81.
Elsevier/North-Holland Biomedical Press, Amsterdam, New York,
Oxford.

Tullar, B.F., Harris, L.S., Perry, R.L., Pierson, A.K., Soria,
A.E., Wetterau, W.F., and Albertson, N.F. (1966). J. Med. Chem.
10, 383.

Verlinde, C.L. and De Ranter, C.J. (1983). Personal communication.

Villarreal, J.E. (1971). In Agonist and antagonist actions of
narcotic analgesic drugs (ed. H.W. Kosterlitz, H.O. Collier,
and J.E. Villarreal), pp 73-94. Macmillan Press.

Villarreal, J.E. and Seevers, M.H. (1972). In Addendum to the
proceedings of the 34th annual scientific meeting of the Committee on Problems of Drug Dependence, p 1047. National Academy of Sciences, Washington, D.C.

Wheland, G. (1955). In Resonance in organic chemistry, p 98.
Wiley & Sons, New York, N.Y.

Woods, J.H., Smith, C.B., Medzihradsky, F., and Swain, H.H.
(1979). In Mechanisms of pain and analgesic compounds (ed.
R.F. Beers, Jr., and E.G. Bassett), pp 429-45. Raven Press,
New York, N.Y.

Woods, J.H., Katz, J.L., Medzihradsky, F., Smith, C.B., Young,
A.M., and Winger, C.D. (1982). In Problems of Drug Dependence
(ed. L.S. Harris), in press. U.S. Government Printing Office,
Washington, D.C.

18 Neuromuscular blocking drugs
D.S. Savage

These drugs are used as adjuncts in anaesthetised patients undergoing
surgical operations in order to block neuromuscular transmission. The
suitably paralysed patients are of course intubated and maintained under
artificial respiration during the operation. The drugs act by associating with cholinergic receptors at the post synaptic cleft of the neuromuscular junction and thereby block the action on muscle of the neurotransmitter acetylcholine. The classical non-depolarising agent which
acts in this manner was of course the naturally occurring d-Tubocurarine.
Unfortunately this drug has its limitations in anaesthetic practice, being
too slow in onset of action, too long in time course of action, having
cumulative properties and producing unwanted side effects such as hypo-
tension and histamine release. Consequently the aim of medicinal chemists in this field was to produce an active drug lacking side effects in
clinical dose range and having a quicker and shorter action without cumulative effect. The object of this chapter is to show how specificity of
chemical structure has contributed to advancing this aim in all respects.

This work commenced about 25 years ago when an industrial company
decided to invest in the philanthropic project of conferring a non-hormonal
activity on a steroidal molecule. This was my first brief on entering
industry and the idea was based on stereoselective introduction of vicinal
1,2-amino alcohol groupings to a steroid nucleus. Such vicinal amino
alcohol derivatives occurring endogenously in the human are for instance
noradrenaline and acetylcholine.

A range of such amino sterols[1,2,3,4] was synthesised over a few years
and proved to be lacking in hormonal activity. Obviously the amino
alcohol groupings were conferring different physical properties on the
total molecule and these compounds were being transported to sites other
than those normally associated with steroids. Some members of the series
were reported[3,5] to have anticonvulsant, interneuronal blocking, local
anaesthetic and neuromuscular blocking activity.

The most potent of the neuromuscular blocking drugs reported[3] is
shown in Fig.18.1 in comparison with acetylcholine. This quaternary ammonio
steroid, a 2β-piperidinio-3α-acetoxy-5α-androstane derivative possessed
$\frac{1}{16}$ the potency[3] of d-Tubocurarine as a non-depolarising neuromuscular
blocking agent.

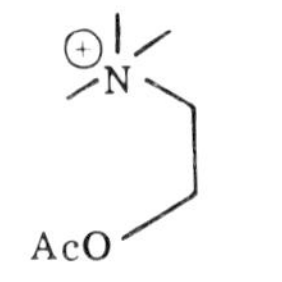

Acetylcholine

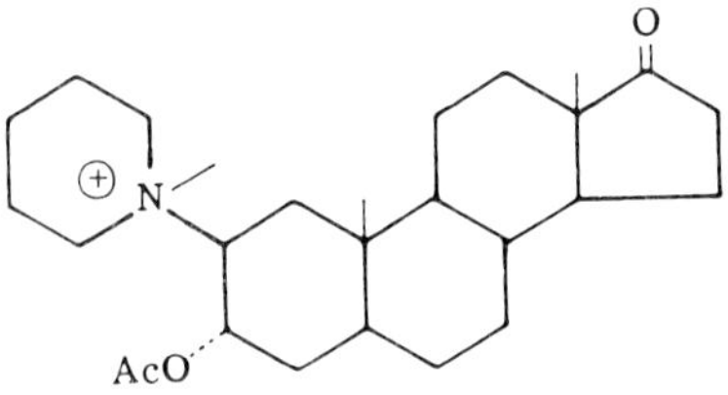

P = 1/16th d-TC

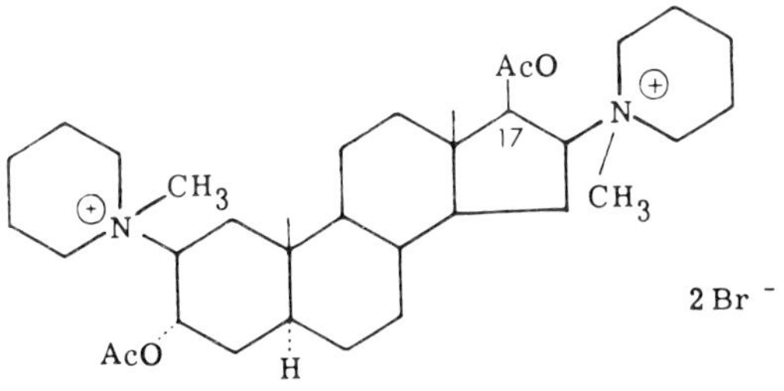

Pancuronium Bromide

Fig. 18.1

Revised Structure of (+) - Tubocurarine Chloride

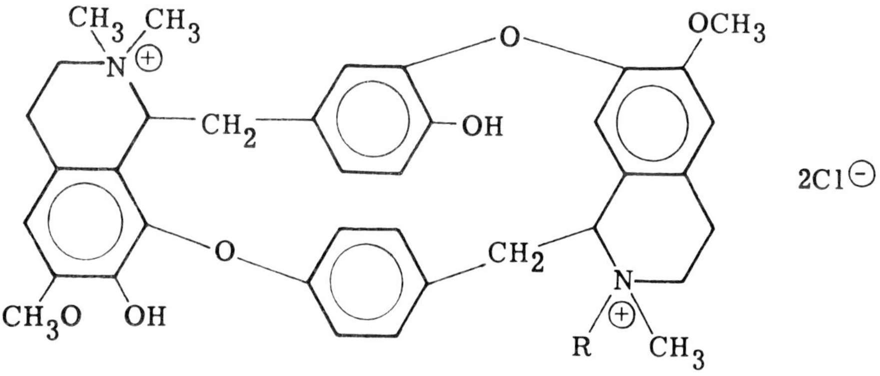

d-TC in 1964, R = CH$_3$

d-TC in 1970, R = H

Fig. 18.2

This novel finding stimulated further research and the structure
was compared with that of d-Tubocurarine which was a bis-quaternary in
1964 although later it was shown[6] to have the monoquaternary structure
shown in Fig.18.2. Hence the obvious deduction was made in 1964 that a
distant second acetylcholine fragment should be added to potentiate the
action of the mono-quaternary steroid shown in Fig.18.1. Further, our
experience[4] of synthetic chemistry suggested the stereoselective introduction of the second fragment at positions 16 and 17 of the androstane
nucleus. This gave rise to our very first bis-quaternary ammonio steroid,
Pancuronium Bromide (Fig.18.1), which is the leading non-depolarising
neuromuscular blocking drug in routine clinical use world-wide. The
structure is shown in Fig.18.3.

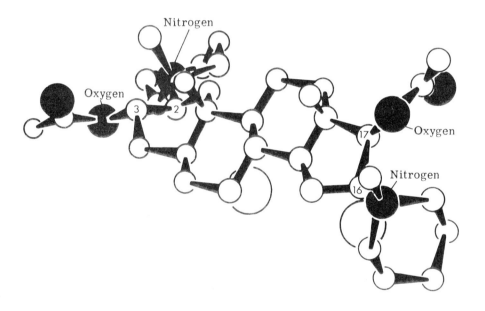

Fig.18.3

The rapid acceptance clinically of this novel drug with its endogenous
nature of two acetylcholine-like fragments incorporated in an androstane
nucleus and its improved selectivity of action again led to a further
research effort in order to discover a similar drug having in addition a
quicker and shorter action.

Indeed two further drugs were developed as far as man but although quick and short-acting in the cat they were long-acting in man.   These drugs were Dacuronium Bromide and Org 6368, differing only from Pancuronium Bromide in their 17β-substituents which were OH and H respectively instead of OAc. This disappointing finding had emerged by 1975 and led to a complete reappraisal of our research approach.

Our experience had shown that the cat was a reliably predictive species for potency[7] and that there was no viable species for predicting duration of action reliably.   However another factor was emerging from the clinical use of all muscle relaxants and that was the desirability of discovering and developing a completely clean drug in terms of lack of cardiovascular side effects.   Even Pancuronium Bromide was associated with limited but significant tachycardia.   Furthermore an animal model was published[8] which showed a good correlation with clinically used agents' abilities to reverse vagally stimulated bradycardia in the cat and the degree of tachycardia which the corresponding drug was liable to induce in man.

This led to a new approach in which we would concentrate on finding a drug lacking side effects and, by examining a range of compounds with in-built differing substituents and physical properties, perhaps also find a quicker and shorter acting drug in man.

This in turn led to a close scrutiny of Pancuronium's properties, of its precursors and also to investigation of some selective chemical reactions.

The latter term can be dismissed most readily since it suffices to say that it proved possible to esterify or quaternise particular amino or hydroxyl groups selectively and in high yield and this has been reported[7]. The fact that structural specificity in these molecules was conferring both a high selectivity of chemical reaction and a source of readily available intermediates suggested that a suitably chosen range of compounds could be prepared and examined to see if 'structural specificity could also give rise to selectivity of pharmacological action'.

In this respect, in order to identify the most important molecular fragments of Pancuronium Bromide, the crystal structure[9], pharmacology and physical properties of this parent compound and closely related analogues were examined.

Firstly the molecular geometry of the two acetylcholine-like fragments were compared.   The essential difference is the torsional angle between the acetoxy O atom and the quaternary N atom on the carbon carbon

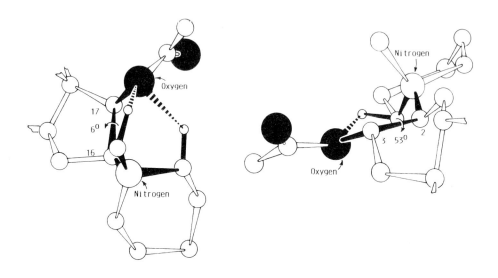

Fig. 18.4

axes. at C_2 C_3 and C_{16} C_{17}.   The structures of these two fragments are
shown in Fig.18.4.   The torsional angle in the ring A fragment is 53°
whereas that involving carbons C_{16} and C_{17} is only 6°.   This smaller
angle gives rise to greater proximity of oxygen and nitrogen substituents
and a unique dual hydrogen bonding system of C - H to O bonding in
contrast to only one such bond in the isomeric A ring system.   Indeed
this is reflected in the abnormally high infrared absorptions for the 3
and 17 acetoxy groups, that for the ring A acetate being 1744 cm^{-1} against
1756 cm^{-1} for the ring D acetate.   Obviously an electron drift from the
carbonyl oxygen is induced more by the dual C - H --- O hydrogen bonding
system in the ring D fragment.

Furthermore it was established by N.M.R. considerations[9] that the
degree of steric interactions among the C_{18} and C_{19} angular methyl groups
and the substituents at positions 2,3,16 and 17 were sufficiently great to
lock the tetracyclically fused molecule constantly in approximately the
single conformation found in the crystal structure.   As a measure of
distance between the two active centres the interonium distance is 11.08 Å.
In contrast it is suggested d-Tubocurarine has a less limited range of

conformation in that crystal structure studies have shown the interonium distance in the dichloride salt[10] to be 9.0 Å and in the dibromide[11] 10.7 Å.

It is pertinent to mention here briefly other neuromuscular blocking drugs in clinical use such as Gallamine and Succinylcholine Chloride which have linear structures allowing greater conformational flexibility.   The unwanted side effects of these drugs are well recognised in clinical practice.

However, to return to specificity of chemical structure and consequent electron topography, it is very interesting to compare the conformations of the two acetylcholine-like fragments, steroidally trapped in their specific conformations, with those found for various salts of acetylcholine itself in different crystal structure studies.   A revealing view is taken along the carbon carbon axis bearing the alcoholic O and quaternary N atoms.

ORG NA97 Pancuronium bromide

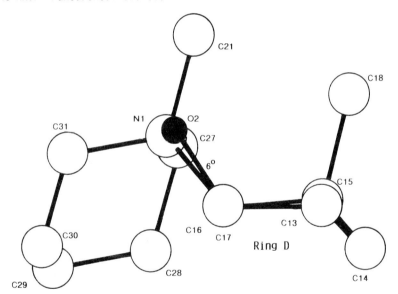

Fig.18.5

For example in the ring D fragment of Pancuronium Bromide the view through C_{16} C_{17} shows an angle of approximately only 6° (Fig.18.5) in contrast to the analogous angle of 53° in the isomeric ring A fragment.

Torsional Angle	N to O Distance (Å)	Source of Ach Structure
78.43	3.2	Br salt
84.64	3.26	Cl salt
73.66	3.13	$CL\,O_4$ salt
76.73	3.19	resorcylate A
95.72	3.21	resorcylate B
53.06	2.96	Pancuronium ring A
6.35	2.92	Pancuronium ring D

Table 18.1

    Comparison with similar cross-sectional views of the ring A fragment and five conformations of acetylcholine salts found in crystallography studies[12-15] shows that this 6° is a uniquely small angle.   In Table 18.1 it is obvious that the second smallest angle of 53° is found in the conformationally restricted ring A fragment.   Correspondingly increasing this torsional angle gives greater N to O interatomic distances.   Obviously these more relaxed conformations of acetylcholine salts do not permit the intimate interatomic relationships which exist within the ring D fragment. The importance of this unique structure will emerge later in this chapter and be discussed further in a biological context.   It suffices to say at this juncture that the ring A and ring D acetylcholine fragments were very much in mind when planning our further synthetic effort.

    Before describing the newer compounds it is pertinent to give one more point of information.   Structure activity studies[16] had shown that for high activity in the Pancuronium series it was essential to have at least two amino groups per molecule at least one of which would be quaternary.

    The standard drug used of course was Pancuronium Bromide and its

ED_{50} (mg/kg) (N) to produce neuromuscular block was compared with its ED_{50} (mg/kg) (V) in the cat to antagonise the bradycardia (approx. 50% decrease in heart rate) produced by vagal stimulation[16]. The ratio $\frac{V}{N}$ of these values was intended to give an approximate indication of a drug's ability to effect neuromuscular blockade without causing tachycardia in humans. This ratio[16] $\frac{V}{N}$ for Pancuronium Bromide was $\frac{62}{18}$, i.e. 3.4; obviously a high value for V and a low value for N is desirable.

Although many compounds were synthesised and tested only a few examples of chemical series are necessary for the purpose of this chapter. These examples are basically selective changes of the 3 and 17 substituents of Pancuronium and of a monoquaternary series[16]. These changes and the appropriate $\frac{V}{N}$ values for each member of the series are given in Tables 18.2-18.5. In Table 18.2 we see a range of diesters in which only the dipropionate $\frac{V}{N}$ = 6.9 offers a little more selectivity over the diacetate. Retaining a 17-ester group in conjunction with a 3-ol (Table 18.3) gives no significant advantage. Retaining the 3-acetate and changing the 17-substituent gives less selective compounds in each case (Table 18.4).

R	$\frac{V}{N}$
(Pancuronium) $CH_3\,CO$	3.4
$CH_3\,CH_2\,CO$	6.9
$CH_3\,(CH_2)_2\,CO$	5
$(CH_3)_2\,CH_2\,CO$	2

Table 18.2

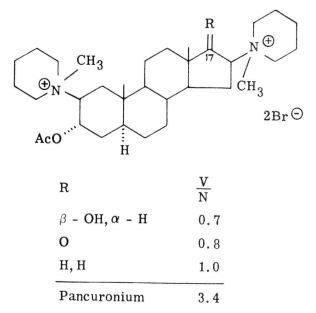

R	$\dfrac{V}{N}$
CH_3CO	4.7
CH_3CH_2CO	0.6
$CH_3(CH_2)_2CO$	0.5
$(CH_3)_2CH_2CO$	0.5
Pancuronium	3.4

Table 18.3

R	$\dfrac{V}{N}$
β - OH, α - H	0.7
O	0.8
H, H	1.0
Pancuronium	3.4

Table 18.4

However a remarkable compound emerged from the diester monoquaternary series (Table 18.5), namely the diacetate 16-N-monomethobromide, with the exceptionally high ratio of 63.   Notice how changing even to dipropionate completely removes this selectivity of action.   This compound, code number Org NC 45, now known as Vecuronium Bromide, differs from Pancuronium Bromide in the chemical nature of its 2-amino substituent as shown in Fig. 18.6.

R	$\dfrac{V}{N}$
$CH_3 CO$ (Org NC 45)	63
$CH_3 CH_2 CO$	0.4
$CH_3 (CH_2)_2 CO$	0.7
$(CH_3)_2 CH_2 CO$	0.2
Pancuronium	3.4

Table 18.5

The important postulate allowed from this structure-activity work[16] with particular respect to the subject of this symposium is that the conformation of the ring D acetylcholine fragment, whilst intrinsically suited to the neuromuscular post-synaptic receptor, is relatively incompatible with the cardiac muscarinic receptor, whereas the isomeric fragment in ring A confers some affinity for this receptor.   This contributes to a theme of particular physiological roles being associated with specific conformations of neurotransmitter.

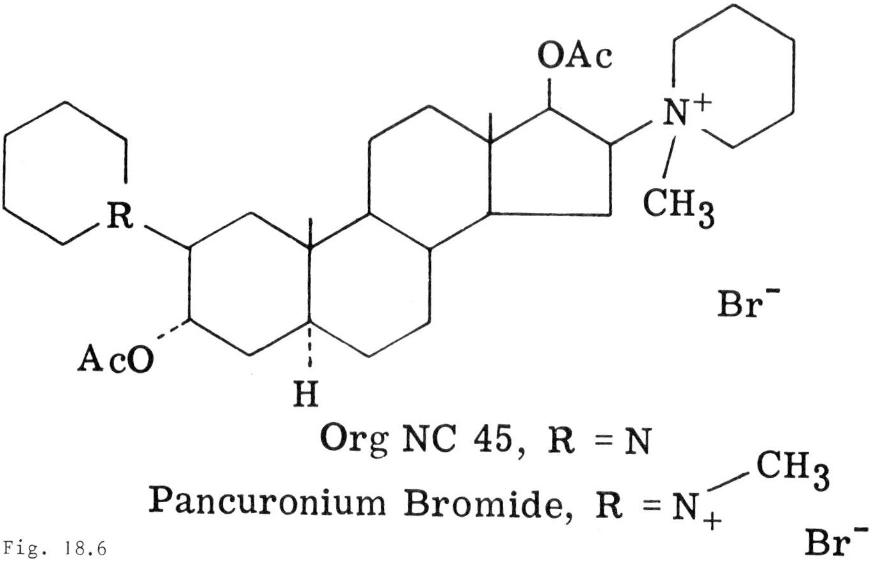

Org NC 45, R = N

Pancuronium Bromide, R = N$_+$ —CH$_3$

Br^-

Fig. 18.6

It is pertinent now to consider Vecuronium Bromide in order to see what it contributes to the theme of this meeting on molecular structure and drug action. Vecuronium Bromide has a vagolytic ED_{50} (V) of 2145 mg/kg and a neuromuscular blocking (N) ED_{50} of 34 mg/kg in the cat. It is the same order of potency as Pancuronium Bromide as a neuromuscular blocking agent both in the cat[16] and human[17,18].

In-depth studies[19] in animal pharmacology have demonstrated further that Vecuronium Bromide has little effect on several mechanisms involved in the control of cardiovascular stability. This further evidence of marked selectivity throws more importance on the conformation of the acetylcholine ring D fragment in this molecule. It must be mentioned here that the conformation of the ring D fragment in Vecuronium is assumed to approximate to that in Pancuronium because of steric compression of substituents.

For instance ring A is in a skew form in both molecules. This

assumption has to be made because Vecuronium Bromide has resisted many
attempts to provide a crystal suitable for crystallographic study by X-ray
analysis.

To continue, it is interesting to remark again on the dissimilarity
of this conformation to that found in the crystal structures of all the
various acetylcholine salts described previously (see Table 18.1). It
poses the question of what is the conformation, if singular, of the
acetylcholine discharged from the vesicles (which are of the order of
500 Å in diameter[20]) presynaptically into the synaptic cleft.   Usually
these vesicles with their quanta of acetylcholine are sited opposite a
fold in the post-synaptic cleft.

Perhaps the conformation of the acetylcholine on leaving the vesicle
is not the important factor in neuromuscular transmission but the energetics involved with the dispatched acetylcholine associating with the
post-synaptic cholinoceptor which in turn is postulated[16] to reflect the
structure of the acetylcholine ring D fragment of Vecuronium Bromide.

It will be interesting to see to what extent this highly selective
drug with its specific structure will contribute to elucidating further
the mechanism of neuromuscular transmission.   When one thinks of the
extraordinary and instant control which the normal human has over the
multitudinous muscle movements at his or her command it does suggest that
the process must be very simple and yet highly selective and well-balanced.

However, to return to the theme of molecular structure and drug
action, our aim stated at the end of the first paragraph of this chapter
should be reconsidered.   It has been shown so far that specificity in
the molecular structure of Vecuronium Bromide has given a highly active
and selective drug.   It remains to consider the time course of action
and the cumulative properties, if any, of this novel agent.

Therefore it is now pertinent to consider the contribution that the
apparently minor, singular molecular change involved in going from Pancuronium to Vecuronium makes to the difference in chemical properties of
these two drugs and the relevance of this difference to the time course
of action and lack of cumulation of the new drug in the human.   Note
the use of the term human because the time course of action of neuromuscular blocking drugs in general is known to differ between animals
and humans.   In particular non-depolarising drugs short acting in animals
tend to be longer acting in man.

The most obvious difference between the two drugs of course is that
Vecuronium Bromide has a tertiary amino function.   Hence the pH values

for Vecuronium Bromide at 25°C for a 1% solution in water is 9.30 in contrast to 6.72 for Pancuronium Bromide. Further in concentration of 4 mg/ml in aqueous solution at 25°C Vecuronium Bromide is known to de-acetylate significantly in 0.5 hr. whereas Pancuronium is relatively stable[21]. This deacetylation is of significance because two of the hydroxy products are 25 and 50 times less potent than the parent drug. So any such metabolism in vivo, especially at an active site, could shorten duration of action.

However in the human plasma at about pH 7.4 of patients given Vecuronium Bromide the drug is found to be relatively stable[22]. This may well be due to the effect of dilution because only 4 to 10 mg of drug will be introduced to several litres of plasma. So only in sites such as the synaptic cleft where local concentrations may gather, can metabolism be expected due to intermolecular interaction.

A more important factor contributing to the significantly shorter duration of action of Vecuronium Bromide is its relatively faster disappearance from the plasma[23], plasma levels being associated with degree of drug activity.

This may also be a factor in the drug's marked lack of cumulative effect. Another factor is possibly the attachment of only the one quaternary ammonium of a Vecuronium molecule to the receptor whereas Pancuronium has two such quaternary heads. Consequently metabolism and pharmacodynamics in the neuromuscular junction could play a very important role in contributing to short duration of action and lack of significant cumulative effect.

The physical properties of Vecuronium Bromide must also give rise to a faster onset of action[24] which means that its complete time course of action fulfils the initial aim.

It is interesting to note that the conformationally restrained molecule, Vecuronium Bromide, has led already to an increased flexibility in anaesthetic practice in that relatively higher doses can be used to give a shorter time to intubation without fear of either prolonged block-ade or appearance of unwanted side effects.

This covers very briefly 25 years of study by many workers in many disciplines, both in academicia and industry. Further the component of serendipity in drug research should not be omitted. On this occasion it is left to the reader to quantify this. You require to put yourself in my position 20 years ago and then to be shown the correct structure of d-Tubocurarine (Fig. 18.2). The monoquaternary, Vecuronium Bromide, is one obvious drug to make but the selective monoalkylating reaction on the precursor to both Pancuronium and Vecuronium Bromides to give Vecuronium Bromide, was not known then.   Ironically this chapter may never have been written had we been presented with the correct structure of d-Tubo-curarine at that point in time.

Fortunately, however, Pancuronium Bromide was developed first which allowed the crystallographical study[9].   This, in conjunction with the advantageous action of the drug clinically, stimulated further research as described above, which in turn led to further refinement in the form of Vecuronium Bromide.

REFERENCES

1.   Hewett, C.L. and Savage, D.S. (1966). J.Chem.Soc. (C). 484-488.

2.   Hewett, C.L. and Savage, D.S. (1967). J.Chem.Soc. (C). 582-588.

3.   Hewett, C.L. and Savage, D.S. (1968). J.Chem.Soc. (C). 1134-1140.

4.   Hewett, C.L. and Savage, D.S. (1969). J.Chem.Soc. (C). 1880-1883.

5.   Hewett, C.L., Savage, D.S., Lewis, J.J. and Sugrue, M.F. (1964). J.Pharm.Pharmacol. 16, 765.

6.   Everett, A.J., Lowe, A. and Wilkinson, S. (1970). Chem.Comm. 1020-21.

7.   Savage, D.S. (1980). In : Curares and Curarisation.   Elsevier, New York, 21-31.

8.   Hughes, R. and Chapple, D.J. (1976). Br.J.Anaesth. 48, 59-67.

9.   Savage, D.S., Cameron, A.F., Ferguson, G., Hannaway, C. and Mackay, I.R. (1971). J.Chem.Soc. (B). 410-415.

10.   Codding, P.W. and James, M.N.G. (1973). Acta Cryst. B29, 935-941.

11.   Reynolds, C.D. and Palmer, R.A. (1976). Acta Cryst. B32, 1431-1436.

12.   Herdklatz, J.L. and Sass, R.L. (1970). Biochem.Biophys.Res.Comm. 40, 583-588.

13.   Svinning, T. and Sørum, H. (1975). Acta Cryst. B31, 1581-1586.

14.   Mahajan, V. and Sass, R.L. (1974). J.Cryst.Mol.Struct. 4, 15-24.

15.   Jensen, B. (1975). Acta.Chem.Scand. 531-537.

16.   Durant, N.N., Marshall, I.G., Savage, D.S., Nelson, D.J., Sleigh, T. and Carlyle, I.C. (1979).   J.Pharm.Pharmacol. 31, 831-836.

17.   Crul, J.F. and Booij, L.H.D.J. (1980) Br.J.Anaesth. 52, Suppl. 1., 49S-52S.

18.   Agoston, S., Salt, P., Newton, D., Bencini, A., Boomsma, P. and Erdmann, W. (1980). Br.J.Anaesth. 52, Suppl. 1, 53S-60S.

19.   Marshall, R.J., McGrath, J.C., Miller, R.D., Docherty, J.R. and Lamar, J-C. (1980). Br.J.Anaesth. 52, Suppl. 1, 21S-32S.

20. Feldman, S.A. (1973). In : Muscle Relaxants. W.B. Saunders & Co.
     Ltd., London.

21. Savage, D.S., Sleigh, T. and Carlyle, I.C. (1980). Br.J.Anaesth.
     52, Suppl. 1, 3S-6S.

22. van der Veen, F. and Bencini, A. (1980). Br.J.Anaesth. 52,
     Suppl. 1, 37S-41S.

23. Fahey, M.R., Morris, R.B., Miller, R.D., Nguyen, T.L. and Upton, R.A.
     (1981). Br.J.Anaesth. 53, 1049-1053.

24. Krieg, N., Crul, J.F. and Booij, L.H.D.J. (1980). Br.J.Anaesth.
     52, 783-788.

Acknowledgement :   To J. Kelder and M.D. van Wendel de Joode for the
computer graphics in this chapter.

19 Structure–activity relationships for quinazoline α-adrenoceptor antagonists

Simon F. Campbell

1. Introduction

Hypertension afflicts between 10–20% of the adult population and is a major risk factor in many forms of cardiovascular disease.[1] However, while elevated blood pressure can be controlled and morbidity/mortality reduced with modern drug therapy, there is still keen clinical interest in novel antihypertensive agents which are free from obtrusive side effects.[2] In the majority of cases, the underlying causes of hypertension are unknown, but considerable attention has focussed on the role of the sympathetic nervous system and α-adrenoceptors located in blood vessel walls.[3,4] The vasomotor centre in the brain determines arteriolar tone by modulating electrical signals in the sympathetic fibres which initiate release of noradrenaline from nerve endings (Fig. 1). This neuro-transmitter traverses the synaptic gap, stimulates postjunctional α-adrenoceptors and the resultant vasoconstrictor response is reflected by an increase in blood pressure. Clearly, overactivity of the sympathetic nervous system could be an important factor in the initiation and maintenance of hypertension, and attempts to develop drugs which act by blockade of the vasoconstrictor action of noradrenaline commenced over 50 years ago.[5] Despite intense efforts, agents of only limited clinical utility were identified, and α-adrenoceptor research waned in the 1960s.

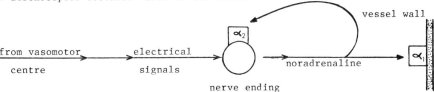

FIG. 1.   Location and role of prejunctional-(α_2) and postjunctional-(α_1) adrenoceptors at a sympathetic nerve synapse.

During the next decade, interest was rekindled when α-adrenoceptors were identified on nerve endings and pharmacological criteria differentiated this new α_2-subtype from the traditional α_1-population in blood vessel walls[6,7,8] (Fig. 1). Moreover, it was proposed that when the concentration of noradrenaline in the synaptic gap was sufficient to elicit appropriate α_1-mediated vasoconstriction, excess transmitter stimulated the prejunctional α_2-receptor and further noradrenaline release was prevented.[9] Thus, arteriolar tone can be regulated by a simple local control mechanism as well as via baroreceptor activation/vasomotor compensation. One consequence of this model is that agents which block prejunctional α_2-receptors will enhance noradrenaline release, and this could overcome any concomitant antagonist activity at postjunctional α_1-sites. As a corollary, useful antihypertensive drugs should only act at the latter receptor, but re-examination of a wide range of α-adrenoceptor antagonists generally indicated poor selectivity for α_1- as opposed to α_2-receptor subtypes.[10] Indeed, several agents showed a preference for prejunctional receptors, and therefore the previous lack of clinical success can be readily understood. More recently, however, selective postjunctional α_1-antagonists have been synthesised and as predicted, such agents as prazosin[11] and doxazosin[12] are clinically effective antihypertensive drugs.[13,14]

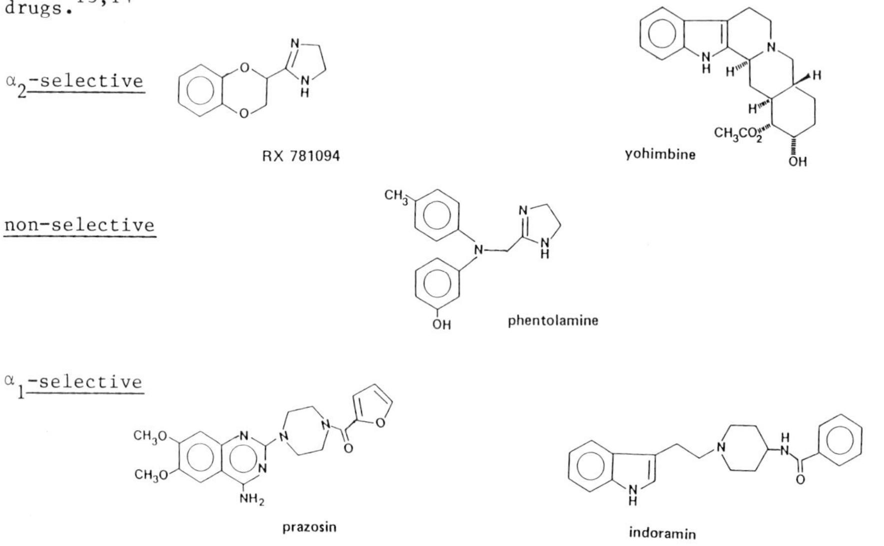

FIG. 2.   Representative α-adrenoceptor antagonists.

It is still surprising that many compounds do not adequately discriminate between α_1- and α_2-adrenoceptors as the two populations are located differently, and their detailed activation mechanisms are quite separate.[15] By analogy with other receptor systems, agents with even limited α_1- or α_2-preferences would be expected to fall into two major structural categories, but in fact α-adrenoceptor antagonists comprise a particularly heterogeneous collection (Fig. 2). For example, the pentacyclic alkaloid yohimbine and the benzodioxan derivative RX781094 are selective for α_2- as opposed to α_1-receptors. Phentolamine shows little preference for either site, whereas such disparate structures as prazosin and indoramin are selective α_1-antagonists. It is not realistic at this stage to rationalise structure activity relationships (SARs) in all these diverse areas, and this paper will therefore concentrate on our own work with 2,4-diamino-6,7-dimethoxyquinazoline derivatives. These compounds are potent, highly selective postjunctional α_1-adrenoceptor antagonists, and two aspects of SARs for this clinically useful series will be discussed:-

a)   a possible mode of interaction of the quinazoline nucleus at the α_1-adrenoceptor;

b)   the role of the quinazoline 2-substituent in influencing both receptor affinity and antihypertensive activity.

In establishing SARs, extensive use has been made of ligand binding methods[16] to determine receptor affinity. α_1- and α_2-adrenoceptors in a rat brain membrane preparation were labelled with tritiated prazosin and clonidine respectively[17,18] and the abilities of test compounds to displace these ligands measured. Thus, an entry, α_1, 2nM; α_2, -, indicates that the Ki for the compound at α_1-adrenoceptors is 2×10^{-9}M while for the α_2-site, the Ki is $>1 \times 10^{-6}$M. Assessment of functional antagonism against noradrenaline at pre- and post-junctional α-adrenoceptors was determined in a rabbit pulmonary artery preparation and results expressed as EC_{40} values (nM.).[6]   Finally, antihypertensive activity was evaluated after oral administration (5mg/kg) to spontaneously hypertensive rats (New Zealand or Okamoto strain). The falls in blood pressure (mm.Hg) during the test period (4.5hr) were measured using an indirect tail cuff method and the maximum value expressed as follows,

$$\% \text{ reduction in hypertension} = \frac{\text{fall in blood pressure}}{(\text{control blood pressure} - 130)} \times 100$$

where 130 represents the systolic blood pressure of normotensive rats. Of course, this test system gives little information relevant to SARs at the receptor level since metabolic, pharmacokinetic and reflex haemodynamic effects may intervene.

2.   A role for the quinazoline nucleus at α_1-adrenoceptors

In a variety of pharmacological preparations prazosin (I), doxazosin and related derivatives competitively antagonise the α_1-mediated responses to noradrenaline (II). Strictly, these observations only require that the binding of the agonist and antagonist be mutually exclusive. However, it is more profitable for the medicinal chemist to assume that these compounds

recognise common receptor sites, and then to identify structural features
which might perform similar functions. The objective of this approach is
to build up a receptor model which allows rationalisation of SARs and which
also stimulates the design of novel, superior agents. In the present case,
the aromatic rings of prazosin and noradrenaline serve as initial reference
points, and the respective 6,7-dimethoxy- and 3,4-dihydroxy-functions could
be similarly accepted by the receptor, but only in the latter case would
activation occur [19] (Fig. 3).

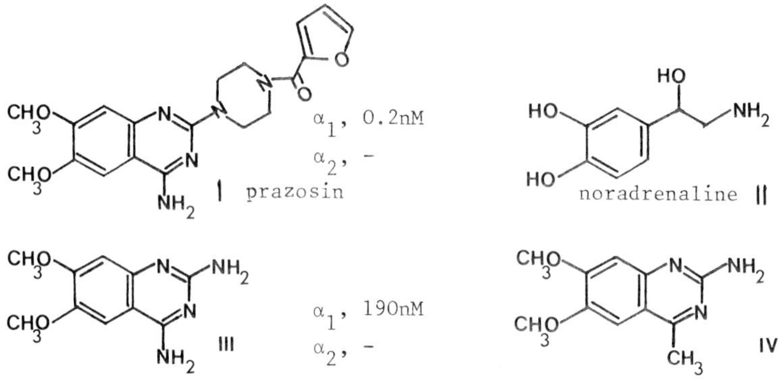

FIG. 3.   Structural/basicity requirements for quinazoline α_1-adrenoceptor
affinity.

By contrast, an obvious structural difference is that noradrenaline
contains a benzylic hydroxyl group but, although important for potency,
this is not essential for agonist activity.[20] Moreover, an hydroxyl moiety
is not present in several other α-antagonists, and there is no reason for
concern over its absence from this quinazoline series. On the other hand,
a basic centre is a common feature of agonist and antagonist structures and
at first sight, the similarly located amino-function of noradrenaline and
the prazosin 2-nitrogen atom could interact with a common receptor site.
If such ideas are valid, then simplified derivatives such as 2,4-diamino-
6,7-dimethoxyquinazoline (III) should also be recognised at the α_1-receptor
and although the compound is substantially less potent than prazosin, the
selectivity characteristic of the whole series is clearly evident. On the
basis of these initial SARs, the 2-amino-4-methyl analogue (IV) should also
have α_1-receptor affinity but, unexpectedly, the compound was inactive, and
factors besides structural similarity must be considered.

The measured pKa of the amino function of noradrenaline is 9.6[21] and at
physiological   pH   (7.4)   the   protonated   species   (V)   will   predominate

(ca. 95%).[22]   This suggests that the α-receptor contains an anionic or
nucleophilic site which interacts with the positive centre in the natural
transmitter, and therefore a similarly charged moiety should also be
present in antagonist structures.   Thus, the different affinities of
compounds (III, pKa 7.8) and (IV, pKa 5.2) can be rationalised in terms of
relative abilities to be protonated at physiological pH.   In the case of
prazosin (pKa 7.2), almost equal amounts of the parent base and the
protonated species will be in equilibrium under physiological conditions,
but the latter should be preferred at the α_1-adrenoceptor.

In contrast to noradrenaline (V), inspection of the 2,4-diamino-
quinazoline system indicates four possible protonation sites, but the
exocyclic nitrogen atoms need not be considered as these electron pairs are
extensively conjugated with the aromatic π-system.   For the ring nitrogen
atoms, molecular orbital calculations indicate that electron density is
highest at N-1 and the N-1 protonated species is also more stable.[23]
Protonation at N-1 (VI) is therefore preferred over N-3 (Fig. 4) and this
is certainly the case in the crystal structure (Fig. 5).   Similar
conclusions have been reached for quinazoline dihydrofolate reductase
inhibitors,[24] while protonation[25] and quaternisation[26] of related
2,4-diaminopyrimidine derivatives also predominate at N-1.

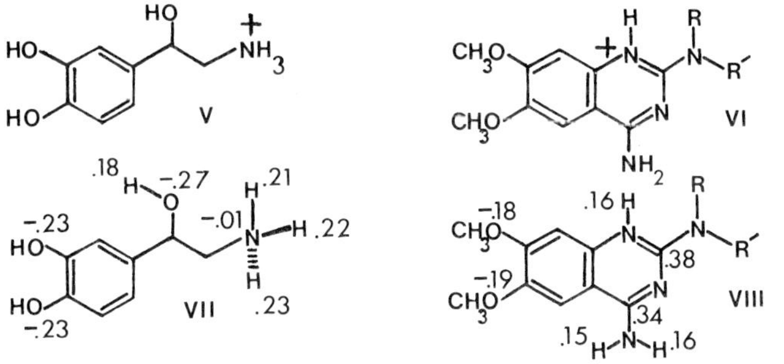

FIG. 4.   Protonated forms of noradrenaline and a quinazoline derivative,
(a) localised; (b) delocalised, CNDO/2 charge distribution by Mulliken
population analysis (only selected centres shown).

Clearly, these results are inconsistent with a receptor model which
envisaged similar roles for the amino-function of noradrenaline and the
nitrogen atom at the quinazoline 2-position.   Major electronic differences
exist between the two centres, and their ability to participate in similar
receptor interactions must be questioned.   However, although convention
places a formal positive charge on the nitrogen atom of a protonated amine,

calculations show that the charge is distributed over the four substituent groups.[27,28] Indeed, the consequences of charge delocalisation in protonated drug molecules have been pointed out on several occasions[29] but with little apparent impact on SAR interpretation.

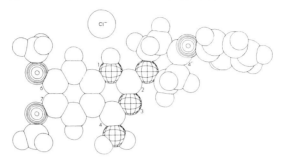

FIG. 5. Space-fill diagram for the X-ray structure of 4-amino-6,7-dimethoxy-2-(4-phenoxy)piperidinoquinazoline hydrochloride salt.[30] Note protonation on N-1.

For noradrenaline (VII), most of the positive charge is distributed over the three hydrogen atoms, and therefore in considering agonist-receptor recognition, there is little reason to emphasize either the location or Coulombic interaction of an essentially uncharged nitrogen centre. Charge distribution is even more extensive in the protonated quinazoline (VIII), and although both the N_1-H and 4-NH_2 systems appear equally suited as focal points for receptor interaction, attention will concentrate on the former, for the present. Thus, charge-reinforced hydrogen bonding involving an anionic/nucleophilic moiety is probably important for agonist and antagonist molecules, but clearly, this negative site must be accessible to both noradrenaline and the quinazoline derivatives. However, as positive charge is symmetrically distributed around the hydrogen atoms, directional flexibility for electrostatic interaction with a negatively charged centre is allowed. Identification of the nucleophilic site, on the other hand, is purely speculative as characterisation of the α_1-adrenoceptor is still in progress[31] and the aminoacid residues which line the active region are unknown. However, the recognition area is part of a complex protein structure, and carbonyl moieties in the polypeptide backbone are effective hydrogen bond acceptors. Moreover, experimental[32] and theoretical studies[33,34] indicate considerable directional latitude for these carbonyl-hydrogen bonds, with little detrimental effect on bonding energy.[35]

Thus, the α_1-adrenoceptor could consist of an hydrophobic area which accommodates the aromatic rings of both noradrenaline or the quinazoline series,[36] and an in-plane carbonyl function which accepts a partially charged hydrogen atom from either structure [37] (Fig. 6). In each case, hydrogen bond lengths are approximately $2A^O$ and as deviation from linearity is only thirty degrees, significant stabilising interactions should be established. Agonist and antagonist can therefore bind at the same sites, but the original formally charged nitrogen atoms do not play direct roles in drug-receptor interactions, and attention is focussed on charge-reinforced hydrogen bonding. Indeed, the high α_1-adrenoceptor affinity of these quinazoline derivatives may reflect the fact that N-1 protonation is particularly appropriate for the nucleophilic site on the receptor protein.

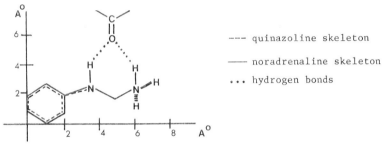

--- quinazoline skeleton

—— noradrenaline skeleton

... hydrogen bonds

FIG. 6.   Composite diagram illustrating interaction of noradrenaline and the quinazoline nucleus with common hydrophobic and hydrogen bond acceptor sites. (Partial structures only).

Although this simple model appeared attractive, and could be further refined, it seemed more appropriate to examine alternative anionic centres for potential advantages over a weakly charged carbonyl function. Therefore, interaction of the protonated form of noradrenaline (VII) with various counterions was assessed using molecular mechanics techniques but first, two provisos were made:

(a)   the anion and the aromatic ring should be essentially coplanar, consistent with the concept of a flat α_1-adrenoceptor;

(b)   simultaneous hydrogen bonding involving the anion, the benzylic hydroxyl group and the positive charged ammonium head should be feasible.

This latter condition was introduced to avoid separate binding sites for each functionality, as, intuitively at least, a single acceptor moiety should be preferred on both enthalpy and entropy considerations. The interaction of various counterions with noradrenaline was studied using

molecular mechanics simulation of non-bonded forces (Van der Waals and
Coulombic) to identify favourable binding positions then full relaxation
energy minimisation was initiated to determine the preferred geometries of
interaction.   Final binding energies (enthalpies of interaction) were
calculated by standard INDO methods.   Naturally, a carbonyl function was
the first ligand evaluated, but concomitant interaction with both hydrogen
bonding moieties could not be accommodated,[38] and this acceptor may be more
appropriate for agonists which do not contain a β-hydroxyl function (e.g.
histamine, dopamine, 5-HT).

By contrast, chloride, phosphate and carboxylate anions satisfied both
criteria, and for the first two cases, similar cyclic hydrogen bonding
arrangements have been observed in crystal structures.[39,40]   These
theoretical and experimental results may be relevant for drug-receptor
interactions as although agonists and antagonists must be transported
through an aqueous milieu, the target site is most likely to be a lipoid
environment.   Moreover, charged counterions may be buried in hydrophobic
pockets [41] rather than being exposed to the outer aqueous phase, and drug-
receptor interactions probably involve desolvated species.   Gas phase
calculations therefore   may be particularly useful although binding
enthalpies should be treated qualitatively rather than quantitatively.
Interestingly, n.m.r. studies support cyclic hydrogen bonded structures for
the hydrochlorides of noradrenaline-like molecules in chloroform, but not
aqueous solution.[42] These results may mirror conformational preferences at
the receptor and in physiological fluids respectively.

Each of the three counterions mentioned previously appears equally
suited for further evaluation,[43] but the carboxylate anion was selected for
detailed attention, principally   because salt bridges involving aspartate
and glutamate residues with protonated heterocycles have been observed in
other enzyme systems, e.g. the dihydrofolate reductase series.[44] Thus for
noradrenaline, a coplanar cyclic hydrogen bonding arrangement (Fig. 7, IX)
is quite stable (binding energy, - 155.93Kcal/mole) and, as this
conformation is less than 2Kcal/mole above the global minimum (phenyl ring
rotated through 60°), it is accessible to the natural α_1-adrenoceptor
agonist.   A similar interactive study was then carried out with a
protonated quinazoline derivative where charge-reinforced hydrogen bonding
occurred, (X, binding energy, - 72.2Kcal/mole) but closer approach ($<2.5A^{\circ}$)
between the carboxylate anion and the N_1-H was prevented by the piperidine

ring.[45] Comparison of structures (IX) and (X) indicates that relative to
the aromatic rings, the carboxylate counterions are differently located
(ca. 4A$^{\circ}$ separation) and appear incompatible with the previous concept of
fixed recognition sites.

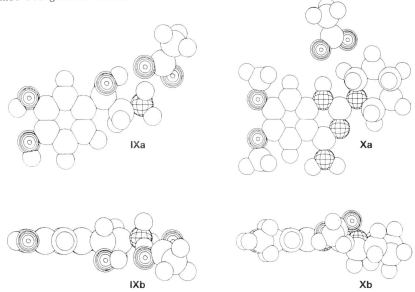

FIG. 7. Interaction of noradrenaline (IX) and a quinazoline derivative (X)
with a carboxylate counterion (see text for provisos). Face-on (a) and
side-views (b) illustrated.

However, the earlier carbonyl model was static and although initial
agonist and antagonist receptor recognition may be similar, the consequences of their binding have nothing in common. By definition, agonists
activate the receptor to produce a physiological response, whereas antagonists exert "squatters' rights", without markedly perturbing the active
site. For example, in the β-receptor series, agonist binding is enthalpy-
driven, implying that strong bonds are required at the receptor in order to
overcome an unfavourable decrease in entropy.[46] By contrast, antagonist
interactions are entropy-driven, with weaker bonds being formed, since the
major contribution to binding affinity results from the entropy increase
associated with released water molecules.[47] Preliminary data also indicate
that prazosin binding to the α_1-adrenoceptor is entropy-driven.[48]

An additional important factor is that even quite simple quinazoline
derivatives have high α_1-binding affinity (see Table I), and these

compounds do not appear to interact with other receptor systems.[49] Thus, the quinazoline nucleus exhibits a high degree of complementarity for the α_1-adrenoceptor, possibly representing a conformationally restricted form of noradrenaline, and its recognition may be highly selective. Therefore, this protonated heterocyclic system could be exquisitely suited to the ground·state of the α_1-adrenoceptor, and the high binding affinity reflects an hydrophobic attraction, charge-reinforced hydrogen bonding and the favourable entropy component associated with release of water molecules.[50] As the drug-receptor complex is also a minimum enthalpy arrangement (X), given the original constraints, the conformational reorganisation required for receptor activation would be energetically unfavourable.

Noradrenaline must also approach this receptor ground state and the carboxylate counterion forms a hydrogen bond with the benzylic hydroxyl function and also participates in medium-range electrostatic interaction with the ammonium head (Fig. 8, XI, binding energy, −74.54Kcals/mole). However, this is not a minimum energy state, and charge-reinforced hydrogen bonding can be optimised by a 4A° movement of the counterion (IXa, binding energy −155.93Kcal/mole). Thus, the conformational change in the protein usually associated with receptor activation could be promoted by the free energy decrease when the initial agonist complex (XI) is transformed into the more stable arrangement (IXa).[51,52]

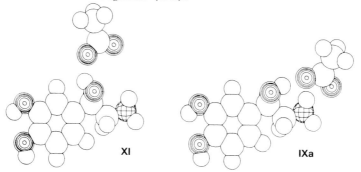

FIG. 8. Interaction of noradrenaline with a carboxylate counterion, α_1-adrenoceptor ground state (XI); activated state, (IXa).

This model therefore provides a simple rationalisation of the different consequences of agonist and antagonist receptor occupancy, and preliminary studies suggest that it may be extended to other structures. Thus, for S(+) noradrenaline the counterion complex is less stable (Fig. 9, binding

energy -141.86Kcal/mole), as repulsive interactions between the hydroxyl
proton and the positively charged ammonium head prevent simultaneous hydrogen bonding. Similar binding energies are also obtained for dopamine (the
deshydroxy analogue), and these results are consistent with pharmacological
evidence that both compounds elicit maximal α_1-mediated responses, but are
considerably less potent (ca. 100 fold) than R(-) noradrenaline.[53]   In
these cases, counterion migration is not steered by the hydroxyl function
but simply results from medium-range Coulombic attraction.

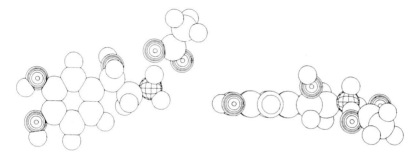

FIG. 9.   Interaction of S(+) noradrenaline with a carboxylate anion on the
α_1-adrenoceptor, activated state.   Counterion location as for (IX).

It should be stressed again that hydrophobic interactions involving
aryl rings have been assumed to be the initial receptor recognition
processes for both agonists and antagonists, and that charge-reinforced
hydrogen bonding is viewed as a secondary event.   This is probably
reasonable, as hydrophobic areas on a receptor surface should be accessible
to drug molecules, while a carboxylic acid moiety buried in a lipoid pocket
would be much less available.   Thus, the initial hydrophobic association
probably not only locks the drug onto the receptor surface but it may also
expose the "naked" counterion for subsequent electrostatic interactions.

Finally, the remarkable α_1-adrenoceptor selectivity of these quinazoline derivatives (Table 1) is consistent with a receptor model having a
coplanar arrangement of both the hydrophobic binding area and the
carboxylate counterion.   By contrast, initial analysis of α_2-adrenoceptor
SARs suggests that these key sites may lie in orthogonal planes [54,55] and
a flat, protonated quinazoline nucleus could not accommodate such
alternative geometry.   Further studies are in progress to determine whether
these proposals can rationalise other SARs observed at each of the
α-adrenoceptor subtypes.[56]

3. The Role of the Quinazoline 2-Substituent

Although the amino-function of noradrenaline and the quinazoline 2-nitrogen atom have quite different electronic characteristics, they should occupy similar regions on the α_1-adrenoceptor and therefore the well-known SARs for the classical catecholamine series may be relevant. It has been known for some time that modification of the nitrogen substituent in the latter compounds has profound effects on receptor selectivities and indeed, Ahlquist's original distinction between α- and β-receptors relied heavily on the observation that isoprenaline had reduced α-adrenoceptor activity. This initially suggested stringent steric requirements for the region around the nitrogen atom, but as the substituent is further elaborated, α-receptor affinity returns (Figure 10).

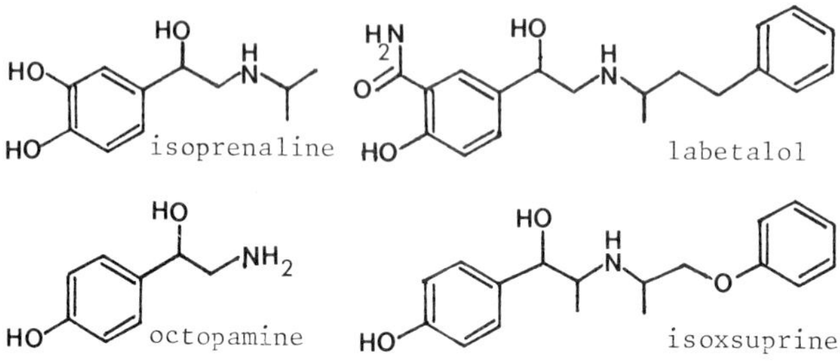

labetalol: α_1, 150nM; α_2, 1,600nM

isoxsuprine: α_1, 26nM; α_2, 1,100nM

FIG. 10. Effect of N-substitution on α-adrenoceptor affinity of "catecholamine" derivatives. (Note octopamine is an α-agonist).

Importantly, labetalol, isoxsuprine and related compounds are antagonists, and their extended nitrogen substituents could occupy the receptor area where the conformational consequences of agonist-induced counterion migration are normally expressed. Consistent SAR information is not available for many of these analogues,[57] but the compounds could occupy the α_1-receptor in a similar fashion, the second aromatic rings capitalising on a distal, hydrophobic site. However, this hydrophobic region is not directly involved in agonist-induced receptor activation, and it may not be an integral part of the receptor *per se* but simply one of the many lipoid areas which are commonly found in protein structures. Even so, it does appear, that in addition to the aromatic and anionic regions, α_1-antagonists can also capitalise on an hydrophobic site centered some 7.5A° from the parent nitrogen atom.

Can these observations be extrapolated to the quinazoline series? (Table I). As mentioned previously, 2,4-diamino-6,7-dimethoxyquinazoline

NO.	R	α_1	α_2	EC$_{(40)}$ POST	EC$_{(40)}$ PRE	% reduction in rat b.p.
1	$-NH_2$	190	—	NT	NT	5*
2	$-N(C_2H_5)(C_2H_5)$	2.0	—	130	1,500	50*
3	piperidine	2.5	—	300	—	18*
4	piperidine–O–CH$_2$CH$_2$–O–	1.5	NT	38	—	83
5	piperidine–CONHC$_4$H$_9$	0.3	—	50	—	100
6	phenylpiperidine	1.0	—	NT	NT	26*
7	tetrahydroisoquinoline, OCH_3, OCH_3	0.02	—	NT	NT	100*
8	piperazine–NH	33	—	1750	—	21*
9	N-phenylpiperazine	3.4	—	NT	NT	45
10	piperazine–CO–furyl	0.2	—	4.5	1,300	70*
11	piperazine–CO–benzodioxane	0.5	—	50	—	77

Table I.   Binding (Ki, nM.), functional (EC$_{40}$, nM.) and antihypertensive activities for selected quinazoline derivatives.   *Okamoto strain.   No. 10, prazosin; No. 11, doxazosin (UK-33,274).

(No. 1) displays moderate binding affinity for α_1-adrenoceptors, similar to more elaborate antagonists such as labetalol and isoxsuprine.   This again highlights the effective receptor fit of the rigid quinazoline nucleus, since more flexible structures require an additional hydrophobic interaction in order to offset the entropy loss associated with conformational restriction.   However, receptor recognition of this

prototype quinazoline derivative (No. 1) may not be typical of the series
as the 2-amino function will also interact with the carboxylate counterion
and so prevent optimum alignment of the aromatic ring. No such problem
exists for trisubstituted derivatives however, and the immediate one
hundred-fold increase in affinity (Nos 2,3) could support the importance of
a single N_1-H-carboxylate interaction (Fig. 7, X) rather than a particular
receptor role for the simple nitrogen substituents. When a carbonyl
function is attached at the piperidine 4-position, potency increases a
further ten fold (No. 5) but the corresponding 4-phenyl derivative, which
also contains a π-system, was unexpectedly less active (No. 6). However,
when the aromatic ring is fused in a bicyclic arrangement (No. 7), a major
increase in affinity resulted (cf. No. 3). Indeed, the exceptional $α_1$
binding affinity of this tetrahydro-isoquinoline derivative (Ki, 0.02nM)
suggests that the compound is capitalising on an additional hydrophobic
area, but as the centre of the aromatic ring is only some 4A° from the
2-nitrogen atom, this may not be the site implicated in the labetalol
series. For the piperazine derivatives, similar fusion of an aromatic ring
is obviously not feasible, but even so, SARs generally correspond to the
piperidine series. In particular, prazosin (No. 10) and doxazosin (No. 11)
are the most potent members of this group, and incorporation of a planar
π-system _via_ a carboxamide moiety appears to be an effective alternative
for a fused aromatic ring.[58] In summary, although the parent diaminoquina-
zoline system (No. 1) is an effective $α_1$-adrenoceptor ligand, binding
affinity can be increased some ten thousand-fold by appropriate
modification of the 2-substituent and compounds of outstanding potency have
been identified. Equally important, these derivatives show little, if any,
propensity to displace clonidine from $α_2$-receptor sites, and exceptional
$α_1$-selectivity is characteristic of this whole quinazoline series.

Although these binding studies provide a measure of affinity for
$α_1$-receptors, evaluation of functional antagonism is also important in
order to determine the ability of a compound to block the action of
noradrenaline. Results from these experiments (Table I) roughly agree
with binding data, but an exact parallel would not be expected as both
receptor affinity and physicochemical properties are important in this test
system. In fact, all of the compounds examined are selective
postjunctional $α_1$-adrenoceptor antagonists with minimal effects at
prejunctional sites, and they should not suffer the detrimental
consequences of non-selective blockade. However, _in vitro_ receptor

affinity is only a guide to in vivo behaviour, and antihypertensive activity in animal models must be established in order to select compounds worthy of secondary pharmacological evaluation. In the spontaneously hypertensive rat, all of these quinazoline derivatives (except No. 1) lower blood pressure although the 2-substituent markedly influences both efficacy and duration of action. For example, the simple piperidine derivative (No. 3) is only weakly active, but incorporation of a 4-substituent (Nos. 4, 5) leads to a substantial improvement. The tetrahydroisoquinoline derivative (No. 7) also displays excellent, long-lasting activity in the rat but more detailed profiling did not justify clinical evaluation. Interestingly, although the desmethoxy analogue (see No. 7, H for OCH_3) has similar binding affinity, (Ki, 0.05nM) it is essentially inactive in vivo. Thus, while substitution of the aromatic ring is not required for α_1-adrenoceptor interaction, it plays a crucial role in stabilising a metabolically vulnerable moiety. Prazosin and doxazosin display similar efficacy in the rat but the compounds do not appear to be the most active of the series. However, direct comparison of these data can be misleading as single-dose experiments may represent quite different points on the respective anti-hypertensive dose-response curves. In man, both compounds are potent, effective antihypertensive agents which appear to lower blood pressure by antagonising the α_1-mediated vasoconstrictor responses to noradrenaline.[12] Importantly, the elimination half-lives of prazosin and doxazosin in man are approximately 2.5 and 11 hrs respectively[59] and the latter compound offers the convenience of once-a-day dosing. The origin of this unexpected advantage appears to be that doxazosin is more resistant to metabolic O-de-methylation than prazosin, and overall molecular behaviour has therefore been markedly influenced by a seemingly minor modification in structure.

4.   Relative Importance of the Quinazoline Nucleus and 2-Substituent

SARs for the quinazoline nucleus and 2-substituent have been discussed separately so far, but structural modifications cannot be viewed as independent events, and the properties of the whole molecule must be considered. For example, elaboration of a quinazoline 2-dimethylamino substituent to the prazosin side chain increases binding affinity only some twenty fold, and this does not suggest a major role for the furoylpipera-zino moiety (Fig. 11). However, the pKa of the former compound (8.1) is almost one log unit higher, and this could be of utmost importance if a protonated quinazoline nucleus is required at the receptor. Thus, the unexpectedly high α_1-adrenoceptor affinity of (XII) may simply result from

increased basicity, and quantification of the contribution of the extended
2-substituent is not possible.

When a simple alkyl substituent is incorporated at the quinazoline
2-position, the compounds are essentially inactive, whereas, by contrast,
the "deaza" prazosin analogue (XIII) displays α_1-affinity similar to
labetalol. At first sight, the more elaborate 2-substituent now appears to
dominate receptor interactions, but no conclusions can be drawn over
relative binding roles when two inactive fragments are combined in a single
molecule.[60] Perhaps the most informative comparison is between structures
(I) and (XIII); prazosin is substantially more basic and the six hundred-
fold increase in α_1-receptor affinity compared to the "deaza" analogue
again reflects the crucial role played by the protonated quinazoline
nucleus.[61] Equally important, (XIII) is only poorly active in the rat
whereas prazosin is particularly effective in this model and in man!

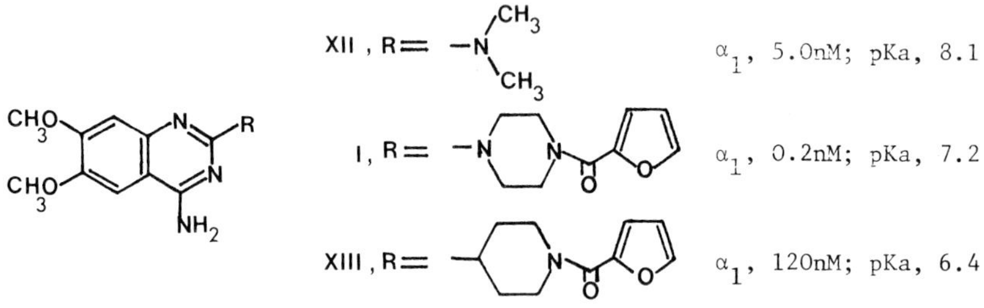

FIG. 11. Importance of nuclear and 2-substituent interactions for
quinazoline derivatives at α_1-receptors.

5. Summary

The α_1-receptor model presented in this paper focusses on the
importance of charge-reinforced hydrogen bonding between protonated agonist
and antagonist structures and a negatively charged carboxylate counterion.
Both noradrenaline and the quinazoline derivatives interact with the
receptor ground state in a similar fashion and for the antagonist, this is
a minimum energy arrangement. For the agonist, however, reorganisation of
the initial complex allows optimisation of charge-reinforced hydrogen
bonding and promotes the conformational change in the receptor protein
associated with agonist-induced activation.

Most of the 2,4-diamino-6,7-dimethoxyquinazoline derivatives described
demonstrate high potency and selectivity for α_1-adrenoceptors. Although

this profile is evident in the parent system, modification at the 2-position can have profound effects on both in vitro α_1-adrenoceptor affinity and in vivo antihypertensive activity. Metabolic stability and pharmacokinetic behaviour can also be influenced as, for example, doxazosin has a markedly longer plasma half-life than prazosin in man.

6. Acknowledgements

The author gratefully thanks his colleagues in the Biology (Drs. V.A. Alabaster, D.A. Cambridge, G.M.R. Samuels) and Chemistry Departments (Dr. J.D. Hardstone, Mr. M.J. Palmer, Miss R.M. Plews) without whose assistance this work could not have been completed. All of the theoretical calculations were performed by Dr. M.S. Tute who also provided invaluable help during informal discussions. Finally, the initial programme which led to the discovery of prazosin was carried out in our Groton laboratories and was directed by Dr. H-J. Hess.

References

1. N.M. Kaplan, Arch. Intern. Med., 143, 255 (1983).

2. R.M. Graham and W.B. Campbell, Federation Proc., 40, 2291 (1981).

3. D.S. Goldstein, Hypertension, 5, 86 (1983).

4. O. Bertel, F.R. Bühler, W. Kiowski and B.E. Lütold, Hypertension, 2, 130 (1980).

5. For example, see, O. Schier and A. Marxer, Progress in Drug Research, 13, 101 (1969).

6. K. Starke, H. Montel, W. Gayk and R. Merker, Naunyn-Schmiedeberg's Arch. Pharmacol., 285, 133 (1974).

7. S. Berthelsen and W.A. Pettinger, Life Sci., 21, 595 (1977).

8. α_2-receptors are now known to be more widely distributed; for a recent review see, P.B.M.W.M. Timmermans and P.A. van Zwieten, J. Med. Chem., 25, 1389 (1982).

9. S.Z. Langer, Pharmacol. Rev., 32, 337 (1981).

10. For a recent review see, K. Starke and J.R. Docherty, J. Cardiovas. Pharmacol., 2, S269 (1980).

11. I. Cavero and A.G. Roach, Life Sci., 27, 1525 (1980).

12. W. Singleton, C.A.P.D. Saxton, J. Hernandez and B.N.C. Prichard, J. Cardiovas. Pharmacol., 4, S145 (1982).

13. R.M. Graham and W.A. Pettinger, N. Eng. J. Med., 300, 232 (1979).

14. P.W. de Leeuw, J.J. Ligthart, A.J.P.M. Smout and W.A. Birkenhäger, Eur. J. Clin. Pharmacol., 23, 397 (1982).

15. J.N. Fain and J.A. García-Sáinz, Life Sci., 26, 1183 (1980); but see also P.B.M.W.M. Timmermans and P.A. van Zwieten, J. Auton. Pharmacol., 1, 171 (1981).

16. H. Glossmann, R. Hornung and P. Presek, J. Cardiovas. Pharmacol., 2, S303 (1980).

17. P. Greengrass and R. Bremner, Eur. J. Pharmacol., 55, 323 (1979).

18. D.A. Greenberg, D.C. U'Prichard and S.H. Snyder, Life Sci., 19, 69 (1976).

19. P. Pratesi and E. Grana, Adv. Drug. Res., 2, 127 (1965).

20. P.N. Patil, D.D. Miller and U. Trendelenburg, Pharmacol. Rev., 26, 323 (1974).

21. F. Mack and H. Bönisch, Naunyn-Schmiedeberg's Arch. Pharmacol., 310, 1 (1979).

22. C.R. Ganellin, J. Med. Chem., 20, 579 (1977).

23. M.S. Tute, unpublished INDO calculations.

24. G.M. Crippen, J. Med. Chem., 22, 988 (1979).

25. D.V. Griffiths and S.P. Swetnam, J.C.S. Chem. Comm., 1224 (1981).

26. D.J. Brown and T. Teitei, J.C.S., 755 (1965).

27. D.H. Aue, H.M. Webb and M.T. Bowers, J. Am. Chem. Soc., 98, 311 (1976).

28. L.J. Saethre, T.A. Carlson, J.J. Kaufman and W.S. Koski, Mol. Pharmacol., 11, 492 (1975).

29. B. Pullman, P. Courrière and H. Berthod, ibid., 11, 268 (1975).

30. X-Ray data kindly provided by Dr. D.J. Williams, Imperial College, London.

31. R.M. Graham, H-J. Hess and C.J. Homcy, J. Biol. Chem., 257, 15174 (1982).

32. N. Sakabe, K. Sakabe and K. Sasaki in Structural Studies on Molecules of Biological Interest, Eds., G. Dodson, J.P. Glusker and D. Sayre, Clarendon Press, Oxford (1981), Chapter 41.

33. P.A. Kollman, J. Am. Chem. Soc., 94, 1837 (1972).

34. A. Johansson, P. Kollman, S. Rothenberg and J. McKelvey, ibid, 96, 3794 (1974).

35. Z. Berkovitch-Yellin and L. Leiserowitz, ibid, 102, 7677 (1980).

36. T. Dipaolo, L.H. Hall and L.B. Kier, J. Theoret. Biol., 71, 295 (1978).

37. See also, C.W. Thornber, Chem. Soc. Reviews, 8, 563 (1979).

38. Use of an uncharged ligand in these calculations is problematical, as the cationic charge is not adequately balanced. Thus, the hydroxyl

hydrogen is repelled by the ammonium head and cannot approach the carbonyl ligand. An additional counterion must therefore be introduced, but as its arbitary location would markedly influence any charge-reinforced hydrogen bonding, this carbonyl model could not be pursued.

39. Noradrenaline hydrochloride, D. Carlström and R. Bergin, Acta. Cryst., 23, 313 (1967).

40. Ephedrine monohydrogenphosphate, R.A. Hearn, G.R. Freeman and C.E. Bugg, J. Am. Chem. Soc., 95, 7150 (1973).

41. L.F. Kuyper, B. Roth, D.P. Baccanari, R. Ferone, C.R. Beddell, J.N. Champness, D.K. Stammers, J.G. Dann, F.E.A. Norrington, D.J. Baker and P.J. Goodford, J. Med. Chem., 25, 1120 (1982).

42. J. Zaagsma, ibid., 22, 441 (1979).

43. Earlier workers have proposed that the α-receptor contains a carboxylate or phosphate anion. For a review see, B.M. Bloom and I.M. Goldman, Adv. Drug Res., 3, 121 (1966).

44. J.T. Bolin, D.J. Filman, D.A. Matthews, R.C. Hamlin and J. Kraut, J. Biol. Chem., 257, 13650 (1982).

45. Calculations for a 2-piperidino substituent should be relevant for a wide range of quinazoline derivatives (see Table 1).

46. G.A. Weiland, K.P. Minneman and P.B. Molinoff, Mol. Pharmacol., 18, 341 (1980).

47. W. Kauzmann, Adv. Prot. Chem., 14, 1 (1959).

48. H. Glossmann and R. Hornung, Naunyn-Schmiedeberg's Arch. Pharmacol., 314, 101 (1980).

49. For example, prazosin has minimal affinity for α_2, 5HT, β_1, β_2, dopamine and cholinergic receptors.

50. For a review of drug-receptor interactions see, P.A. Kollman in Burger's Medicinal Chemistry, 4th Ed., Part 1, M.E. Wolff, Ed., Wiley, New York, 1980, p. 313.

51. T.J. Franklin, Biochem. Pharmacol., 29, 853 (1980).

52. A.S.V. Burgen, Federation Proc., 40, 2723 (1981).

53. R.R. Ruffolo, E.L. Yaden and J.E. Waddell, J. Pharm. Exp. Ther., 222, 645 (1982).

54. A. Carpy, J.M. Leger, G. Leclerc, N. Decker, B. Rouot and C.G. Wermuth, Mol. Pharmacol., 21, 400 (1982).

55. S.F. Campbell, unpublished observations.

56. The important role of aromatic substituents in influencing α_1- vs α_2-selectivity will not be discussed in this paper.

57. M. Aggerbeck, G. Guellaën and J. Hanoune, Br. J. Pharmacol., 65, 155 (1979).

58. G.L. Olson, H-C. Cheung, K.D. Morgan, J.F. Blount, L. Todaro, L. Berger, A.B. Davidson and E. Boff, J. Med. Chem., 24, 1026 (1981).

59. H.L. Elliott, P.A. Meredith, D.J. Sumner, K. McLean and J.L. Reid, Br. J. Clin. Pharmacol., 13, 699 (1982).

60. M.I. Page, Angew. Chem. Int. Ed. Engl., 16, 449 (1977).

61. L. Cocco, C. Temple, J.A. Montgomery, R.E. London and R.L. Blakley, Biochem. Biophys. Res. Comm., 100, 413 (1981).

20 Structure–activity relationships in antitumor anthracyclines

Federico Arcamone

Doxorubicin (Adriamycin), a clinically useful anticancer antibiotic (1), was discovered in 1967 in cultures of strains derived from *Streptomyces peucetius*. Also on the basis of previous work done on the related compound daunorubicin, structure Ia was established for the new antibiotic, soon made available for clinical trials when a semisynthetic procedure of Ia from the more readily available daunorubicin Ib was developed. The efficacy of doxorubicin in the medical treatment of different human·cancers is however accompanied by undesirable side effects including a cumulative dose limiting cardiotoxicity. A considerable effort has been made in our laboratory as well as in those of others, aimed at the synthesis of related compounds with the hope of finding structural analogues with reduced toxicity and/or wider spectrum of activity. The establishment of structure-activity relationships is an essential part of this programme.

Examination of the spatial distribution of atomic groups in N-bromoacetyldaunorubicin (Figure 1) allows one to separate the anthracycline molecule into different parts. First, the C-9 side chain, equatorially oriented both in the crystals and in solution, typically constituted by a two carbon atom residue in most biosynthetic anthracyclines. Second, the appendage represented by the aminosugar moiety, whose orientation is almost perpendicular to that of the planar chromophore. This moiety is an essential part of the anthracycline antibiotics, and in doxorubicin and related biosynthetic glycosides is invariably represented by the unique aminosugar, daunosamine. The third part of the molecule is the tetracyclic aglycone moiety,

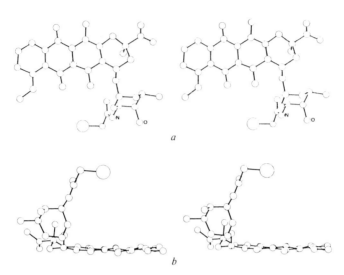

FIGURE 1. Stereo views of N-bromoacetyldaunorubicin: (a) approximately perpendicular to the chromophore plane; (b) parallel to the least squares plane of the chromophore. Reproduced from Nature (London), New Biol. 234, 78 (1971) with the
permission of the Publisher.

represented by the hydroxylated anthraquinone chromophore. A
typical feature of this chromophoric system is the chelation
of quinone oxygens with *peri* hydroxyl groups. Finally, alicyclic
ring A, whose conformation determines the spatial arrangement
of the anthracycline molecules, also bearing the two chiral
centers of the aglycone moiety (Figure 2).

   In this presentation I shall deal with the effects of structural variations of the different parts of the anthracycline
molecule in order to draw some conclusions concerning the molecular requirements for bioactivity. To this end, a brief
summary of available evidence concerning the mechanism of action of antitumor anthracyclines seems appropriate.

   Figure 3 shows a simplified picture of the currently accepted
views concerning the cytotoxicity of antitumor anthracyclines. The factors involved in the action of the drugs at cell level are (a) the uptake of the compound into the cell, (b) binding of the drug to cell constituents, including enzymes involved in the metabolic transformations of the same and (c)

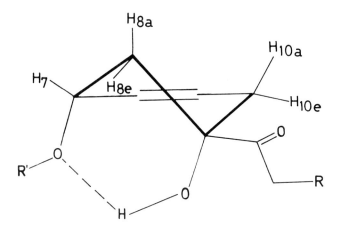

FIGURE 2. Conformation of ring A in daunorubicin (R = H) and doxorubicin (R = OH). R' is the daunosaminyl residue.

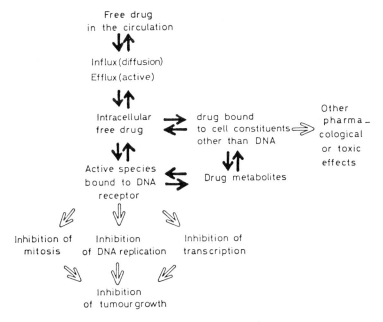

FIGURE 3. Mode of action of antitumour anthracyclines at the cell level.

the affinity for nuclear DNA, apparently the major site of
action of the antitumor anthracyclines. Binding to DNA results in the inhibition of DNA functions such as the template activity in DNA dependent DNA and RNA synthesis and the
mitotic process of cell duplication (1).

A large body of evidence indicates that the main type of anthracycline-DNA interaction is represented by the formation
of an intercalation complex (2). Values of the formation constant, K_{app}, for the doxorubicin-DNA complex reported in the
literature show variations according to different techniques,
experimental conditions and evaluation procedures (2,3). In
more recent studies different values of K_{app} were also found.

Results of equilibrium dialysis experiments with [^{14}C]doxo-
rubicin, performed recently in our   laboratory, were fitted
according to a two classes of binding site model and $K_{app} = 10^7$
was found for the strongest binding site of calf-thymus DNA (3).
On the basis of spectrophotometric measurements, Stutter et al.
(4) found $K_{app} = 2.32 \cdot 10^6$ and $K_{app} = 1.73 \cdot 10^6$ when the size of the
binding site was restricted to three base pairs or the coopera-
tivity constant was kept $\sigma = 1$, respectively. According to the results of the latter authors doxorubicin was able to bind DNA
twice as strongly as daunoribicin. The binding constants
of different anthracyclines to native calf-thymus DNA as determined by the fluorescence quenching method and Scatchard
linear evaluation were the following: doxorubicin, $1.18 \cdot 10^7$;
carminomycin, $5.3 \cdot 10^6$; pyrromycin, $5.1 \cdot 10^6$; musettamycin 7.4 $\cdot$
$\cdot 10^6$; marcellomycin, $1.23 \cdot 10^7$; aclacinomycin $3.4 \cdot 10^6$ (5). Using
different synthetic polydeoxynucleotides as templates, the in-
hibitory effect of daunorubicin on E. coli polymerase I correlated
much better with the relative affinity for the templates than
with the dissociation rate constants of the corresponding drug-
DNA complexes as determined by stopped-flow kinetics. According
to some authors (6), the mode of action of daunorubicin at the
DNA level is presumably related to the consequences of a dis-
ruption of the DNA secondary structure in regions containing
the intercalated drug. Conformational changes of the DNA super-
helix were deduced from the staining patterns after agarose gel
electrophoresis of anthracycline-DNA complexes (7). Variations
in the DNA binding strength of different anthracyclines was

however ascribed to different values of the dissociation rate constants of the complex by Bauer et al. (8).

A good correlation was found between the affinity of 36
daunorubicin and doxorubicin analogues for calf-thymus native DNA or the ability of the same to increase the viscosity
of DNA and cytotoxicity on *in vitro* cultured HeLa cells or
toxicity in the *in vivo* tests (9,10). On the other hand the
antitumor efficacy at optimal non-toxic doses, indicative of
pharmacological selective properties in the animal systems,
appeared to be related with other determinant factors. These
were represented by structural features, such as the hydroxy-
methylketone side chain of doxorubicin and its analogues, or
by other properties such as tissue distribution and metabolism
in vivo (9,10).

Other molecular interactions may contribute to the pharmacological and   toxic effects of the anthracyclines. These
compounds are able to bind to proteins such as non histone
protein fractions from rat liver chromatin (11), spectrin, a
protein present in the membrane of erythrocyces (12), tubulin (13),
G-actin and heavy meromyosine (14), proteins of human erythrocyte ghosts (15) and of different calf tissues (16), heart iso-
citrate dehydrogenase (17), cardiac contractile proteins (18).
Phospholipids bind doxorubicin or its analogues (19-24), the
main interaction being that exhibited with cardiolipin with an
association constant of the order of $10^5 M^{-1}$ (23) for the daunorubicin complex and of $1.8 \cdot 10^6 M^{-1}$ for the doxorubicin complex
(25). A considerable affinity of doxorubicin for the mucopoly-
saccharides has been found in our laboratory (26). Different
investigators have been concerned with the formation of semiquinone radical species upon one-electron reduction of the anthracyclines and their role in the biological effects exhibited by these compounds (27-31). There are two possible modes
in which the semiquinone radicals could interact with biological
systems. Firstly the semiquinone appears to be able to transfer
the unpaired electron to oxygen giving the superoxide anion $O_2^{-\cdot}$
that would be the source of highly reactive hydroxyl radicals
at different cellular sites including the nuclei themselves
(32). Secondly the radicals might be considered as reactive
agents which might give rise to covalent linkages with intra-

cellular macromolecules. The different aspects of this however still unproved hypothetical mode of action of antitumor anthracyclines have been discussed in a recent symposium
(33,34).

The C-9 Side Chain

The C-13 carbonyl group is, within the large family of biosynthetic anthracycline glycosides, a unique feature of doxorubicin (Ia) and its congeners among which daunorubicin (Ib)
and carminomycin (Ic) are the most important. The corresponding C-13 secondary alcohols are also known as biosynthetic
products (IIb, IIc) and as metabolites (IIa, IIb) (1).

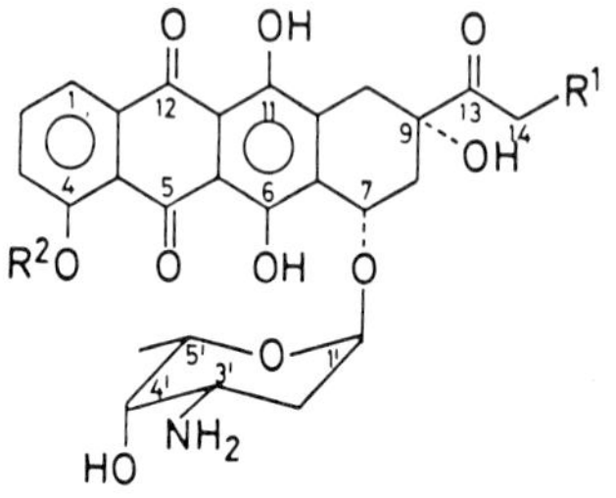

Ia : R^1 = OH , R^2 = Me
Ib : R^1 = H , R^2 = Me
Ic : R^1 = R^2 = H

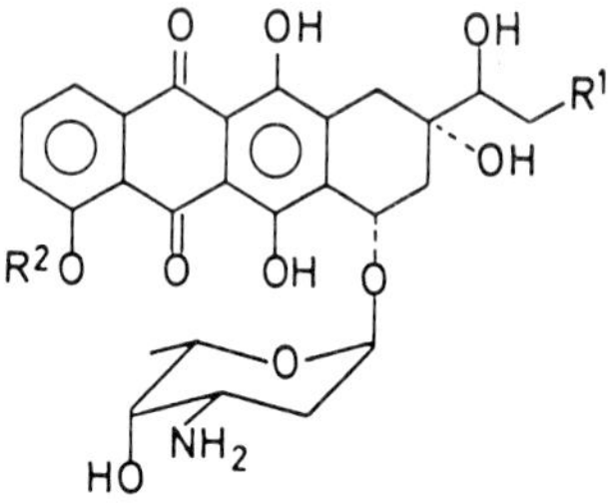

IIa : R^1 = OH , R^2 = Me
IIb : R^1 = H , R^2 = Me
IIc : R^1 = R^2 = H

Other side chain derivatives of the daunorubicin-doxorubicin
group are the 13-deoxo analogues IIIa, IIIb, the derivatives
of the ketone function such as rubidazone (daunorubicin ben-
zoylhydrazone)IIIc, the 14-esters IIId, thioesters IIIe, and
ethers IIIf of doxorubicin, the 14-aminoderivatives IIIg, the
14-C-alkyl derivatives such as IIIh and compounds IIIi-p (1).

    The 13-dihydro derivatives show biological activity similar
to that of the corresponding 13-ketones when tested against murine
leukemias and substantial activity was also retained by the
13-deoxo analogues IIIa and IIIb (1). The antitumor activity of
IIIc was noteworthy, and the compound entered clinical use for
acute leukemias. However the compound was shown to be converted
to the parent drug *in vivo* (35), as also was found to be the case with

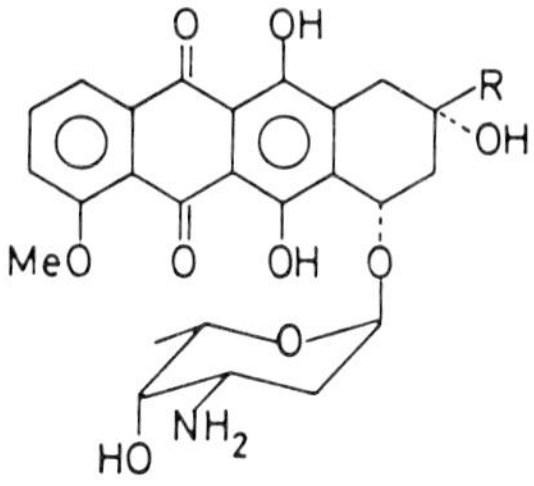

a : R = C_2H_5

b : R = CH_2CH_2OH

c : R = $C(CH_3)$ = $NNHCOPh$

d : R = $COCH_2O$ acyl

e : R = $COCH_2S$ acyl

f : R = $COCH_2O$ alkyl

g : R = $COCH_2N\underset{R^1}{\overset{R^1}{<}}$

h : R = $COCH_2CH_3$

i : R = CO_2H

l : R = CO_2Me

m : R = CH_2OH

n : R = CHO

o : R = $CH-CH$ (with O bridge)

p : R = H

III

doxorubicin esters of general structure IIId (36). Thioesters IIIe, ethers IIIf and amino derivatives IIIg were found to be less effective than doxorubicin in P388 mouse leukemia (1). Similarly all compounds of structures IIIh-IIIp were less potent than Ia or Ib, and, with the exception of IIIm and IIIp, exhibited considerably lower activity, if any, against experimental tumors. A clear relationship was found within this group between antitumor activity and DNA binding ability, when the above mentioned metabolic transformations was taken into account (37).

The Sugar Moiety

Daunosamine, 3-amino-2,3,6-trideoxy-L-_lyxo_-hexopyranose, is the unique sugar constituent of doxorubicin and related biosynthetic anthracycline glycosides. The synthesis of glycosides containing stereochemical or structural variants of this aminosugar has been carried out during the last decade in our, as well as in other laboratories. Stereoisomers of daunosamine were already known constituents of other antibiotics. Acosamine, the L-_arabino_ isomer, is a constituent of actinoidin and ristosamine, the L-_ribo_ isomer, of ristocetin (1). In comparing the antitumor activity of configurational analogues (structures IV to VIII) with Ia and Ib the absolute configuration of the sugar moiety appears to be critical for exhibition of optimal activity (Tables 1 and 2). Whereas the L-_arabino_ ana-

logues behaved similarly to the parent compounds, the antitumor activity in murine experimental leukemias of analogues
possessing an axial amino group or the β-glycosidic structure
was considerably reduced in terms of potency and/or efficacy.

TABLE 1

*Antitumor activity of daunorubicin configurational analogues against
mouse ip inoculated experimental leukemias (38,39)*

Compound	Configuration	Tumor System	O.D.[a]	T/C%[b]
Ib	α-L-*lyxo*	L 1210[c]	4	150
		P 388[d]	2	127
IVa	α-L-*arabino*	L 1210	4	143
Va	α-L-*ribo*	L 1210	50	127
VI	α-L-*xylo*	L 1210	33	120
VII	α-D-*ribo*	P 388	50	125
VIIIa	α-D-*arabino*	P 388	100	125
VIIIb	β-D-*arabino*	P 388	50	no act.

(a) Optimal ($\leq$ LD$_{10}$) dose in mg/Kg body weight. (b) Survival
time of treated animals expressed as per cent of untreated,
tumor bearing mice. (c) Treatment ip on day 1. (d) Treatment
ip on days 5,9,13.

TABLE 2

*Antitumor activity of doxorubicin configurational analogues against
mouse experimental leukemias (38)*

Compound	Configuration	Tumor System	O.D.[a]	T/C%[b]
Ia	α-L-*lyxo*	L 1210[c]	5	166
IVb	α-L-*arabino*	L 1210	5	150
Vb	α-L-*ribo*	L 1210	60	161

(a), (b), (c) see footnotes in Table 1.

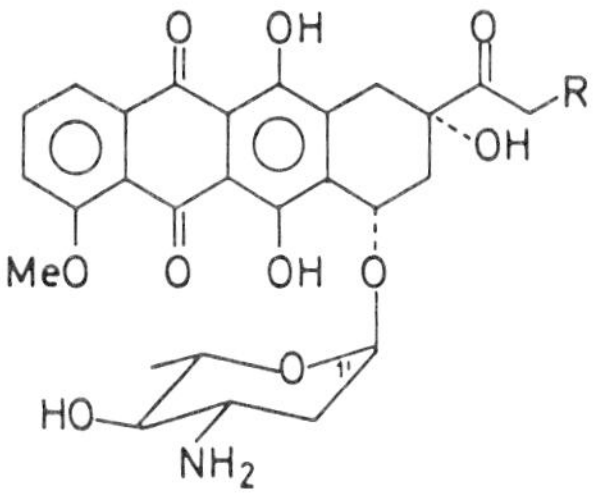

IV a : R = H
IV b : R = OH

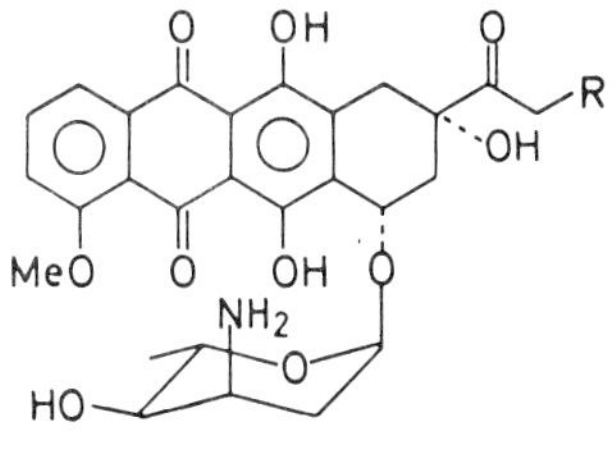

V a : R = H
V b : R = OH

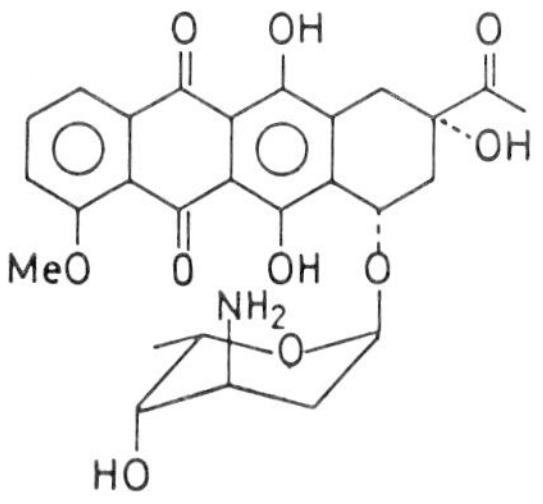

VI

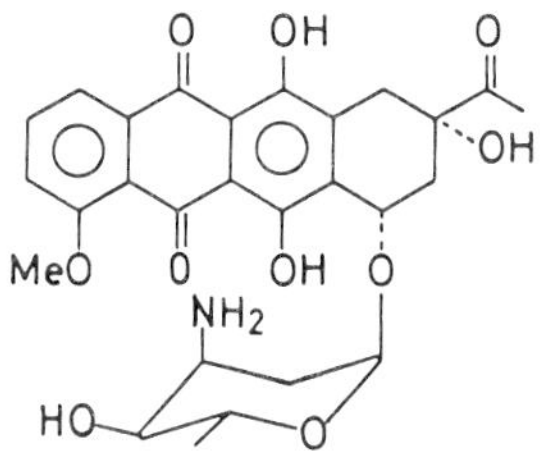

VII

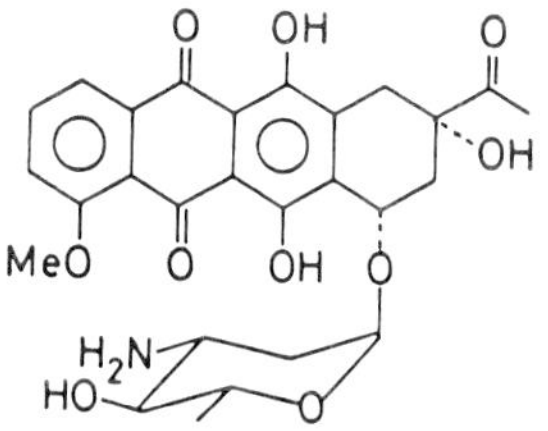

VIII a
VIII b (1¹ - epi)

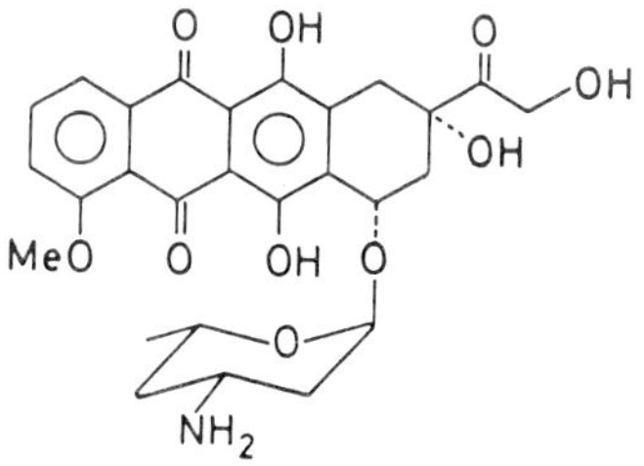

IX

The biological activity of configurational analogues was in good agreement with the DNA binding data (37). The L-*arabino* analogue 4'-epidoxorubin compared well with doxorubicin against different experimental murine tumors such as mammary carcinoma, Moloney sarcoma virus induced sarcoma, metastatic Lewis lung and MS-2 sarcomas because of a generally more favourable therapeutic index (1). The compound was also found to be less cardiotoxic than doxorubicin (40) and to behave differently from the parent drug as regards metabolism and pharmacokinetics in laboratory animals (unpublished work from this laboratory). Extended clinical trials have confirmed the preclinical evaluation of 4'-epidoxorubicin which appears to be a promising new drug for the treatment of human cancer (41).

In addition to 4'-epidoxorubicin, other analogues modified at C-4' are of interest. First, 4'-deoxydaunorubicin (IX), a compound showing, *inter alia*, outstanding inhibitory properties against human colorectal tumors transplanted in nude mice (42) and in the murine colon 38 system (43), together with an absence of cardiotoxic effects at tolerated doses in experimental animals. The more favourable pharmacologica properties in the mouse systems may be related to a different pharmacokinetic behaviour (44). The compound is now undergoing phase 1 clinical trials.

The 4'-O-methyl derivative X showed higher efficacy than doxorubicin in murine L 1210 leukemia but it was not so in other test systems including a doxorubicin resistant P 388 tumor line (1). The presence of a C-methyl group at position 4' was also compatible with a high level of antitumor activity (Table 3). The activity of the compounds against P 388 leukemia in mice was correlated with *in vitro* inhibition of cultured HeLa cells and with DNA binding (46).

Modifications at C-6' however, produced different results as exemplified by the behaviour of 6'-hydroxylated analogues whose affinity for DNA and antitumor activity were lower than those of the corresponding 6-deoxy parent compounds (1). On the other hand the markedly low potency (although accompanied by a comparable antitumor efficacy in currently used mouse leukemia test systems) of the N-acylated analogues of daunorubicin and doxorubicin was more likely explained by a metabolic conver-

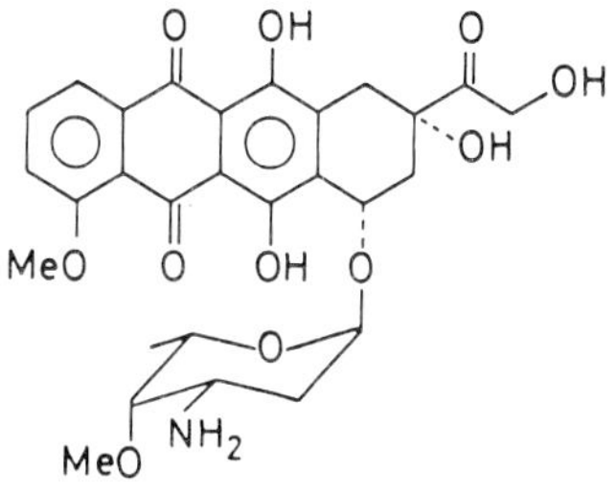

X

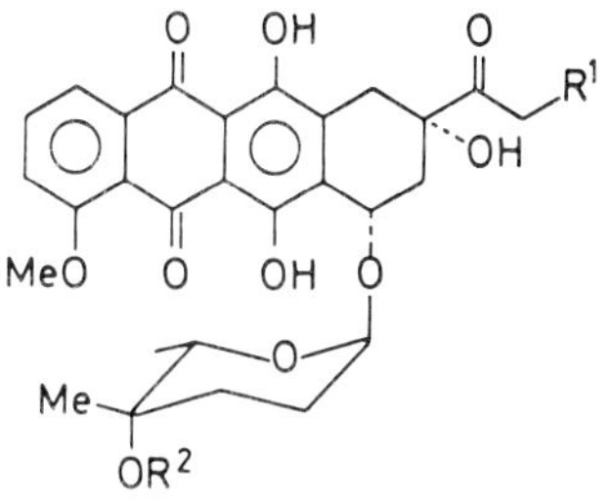

XI a : $R^1 = R^2 = H$
XI b : $R^1 = H ; R^2 = Me$
XI c : $R^1 = OH ; R^2 = H$
XI d : $R^1 = OH ; R^2 = Me$

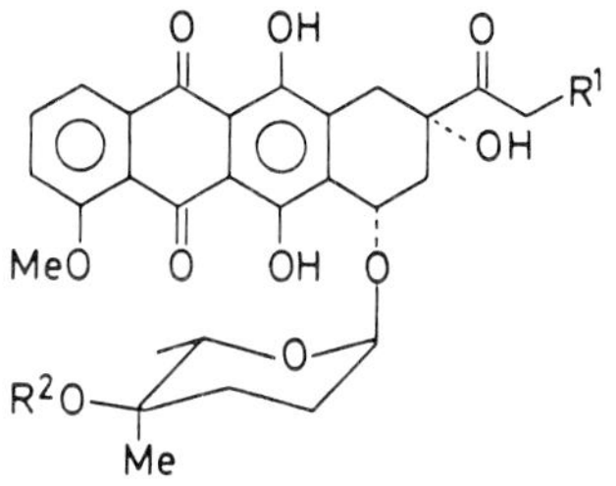

XII a : $R^1 = R^2 = H$
XII b : $R^1 = H ; R^2 = Me$
XII c : $R^1 = OH ; R^2 = Me$
XII d : $R^1 = OH ; R^2 = Me$

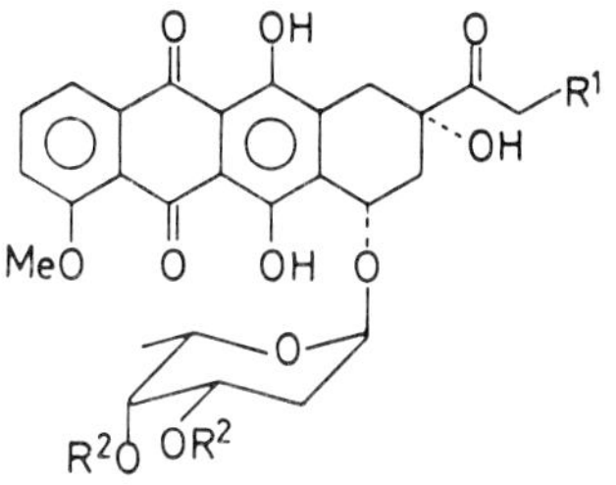

XIII a : $R^1 = R^2 = H$
XIII b : $R^1 = H ; R^2 = Ac$
XIII c : $R^1 = OH ; R^2 = Ac$

sion to the parent free aminoglycosides than by a different mode of action (1,37).

TABLE 3

Activity of 4'-C-methyl derivatives of daunorubicin and doxorubicin against P 388 murine leukemia. Treatment ip on day 1 after tumor inoculation (45)

Compound	Structure	O.D.[a]	T/C%[b]
Daunorubicin	Ib	2.9	175
-, 4'-C-methyl	XIa	20.0	155
-, 4'-epi-4'-C-methyl	XIIa	0.44	163
Daunorubicin	Ib	2.9	163
-, 4'-C-methyl-4'-O-methyl	XIb	33.7	160
-, 4'-epi-4'-C-methyl-4'-O-methyl	XIIb	22.5	150
Doxorubicin	Ia	6.6	193
-, 4'-C-methyl	XIc	7.7	172
-, 4'-epi-4'-C-methyl	XIIc	2.0	156
Doxorubicin	Ia	6.6	201
-,4'-C-methyl-4'-O-methyl	XId	6.6	172
-,4'-epi-4'-C-methyl-4'-O-methyl	XIId	10.0	205

(a), (b) see footnotes in Table 1

The results reported on 3'-deamino-3'-hydroxyderivatives of Ia and Ib are of importance in the establishment of molecular requirements for biological activity. The 3'-deamino-3'-hydroxyderivative of daunorubicin XIIIa, the corresponding 3',4'-diacetate XIIIb (47) and 3'-deamino-3'-acetoxy-4'-O-acetyldoxorubicin XIIIc (48) displayed significant antitumor activity, comparable to that of the parent aminoglycosides, albeit at higher dosages. This finding is in agreement with the general observation of the higher binding affinity shown by the compounds possessing a 3'-amino group.

Consideration of all known modifications of the sugar moiety in Ia and Ib is beyond the scope of this presentation. However, the furanose analogues XIV and XVa-d should be mentioned, because

also in this case significant bioactivity, although at relatively high dosages, was recorded (45.49).

XIV

XVa : R^1 = H ; R^2 = CHO
XVb : R^1 = OH ; R^2 = CHO
XVc : R^1 = H ; R^2 = CH$_2$OH
XVd : R^1 = H ; R^2 = CH$_2$NH$_2$

Modification of the Tetracyclic Moiety

Biosynthetic anthracyclines exhibit different hydroxylation
patterns at positions *peri* to the quinone groups but relationships between the said structural feature and bioactivity are
presently not known. As far as ring D is concerned, the high
biological activity of 4-demethoxy analogues of daunorubicin
and doxorubicin XVIa and XVIb and of derivatives thereof is an
established notion. In fact, 4-demethoxydaunorubicin is presently undergoing enlarged phase 2 clinical trials because of
its lower cardiotoxicity when compared with the parent drug,
higher potency and activity when administered by the oral route.
The reasons underlying the exhibition of higher potency by the
said analogues are still uncertain, both a more pronounced effect on DNA double helix stabilization and a higher intracellular uptake in cultured cells having been related with the
higher biological potency (1).

4-Demethoxyderivatives possessing modifications at C-4',
such as XVIc,d and XVIIa-d maintained the high potency shown
by XVIa and XVIb together with outstanding antitumor activity
in the murine screen systems. Within this series the activity
of XVId against iv inoculated L 1210 leukemia and colon 26 tumor was markedly superior than that of Ia (50).

Methylation of daunorubicin Ib either at C-6 or at C-11
resulted in a loss of potency and in a lower affinity for DNA
when compared with the parent (51) and similar behaviour was

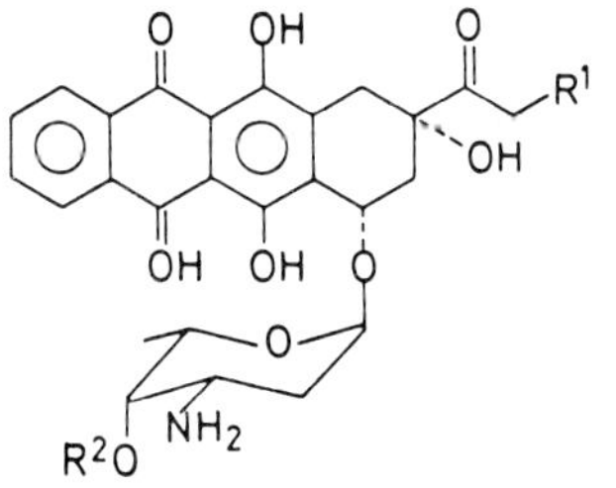

XVI a : $R^1 = R^2 = H$
XVI b : $R^1 = OH$; $R^2 = H$
XVI c : $R^1 = H$; $R^2 = Me$
XVI d : $R^1 = OH$; $R^2 = Me$

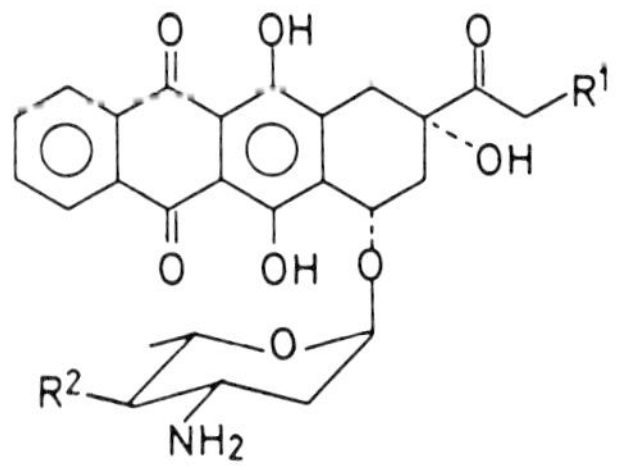

XVII a : $R^1 = H$; $R^2 = OH$
XVII b : $R^1 = R^2 = OH$
XVII c : $R^1 = R^2 = H$
XVII d : $R^1 = OH$; $R^2 = H$

shown by the same derivatives of carminomycin. Significant
bioactivity was however shown by doxorubicin isomer XVIII
(37,45).

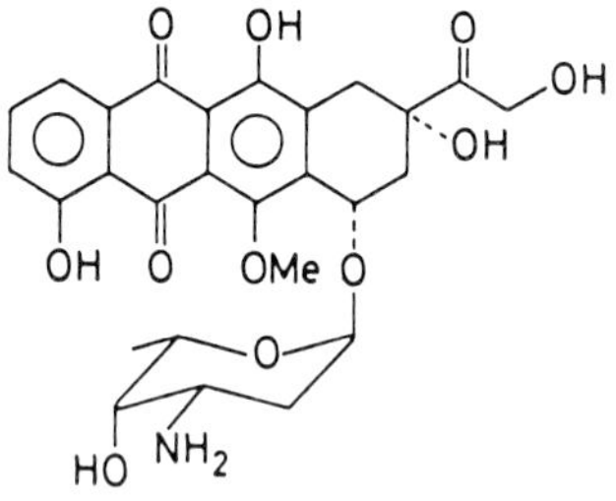

XVIII

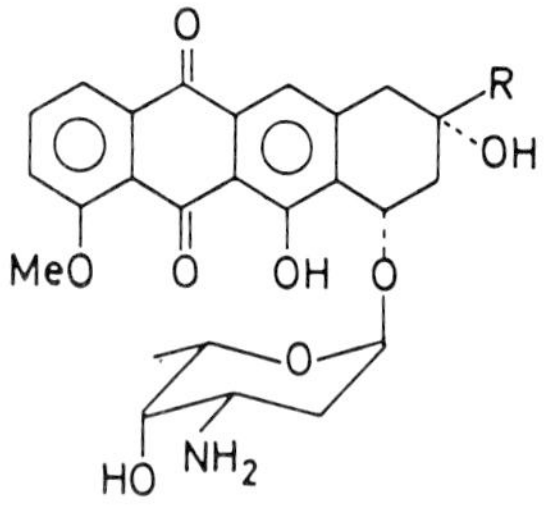

XIX a : $R = COCH_2OH$
XIX b : $R = CHOHCH_3$
XIX c : $R = COCH_3$
XIX d : $R = C_2H_5$

Biosynthetic 11-deoxyderivatives XIXa-d (52) exhibited
significant antitumor activity in mouse experimental leukemias albeit at higher dose levels when compared with 11-hydroxy-
lated congeners (Table 4). Compounds XXa-d, also isolated from
a S. peucetius derived strain (53) showed higher potency, in
agreement with the higher potency of Ic in respect to Ia and
Ib, but not higher antitumor efficacy when compared with
doxorubicin (Table 4). On the other hand compound XXI, obtained by total synthesis (54) appeared to be endowed with biological activity both *in vitro* and *in vivo* very similar to that

TABLE 4

Activity of 11-deoxy derivatives in the P 388 leukemia test (52,53)[a]

Compound	O.D.[b]	T/C%[c]
Ia	10	233
XIXa	66	232
Ib	2.9	183
XIXc	29	161
Ia	10	355
XXa	3.4	190
XXb	2	140
XXc	1.9	159

(a) Treatment ip on day 1. (b), (c) see footnotes in Table 1

exhibited by Ia. The 6-deoxy derivative XXII   recently
obtained by a synthetic procedure in our laboratory (55), showed cytotoxic properties in HeLa cell cultures
at concentrations comparable with those of Ib in the same
in vitro system.

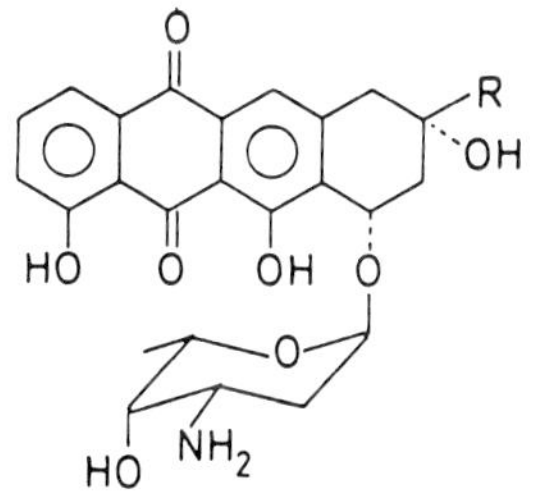

XX a : R = COCH$_2$OH
XX b : R = CHOHCH$_3$
XX c : R = COCH$_3$
XX d : R = C$_2$H$_5$

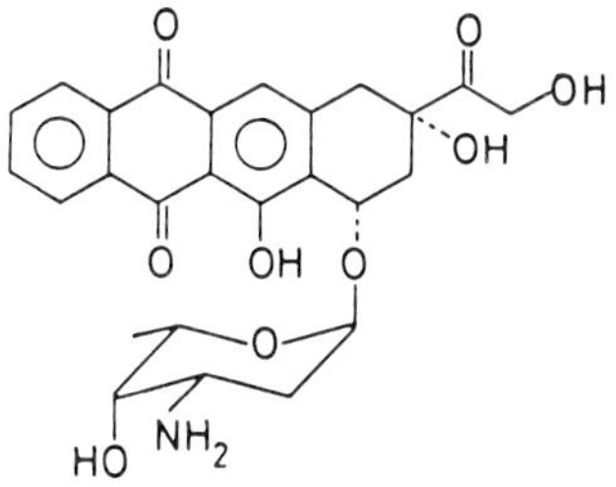

XXI

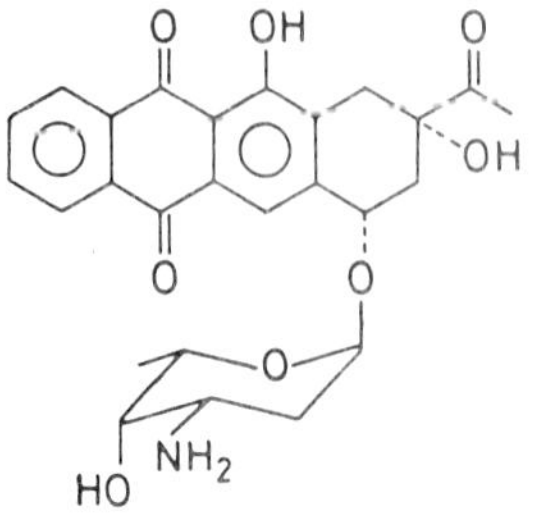

XXII

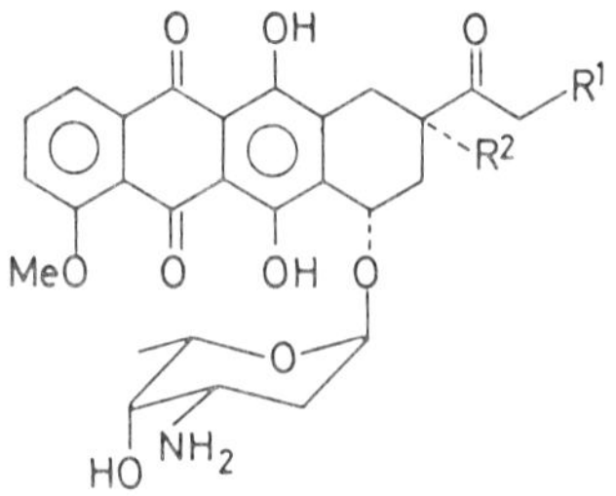

XXIIIa : $R^1 = R^2 = H$
XXIIIb : $R^1 = OH$; $R^2 = H$
XXIIIc : $R^1 = H$; $R^2 = Me$

The importance of the C-9 hydroxyl group is shown by the
lower potency of XXIIIa when compared with Ib and by the lower
potency and antitumor efficacy of XXIIIb when compared with
Ia in the P 388 screen system in agreement with the reduced
affinity for DNA. When the C-9 hydroxyl was substituted with
a methyl group as in XXIIIc, no activity was demonstrated in
the same system (1). On the other hand 10(R)-methoxydaunoru-
bicin (XXIVa), possessing the same conformation in solution
as the parent daunorubicin, retained significant antitumor
activity, whereas the 10(S) epimer XXIVb, having a different
conformation, was inactive at the maximum doses tested   (56). The
conformation of XXIVa and XXIVb was determined to be as in
Figure 4 on the basis of PMR measurements (56). In compound

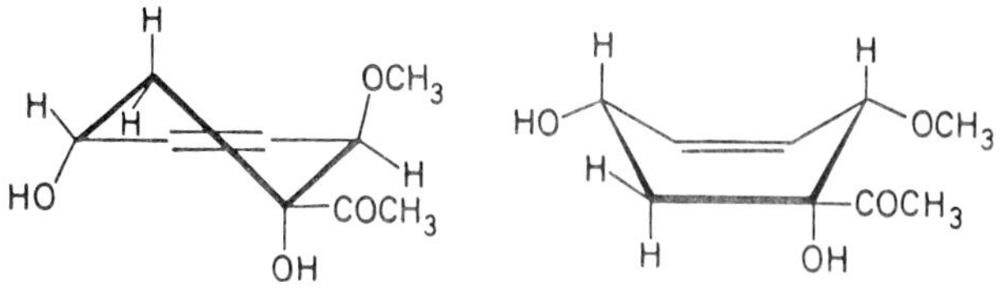

XXIV a   XXIV b

FIGURE 4. Ring A conformation of XXIVa and of XXIVb as
determined by PMR.

XXIVa the long range coupling typical of daunomycinone derivatives $J_{H-8eq,H-10eq}$ = 1 Hz together with $J_{H-7,H-8}$= 5.2 Hz
indicated the half-chair conformation whereas in XXIVb
$J_{H-7,H-8}$= 6.6 He and the absence of coupling between H-8 and
H-10 suggested a boat-like conformation of the A ring. Similarly, 9,10-anhydrodaunorubicin (XXV) and 9,10-methane derivative XXVI appeared devoid of antitumor activity in the
same system (1).

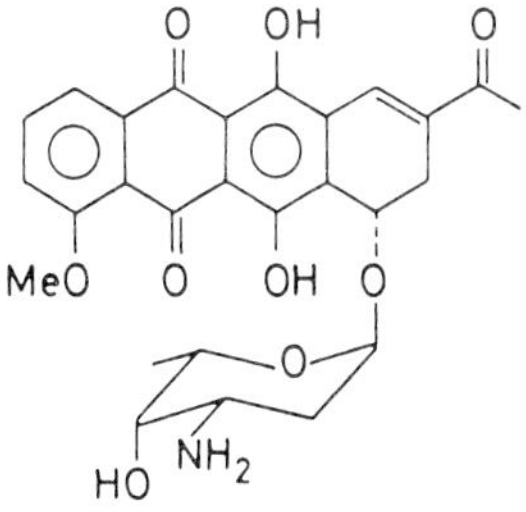

XXV 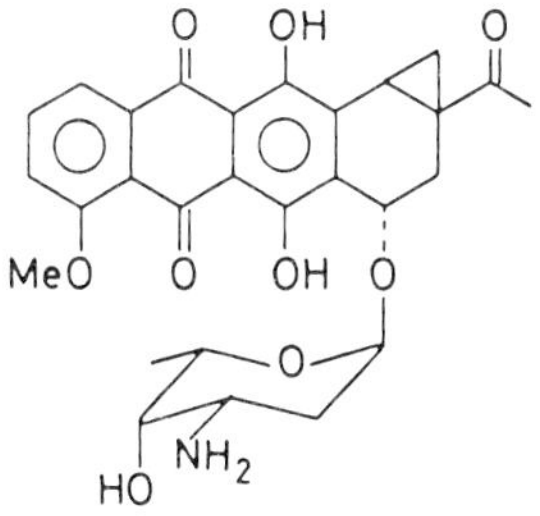   XXVI

As far as the substitution at C-8 (of importance in view
of a potential effect on the metabolic cleavage of the glycosidic bond) is concerned, the two known examples indicate a
substantial loss of bioactivity. The 8-methoxy derivative XXVII
(57) and the 8-C-methyl derivative XXVIII did not show activity superior to that of Ib, the latter being practically
ineffective in the P 388 test (58).

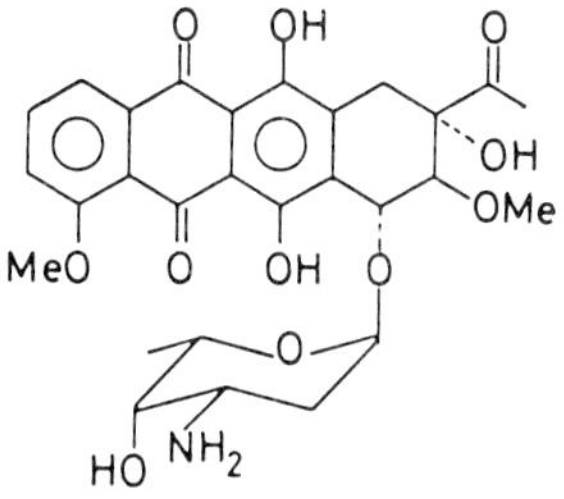

XXVII 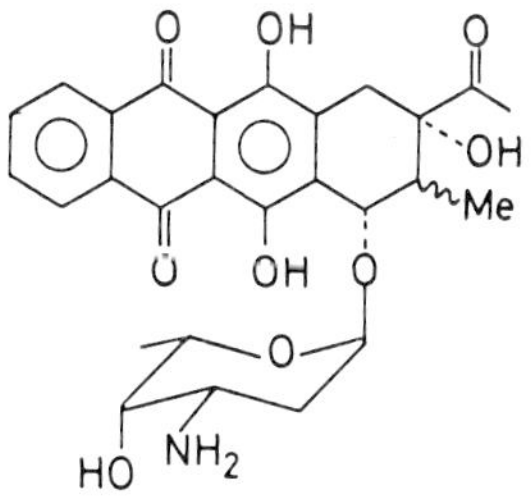   XXVIII

The strict requirement represented by the need for an integral alicyclic ring A is also shown by the complete loss of activity in XXIXa and XXIXb (1).

XXIXa : R = H
XXIXb : R = Me

DISCUSSION

There is no doubt that the pharmacological properties of chemo-therapeutic agents are the consequence of (a) affinity for a given biological receptor and (b) their behaviour as foreign compound (xenobiotics) in the living organism. The former pro-perty, that is intrinsic biological activity or potency, is a function of three-dimensional shape of the drug molecule, of electric charge distribution and of lipophilicity , clearly factors determining the affinity for the main receptor of the drug. The second determinant of pharmacologic behaviour is a function of tissue concentrations of the drug and/or its ac-tive metabolites. These properties, dictating the selectivity of the drug *in vivo*, are related to the distribution of the compound in the polyphasic biological systems and with the interaction of the drug with tissue components. All available experimental evidence supports the current view indicating nuclear DNA as the main biological receptor of antitumor an-thracyclines. The interaction with this macromolecule should therefore be considered as the basic molecular phenomenon underlying the biological activity of doxorubicin and its analogues.

Although not directly involved in the DNA intercalation process according to reported models, the C-9 side chain ap-

pears to be of importance in the stabilization of the drug-
DNA complex as even minor modifications of this moiety
strongly affect either potency or, both potency and antitumor activity. Favourable features of this moiety seem
to be (1) small size and (2) the presence of one or two
oxygen atoms, in the latter case the two oxygen atoms are
preferably linked to different carbon atoms. In active molecules the side chain is $COCH_3$, $COCH_2OH$, $CHOHCH_3$, $CHOHCH_2OH$,
$COCH_2CH_3$, CH_2CH_3, CH_2CH_2OH, CH_2OH, H, whereas groups like
CO_2H, CO_2Me, CHO, $CH{-}CH_2$, $COCH_2CH_2CH_3$, $COCH(CH_3)_2$, $COCH_2OR$,
$COCH_2SR$ (R = alkyl or aryl), $COCH_2NR_2$ were incompatible with
bioactivity. The above requirements   can be understood in
terms of the necessity not to perturb the aqueous polar coat
of the DNA double helix and the possibility of hydrogen bonding of water molecules constituting the same.

Substitution at C-4' in the sugar moiety does not affect
affinity for the DNA receptor and different modifications
appear to be compatible with high biological potency. On the
other hand inversion of configuration at C-3' or the presence
of an hydroxyl group at C-6' reduces the affinity for DNA and
therefore higher dosages are needed for optimal tumor inhibition in the screening systems. The same occurs when the amino
group is replaced   by   an hydroxyl. The role of the sugar
moiety in the intercalation complex is still not established
because the findings of Quigley et al. (59)   have changed
former deductions concerning the structure of anthracycline-
DNA complex (60).

Taking into account the marked effects of the previously mentioned
stereochemical and structural variations of the sugar moiety on affinity for the DNA receptor and consequently on bioactivity
it must be deduced that this moiety plays an important role
in the stabilization of the intercalation complex.

A wide variation in toxicity is shown by the modified anthraquinone moieties. However, details in the substitution at
C-1 to C-4, C-6 and C-11, namely the chelation of the quinone
carbonyl by the *peri* hydroxyl groups, the pattern of hydroxylation or the substitution on ring D do not represent absolute
requirements for bioactivity. Although these   variations
may well be related with a diminished affinity for DNA (as

is the case with the 11-deoxyanalogues) or with enhanced
lipophilicity (as it is the case of the 4-demethoxyanalogues),
no evidence has been provided for hypothetical mechanisms of action
involving metal chelation or, in some way, a quinol structure
of ring B. A markedly different picture arises from the results of ring A modifications. Here we have complete loss of
bioactivity as a consequence of conformational changes, not
to say of changes in chirality at the C-7, C-9 centres (1).
It is clear that the orientation of the sugar moiety at C-7
in the half-chair conformers, resulting in an optimal spatial
arrangement of the molecule, is a strict requirement for cytotoxicity and antitumor activity.

Of great importance are the structural requirements for
selectivity of action, namely optimal antitumor efficacy at
non-toxic doses. When the standard tests in the mouse are
considered, doxorubicin appears to be superior to daunorubicin (61) and all analogues possessing the ketol side chain
typical of doxorubicin appear to be superior to the corresponding methylketone derivatives (10). Different explanations
have been proposed for this, and all are based on a different
behaviour (metabolism, pharmacokinetics, effect on immunolo-
gical responses of the host) at the organism level. Similar
factors should be involved in the more favourable pharmacological properties of analogues presently undergoing clinical
trials as second generation antitumor anthracyclines, namely
4'-epidoxorubicin, 4'-deoxydoxorubicin and 4-demethoxydauno-
rubicin. These compounds, although not differing from the
parent drugs as far as the affinity for DNA is concerned,
show important differences in the metabolism and pharmacokinetic behaviour (unpublished work from   this   laboratory)
that might well be responsible for the more favourable therapeutic index and even for improved antitumor activity.

REFERENCES

1. ARCAMONE, F. (1981) "Doxorubicin" Medicinal Chemistry Series, G.
   Stevens Ed., Vol. 17, Acad. Press, New York.
2. NEIDLE, S. (1978) in "Topics in Antibiotic Chemistry", P. Sammers Ed.,
   part D, pp 240-278, Ellis Horwood Publ., Chichester.
3. ARCAMONE, F., ARLANDINI, E., MENOZZI, M., VALENTINI, L., and VANNINI,
   E. (1982) Anthracycline Antibiotics in Cancer Therapy. Developments in
   Oncology, Vol. 10, F.M. Muggia, C.W. Young and S.K. Carter Eds.

4. STUTTER, E., GOLLMICK, F.A., and SCHÜTZ, H. (1982). Studia Biophys. 88, 131.
5. PACHTER, J.A., HUANG, C., DUVERNAY Jr., V.H., PRESTAYKO, A.W., and CROOKE, S.T. (1982). Biochemistry 21, 1541.
6. GRANT, M. and PHILLIPS, D.R. (1979). Mol. Pharmacol. 16, 357.
7. MONG, S., DUVERNAY, V.H., STRONG, J.E., and CROOKE, S.T. (1980). Mol. Pharmacol. 17, 100.
8. BAUER, E., FÖRSTER, W., GOLLMICK, F.A., SCHÜTZ, H., STUTTER, E., WALTER, A., and BERG, H. (1982). Studia Biophysica 87, 207.
9. ARCAMONE, F. Paper presented at the Symposium on Anthracyclines, Paris, June 24-25, 1981.
10. ARCAMONE, F., CASAZZA, A.M., CASSINELLI, G., DI MARCO, A., and PENCO, S. (1982). see reference 3.
11. KIKUCHI, H. and SATO, S. (1976). Biochim.Biophys.Acta 434, 509.
12. NIKKELSON, R.B. and LIN, P.S. (1977). J. Mol. Med. 2, 33.
13. NA, C. and TIMASHEFF, S.N. (1977). Arch. Biochem. Biophys. 182, 147.
14. SOMEYA, A., AKIYAMA, T., MISUMI, M., and TANAKA, N. (1978). Biochem. Biophys. Res. Comm. 85, 1542.
15. SINHA, B.K. and CHIGNELL, C.F. (1979). Biochem. Biophys. Res. Comm. 86, 1051.
16. LUCACCHINI, A., MARTINI, C., SEGNINI, D., and RONCA, G. (1979). Experientia 35, 1148.
17. YASUMI, M., MINAGA, T., NAKAMURA, K., KIZU, A., and IJICHI, H. (1980). Biochem. Biophys. Res. Comm. 93, 631.
18. LEWIS, W., KLEINERMAN, J., and PUSZKIN, S. (1982). Circul.Res. 50, 547.
19. DUARTE-KARIM, M., RUYSSCHAERT, J.M., and HILDEBRAND, J. (1976). Biochem. Biophys. Res. Comm. 71, 658.
20. ANGHILERI, L.J. (1977). Arzneim.-Forsch. 27, 1177.
21. SCHWARTZ, H.S., SCHIOPPACASSI, G., and KANTER, P.M. (1978). In: Antibiotics and Chemotherapy (Schönfeld, H. et al. Eds.) 23, 247.
22. TRITTON, T.R., MURPHREE, S.A., and SARTORELLI, A.C. (1978). Biochem. Biophys. Res. Comm. 84, 802.
23. SCHWARTZ, H.S. and KANTER, P.M. (1979). Eur. J. Cancer 15, 923.
24. KARCZMAR, G.S. and TRITTON, T.R. (1979). Biochim. Biophys. Acta 557, 306.
25. GOORMAGHTIGH, E., CHATELAIN, P., CASPERS, J., and RUYSSCHAERT, J.M. (1980). Biochim. Biophys. Acta 597, 1.
26. MENOZZI, M. and ARCAMONE, F. (1978). Biochem.Biophys.Res.Comm. 80, 313.
27. SATO, S., IWAIZUMI, M., HANDA, K., and TAMURA, Y. (1977). Gann 68, 603.
28. BACHUR, N.R., GORDON, S.L., and GEE, M.V. (1978). Cancer Res. 38, 1745.
29. MYERS, C.E., McGUIRE, W.P., LISS, R.H., IFRIM, I., GROTZINGER, K., and YOUNG, R.C. (1977). Science 197, 165.
30. GOODMAN, J. and HOCHSTEIN, P. (1977). Biochem. Biophys. Res. Comm. 77, 797.
31. LOWN, J.W., SIM, S.K., MAJUMDAR, K.C., and CHANG, R.Y. (1977). Biochem. Biophys. Res. Comm. 76, 705.
32. BACHUR, N.R., GEE, M.V., and FRIEDMAN, R.D. (1982). Cancer Res. 42, 1078.
33. PIETRONIGRO, D.D. (1982) see reference 3.
34. BACHUR, N.R. (1982) see reference 3.
35. BENJAMIN, R.S., KEATING, M.J., McCREDIE, K.B., LUNA, M.A., LOO, T.L., and FREIREICH, E.J. (1976). Proc. Am. Ass. Cancer Res. 17, 72.
36. ARCAMONE, F., FRANCESCHI, G., MINGHETTI, A., PENCO, S., REDAELLI, S., DI MARCO, A., CASAZZA, A.M., DASDIA, T., DI FRONZO, G., GIULIANI, F., LENAZ, L., NECCO, A., and SORANZO, C. (1974). J. Med. Chem. 17, 335.
37. ARCAMONE, F. (1982) . Structure-Activity Relationships of Antitumor Agents. Developments in Pharmacology, Vol. 3, REINHOUDT,T.A. et al. Ed.

38. ARCAMONE, F. (1980). "The Development of New Antitumor Anthracyclines" Medicinal Chemistry Series, Vol. 16, Acad. Press, New York, p. 1-40.
39. HORTON, D., NICKOL, R.G., WECKERLE, W., and WINTER-MUHALY, E. (1979). Carb. Res. 76, 269.
40. CASAZZA, A.M., DI MARCO, A., BONADONNA, G., BONFANTE, V., BERTAZZOLI, C., BELLINI, O., PRATESI, G., SALA, L., and BALLERINI, L. (1979). in "Anthracyclines" (Proc. Workshop) , 403, Crooke, S.T. and Reich, S.D. Eds., Acad. Press, New York.
41. GANZINA, F. (1982). International Symposium of 4'-epi-adriamycin, Tokyo, Japan, Nov. 12, 1982.
42. GIULIANI, F., COIRIN, A.K., RENE RICE, M., and KAPLAN, N.O. (1981). Cancer Treat. Rep. 65, 1063.
43. CASAZZA, A.M. and SOAVI, G. (1982). in "Adriamycin and analogs in gastrointestinal cancer" (Proc. Workshop), Basel. Feb. 12. 1982.
44. FORMELLI, F., POLLINI, C., CASAZZA, A.M., DI MARCO, A., and MARIANI, A. (1981). Cancer Chemother. Pharmacol. 5, 139.
45. ARCAMONE, F., CASSINELLI, G., and PENCO, S. (1982). in "Anthracycline Antibiotics", El Khadem, H.S., Ed., p. 59-74, Acad. Press, New York.
46. BARGIOTTI, A., CASSINELLI, G., PENCO, S., VIGEVANI, A., and ARCAMONE, F. (1982). Carbohydr.Res. 100, 273.
47. FUCHS, E.F., HORTON, D., and WECKERLE, W. (1977). Carbohydr. Res. 57, C36.
48. HORTON, D., PRIEBE, W., and TURNER, W.R.(1981). Carbohydr. Res. 94, 11.
49. ISRAEL, M., AIREY, J.E., MURRAY, R.J., and GILLARD, J.W. (1982). J. Med. Chem. 25, 28.
50. BARBIERI, B., BELLINI, O., SAVI, G., BERTAZZOLI, C., PENCO, S., and CASAZZA, A.M. (1982). Proc. 5th Int. Symp. on Future Trands in Chemother., Tirrenia (Italy), May 24-26, 1982.
51. ZUNINO, F., CASAZZA, A.M., PRATESI, G., FORMELLI, F., and DI MARCO, A. (1981). Biochem. Pharmac. 30, 1856.
52. CASSINELLI, G., DI MATTEO, F., FORENZA, S., RIPAMONTI, M.C., RIVOLA, G., ARCAMONE, F., DI MARCO, A., CASAZZA, A.M., SORANZO, C., and PRATESI, G. (1980). J. Antib. 33, 1468.
53. CASSINELLI, G., RIVOLA, G., RUGGIERI, D., ARCAMONE, F., GREIN, A., MERLI, S., SPALLA, C., CASAZZA, A.M., DI MARCO, A., and PRATESI, G. (1982). J. Antib. 35, 176.
54. UMEZAWA, H., TAKAHASHI, Y., NAGANAWA, H., TATSUTA, K., and TAKEUCHI, T. (1980). J. Antib. 33, 1581.
55. PENCO, S., ANGELUCCI, F., ARCAMONE, F., BALLABIO, M., BARCHIELLI, G., FRANCESCHI, G., SUARATO, A., and VANOTTI, E. (1983). J. Org. Chem. to be published.
56. PENCO, S., GOZZI, F., VIGEVANI, A., BALLABIO, M., and ARCAMONE, F., (1979). Heterocycles 13, 281.
57. PENCO, S., ANGELUCCI, F., BALLABIO, M., VIGEVANI, A., and ARCAMONE, F. (1980). Tetrahedron Lett. 21, 2253.
58. NAFF, M.B., PLOWMAN, J., and NARAYANAN, V.L. (1982). see ref. 45, p.1.
59. QUICLEY, G.J., WANG, A.H.-J., UGHETTO, G., VAN der MAEL, G., VAN BOOM, J.H., and RICH, A. (1980). Proc.Natl.Acad.Sci. U.S.A. 77, 7204.
60. PIGRAM, W.J., FULLER, W., and HAMILTON, L.D. (1972). Nature New Biol. 235, 17.
61 CASAZZA, A.M. (1982). see reference 3.

21 β-lactam antibiotics and ansamycins

Mario Brufani and Luciana Cellai

This chapter will deal with the contribution made by X-ray crystallography to the research on antibiotics with particular reference to the formulation of structure-activity relationships. This discussion is limited to two classes of antibiotics, namely β-lactam antibiotics and ansamycins. The former are of great scientific, therapeutic and commercial interest, the latter have been for several years one of the main research subjects of the Rome group to which the present authors belong.

In 1964 the Rome group published an X-ray study on a rifamycin which permitted the elucidation of the three-dimensional structure of this kind of compound for the first time. This provided the basis for further work. In the following years many related semisynthetic compounds were synthesized, their structures were studied by spectroscopic methods, in some cases by X-ray diffraction, and their activities were tested. All this work, together with the results published by other groups, above all by the Research Laboratories of Lepetit and Ciba-Geigy, lead to the formulation of a hypothesis on the mode of action of these antibiotics at a molecular level. This brief reference to the work of the Rome group is meant to outline the sequence of stages which generally takes place in research on antibiotics, that is: the discovery of a new compound, the testing of its activity, the structure determination, the chemical modification, the formulation of structure-activity relationships and the preparation of therapeutically useful derivatives. The word "structure" in this scheme can be assumed to be synonimous to "crystal structure". This is not intended to deny or diminish the contribution made by other spectroscopic methods in solving structural problems, nevertheless the size and complexity of most of the natural products of biological interest require the use of X-ray crystallography in order to obtain a complete and unequivocal molecular structure determination, at least in the early phase. Further to the initial, merely structural application, a sophisticated technique such as X-ray crystallography can be of great value in the definition of structure-activity relationships provided that a few complementary conditions are respected: 1) that the chemical details of activity have been sufficiently investigated and defined by the preparation and testing of an adequate number of derivatives; 2) that the studies on the mechanism of action already allow for the formulation of a hypothesis at the molecular level; 3) that the structural study in the solid-state is supported by conformational studies in solution performed by different spectroscopic techniques. When these conditions are not respected the crystallographic work risks remaining sterile. The following examples are cases in point.

β-Lactam Antibiotics

β-lactam antibiotics include, other than the classical penicillins and cephalosporins (Fig.1), a series of more recently discovered substances among which are the cephamycins, nocardicins, clavulanic acid, and thienamycin, all characterized by the presence of a sterically constrained β-lactam ring (Ghuysen 1981).

According to the first hypothesis of Tipper and Strominger (1965), penicillins and cephalosporins inhibit the synthesis of the bacterial cell-wall by inhibiting the peptidoglycan DD-transpeptidase. This enzyme operates the cross-linking of peptidoglycan, the main component of the bacterial cell-wall by a reaction of transpeptidation between the endings of two peptide-chains. One of the endings is always D-Ala-D-Ala. In the absence of β-lactam antibiotics a D-Ala residue is released and an interchain linkage is formed contributing to the building up of a fixed three--dimensional structure of the cell-wall. When instead β-lactam antibiotics are present, they inhibit the transpeptidation by irreversible acylation of the enzyme and this can eventually cause the lysis of the cell.

In order to explain the inhibition mechanism, the hypothesis has

been put forward that irreversible acylation may occur since β-lactam antibiotics mimic the natural substrate in a certain conformation, and are
thus allowed to reach the transpeptidase. This explanation is certainly
over-simplified especially when one bears in mind the fact that the peptidoglycan cross-linking process is regulated by several other enzymes be-
side the DD-transpeptidase, and that some of these enzymes are also inhibited by β-lactam antibiotics, each to a different extent. However, the
hypothesis bears out that the configuration of the asymmetrical centers of
β-lactam antibiotics is such, that they can only be attacked by enzymes
which act on D-aminoacids. The high specificity of these antibiotics is
probably due to this characteristic.

In X-ray crystallographic studies, carried out on both active and
inactive compounds, have contributed greatly to the formulation of structure-activity relationships in these antibiotics. In order to compare
the relative differences, their structure can be divided into three main
sectors: the thiaring, the β-lactam ring and the side-chain (Fig.1).

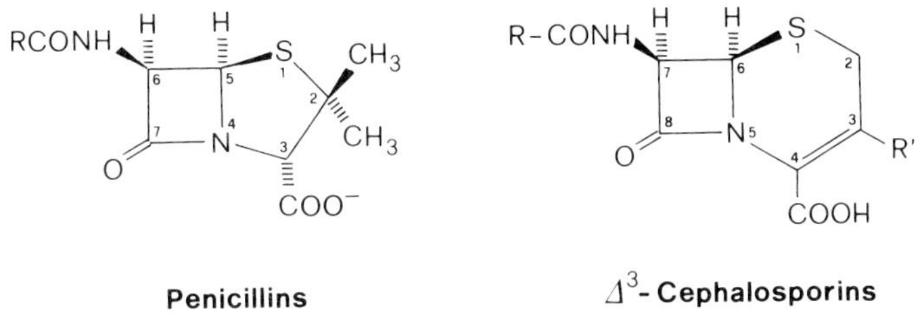

Penicillins Δ^3-**Cephalosporins**

FIG.1. Structural formulas of penicillins and cephalosporins.

For the first sector X-ray studies have shown that penicillins may
display two conformations of the thiazolidine ring, one, the so called
C(3) conformation, with C(3) below the plane of S(1), C(2), N(4) and C(5),
and the other, the so called S(1) conformation, with S(1) above the plane
of C(2), C(3), N(4) and C(5) (Dexter and van der Veen 1978). N.m.r. studies showed that, in solution penicillins preferentially adopt the S(1)
conformation (Dobson et al. 1975), (Fig.2). Only one conformation was

Ansamycins

Ansamycins are a group of antibiotics characterized by a structure in which an aliphatic bridge (the so called onsa-bridge) spans two non-adjacent positions of an aromatic system (benzene or naphthalene). All ansamycins display interesting biological properties, some are antibacterial, others are antiprotozoal and some are antitumor agents. Only for a few of them has a mechanism of action been formulated in some detail or structure-activity relationships been well defined. In particular, among these, only the rifamycins are normally used in therapy.

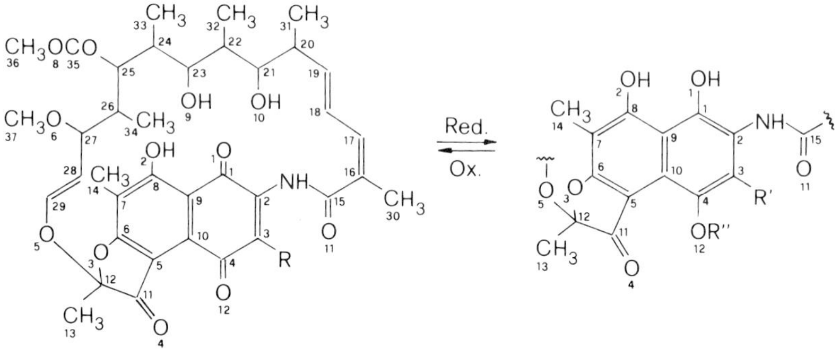

R=H : Rifamycin S   R'=R"=H : Rifamycin SV

R=COOCH₃ : 3-Methoxycarbonyl   R'=H
             rifamycin S   R"=CH₂COOH : Rifamycin B

R=H   Rifamycin S   R'=H
CN:C(OCH₃)C : iminomethyl   R"=CH₂COOH
 2 15   3 16   ether   C(CH₃)(OH)CO : Rifamycin Y
                                       20   21

R=H   R'=CH:NN̄ ̄NCH₃
COCH–CH : Tolypomycinone   R"=H   : Rifampicin
18 19\ /20
    CH
    31 2

FIG.3. Structural formulas of some ansamycins.

Rifamycins are naphthalenic ansamycins with high antibacterial activity, especially against gram-postive bacteria and mycobacteria. Their activity is due to the specific inhibition of the bacterial DNA-dependent
RNA polymerase (DDRP), the enzyme operating the transcripition of DNA (Brufani 1977, Lancini and Zanichelli 1977).

Rifamycin B (Fig.3), which was isolated in the Research Laboratories
of Lepetit, Milan, was the first (Sensi et al. 1959). From it, by mild
oxidation and hydrolysis, rifamycin S has been obtained, and this gives,
by mild reduction, rifamycin SV (Sensi et al. 1961),(Fig.3). Rifamycin B
dimethylamide was used for some time in therapy. Rifamycin SV is still in
use.

The structure of rifamycin S was determined by Prelog and collaborators (Oppolzer et al. 1964) through chemical degradation and spectroscopic
studies and was confirmed through an independent determination of the
crystal structure of rifamycin B p-iodoanilide, (Brufani et al. 1964),
(Fig.4).

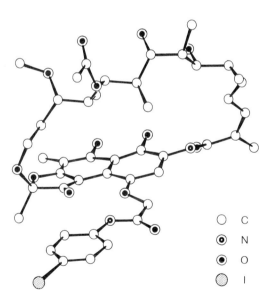

○　C

◉　N

◉　O

◉　I

FIG.4. Molecular structure of rifamycin B p-iodoanilide (data from Brufani
       et al. (1964)).

This crystallographic analysis, furthermore, gave the configuration of the nine asymmetric atoms and provided the first three-dimensional view of the structure of these antibiotics. The structure consists of a planar part, formed by a naphthohydroquinone nucleus fused to a five-membered ring, and of a 17 membered chain (the ansa-bridge), which is connected to atoms C(2) and C(12). The chromophoric group is quite planar and the best plane through the 17 atoms of the ansa-bridge makes a dihedral angle of 109° with it. With respect to the skeleton of the ansa-bridge, the carbonyl oxygen O(11), the hydroxyl groups at C(21) and C(23), and the methoxyl group at C(27) are on the same side as the two phenolic hydroxyls, whereas the acetoxyl at C(25) is on the opposite side.

The second crystal structure of a rifamycin to be determined was that of rifamycin Y p-iodoanilide, (Brufani et al. 1967), an inactive compound which differs from rifamycin B in having a hydroxyl replacing a hydrogen atom at C(20), with retention of configuration, and C(21) as a carbonyl group (Fig.3). In comparing the molecular structures of the two compounds it can be seen that the differences in costitution cause a notable difference in the spatial arrangement of the ansa-bridge. In particular, the C(21)=O(10) double bond is oriented almost parallel to the best plane through the skeleton of the ansa-bridge, while in rifamycin B, the C(21)-O(10) single bond is oriented almost parallel to the best plane of the chromophore-rings. In rifamycin Y the best plane through the skeleton of the ansa-bridge and that through the chromophore-rings make an angle of 103°.

These two structures were first used in the formulation of structure-activity relationships in rifamycins (Brufani et al. 1974), together with the structures of a derivative of tolypomycinone (Kamiya et al. 1969) and a derivative of streptovaricin C (Wang et al.1971, Wang and Paul 1976). These latter compounds are members of two classes of ansamycins, both strictly related to rifamycins, and both inhibitors of bacterial DDRP. In particular, the two compounds used for the crystallographic analysis are inactive due to their derivatization, but display a molecular geometry compatible with activity, as is discussed later. These results were later

supported by the study of the structures of other active and inactive compounds. First of all mention must be made of the study of the structure of
rifampicin (Fig.3), a drug of choice in the therapy of tubercolosis (Maggi
et al. 1966). The crystal structure of rifampicin (Gadret et al. 1975)
shows that the four C-O bonds at 1, 8, 21 and 23 are arranged similarly to
those in rifamycin B. The best planes through the skeleton of the ansa-
bridge and through the chromophore-rings make an angle of 98°.

Another interesting rifamycin recently studied by X-rays is 3-metho-
xycarbonylrifamycin S. This compound, prepared by Alfa Ricerche, Bologna
(Bellomo et al. 1981), was unexpectedly inactive in bacteria. The X-ray
analysis (Cellai et al. 1982a) showed that the conformation of the molecule is not substantially changed with respect to those of the above examined active compounds and that, therefore, its inactivity in bacteria must
be attributed to factors affecting the penetration of the bacterial cell-
wall. In fact, the compound was very active on isolated bacterial DDRP
(Brufani et al. 1982).

Another example of an inactive rifamycin studied by X-rays is the
iminomethylether of rifamycin S (Fig.3). The study of the structure of
this compound (Arora 1981) revealed that in this case the chemical modification causes a dramatic variation in the conformation of the ansa-bridge, so that the C-O bonds at 21 and 23 were almost parallel to the best
plane containing the skeleton of the ansa-bridge, instead of being parallel to that containing the chromophore-rings, as was found in active compounds. This compound is inactive on isolated bacterial DDRP.

An example of merely reduced activity is represented by tolipomyci-
none (Fig.3). The crystal structure of tolypomycinone 8,21,23-tri-m-bromo-
benzoate, mentioned above, was already known (Kamiya et al. 1969). This
compound, although displaying a conformation of the ansa-bridge similar to
that of the active rifamycins, was totally inactive due to its modification. Nontheless the non-derivatized tolypomycinone was less active than
the equivalent rifamycin S. The solution of the crystal structure of tolypomycinone (Brufani et al. 1978), (Fig.5) revealed that the differences in
the constitution of the ansa-bridge, although limited to atoms C(18),

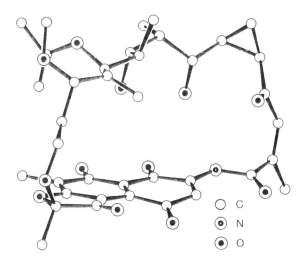

FIG.5. Molecular structure of tolypomycinone (data from Brufani et
al. (1978)).

C(19) and C(20), causes a conformational variation leading the C-O bonds
at 21 and 23 in a position intermediate between that found in active compounds and that found in the inactive iminomethylether of rifamycin S. The
best plane through the skeleton of the ansa-bridge and that through the
chromofore-rings make an angle of 61°. Therefore, in tolypomycinone 8,21,
23-tri-m-bromobenzoate, the ansa-bridge must be constrained in the "active" conformation by the presence of the three bulky substituents, while
in the non-derivatized molecule the conformation of the antibiotic changes. Furthermore the crystal structure showed that, in tolypomycinone, the
amide-carbonyl C(15) is cisoid with respect to the C(2)- C(3) bond, while
it is transoid in rifamycins B and Y, and in rifampicin. This observation
has led to the preparation of several 3-monoalkylamino derivatives of tolypomycinone (Bellomo et al. 1977), with the aim of engaging the amide-
carbonyl in an intra-molecular hydrogen bond with the N-H of the substituent. This should bring about a conformational effect transmitted along
the ansa-bridge up to the C-O bonds at 21 and 23 inducing them to assume
the "fully active" disposition. In fact, these derivatives were more ac-

tive than the parent compound; some are as active as rifampicin.

In conclusion, the results of the chemical (Wehrli and Staehelin 1969, Ghisalba et al. 1980) and structural studies with respect to activity can be summarized as follows:

1) the naphthohydroquinone (rifamycin SV) and naphthoquinone(rifamycin S) forms are equally active;

2) the elimination of the phenolic hydroxyl-groups on C(1) and/or C(8) causes the loss of all activity;

3) the acetylation or methylation of the phenolic hydroxyl groups on C(1) and/or C(8) gives inactive derivatives;

4) the opening of the ansa-bridge causes the loss of all activity;

5) the hydrogenation of the double bonds of the ansa-bridge decreases activity;

6) the acetylation of the hydroxyl-groups on C(21) and C(23) leads to inactive derivatives;

7) the oxidation of the same hydroxyl-groups to carbonyl-groups leads to inactive derivatives;

8) the hydrolysis of the acetoxyl-group on C(25) involves no loss of activity;

9) the hydroxyl-group on C(4) is not essential for activity.

In the light of previous data, it is possible to formulate the hypothesis that, for activity, it is essential to have the presence of the aromatic nucleus with oxygen atoms on C(1) and C(8) either in the quinone form or as free hydroxyl-groups, and of two free hydroxyls on C(21) and C(23). A certain conformation of the ansa-bridge leading to a definite geometric relationship between the two hydroxyls and the aromatic nucleus is also necessary. That is, the interaction with the enzyme DDRP takes place through the formation of hydrogen bonds with oxygens O(1), O(2), O(9) and O(10) which must all be unhindered and free from substituent. The interaction is more efficient when all the four C-O bonds point in the same direction. The geometrical parameters discussed above are summarized in Table 1.

It has also been hypothesized that the naphthalenic-system of the antibiotic contributes to the binding as an acceptor in a π-π interaction

TABLE 1.

Distances in Å between O(1), O(2), O(9), O(10), and dihedral angles Φ° between the best planes through ansa-bridge and chromophore-rings in ansamycins.

	RIF B	RIF Y	BRTOL	RIFMP	TOL	IME-RS*		MCRS
O(1)-O(2)	2.6	2.4	2.8	2.48	2.55	2.54	2.55	2.54
O(1)-O(9)	6.7	6.5	6.0	6.16	3.23	6.18	6.64	7.00
O(1)-O(10)	5.7	8.1	5.4	5.41	3.01	4.11	4.70	5.76
O(2)-O(9)	7.8	7.3	6.9	6.82	3.31	4.76	4.94	7.42
O(2)-O(10)	7.5	9.6	7.3	6.93	4.77	3.81	4.06	7.03
O(9)-O(10)	2.7	3.6	2.8	2.72	2.56	2.70	2.77	2.73
Φ°	109	103	71	98	61	44	41	97

RIF B: Rifamycin B p-iodoanilide

RIF Y: Rifamycin Y p-iodoanilide

BRTOL: Tolypomycinone
8,21,23-tri-m-bromobenzoate

RIFMP: Rifampicin

TOL:Tolypomycinore

IME RS:Rifamycin S iminomethyl ether

* Two molecules per asymmetric unit

MCRS: 3-methoxycarbonylrifamycin S

with an aromatic residue of the enzyme. In fact, the inhibition was increased when rifamycin S carries an electron-withdrawing group as the 3-
substituent (Dampier and Whitlok 1975).

The above structure-activity relationships have been put forward on
the basis of the solid-state data only. Nontheless a n.m.r. conformational
study in solution, performed on a large number of derivatives of both rifamycin S and SV (Cellai et al. 1982b), have shown that, in spite of the
apparent flexibility, rifamycins give rise to only four kinds of isomers,
three of which have also been observed in the solid-state, and all of
which are generated by the combination of two rotations, one relative to
the amide plane and the other relative to the plane containing the C(28)=
C(29) double bond. Apart from these rotations, the remaining parts of the
ansa-bridge between C(17) and C(27) display only small variations. As a
consequence the spatial relationships between O(1), O(2), O(9) and O(10)
remain the same whatever the oxidation state at O(1), the kind of C(3)
substituent, the nature of the solvent or whether the molecules are present in the solid state or in solution. The conclusions derived from the
crystal data are thus confirmed.

The most recent application of X-ray crystallography to a rifamycin
is the study of the structure of rifamycin L 105, (Cerrini et al. unpublished results), (Fig.6).This new antibiotic, prepared by Alfa Ricerche, Bologna, (Marchi, E. et al. 1982), is under clinical evaluation as an
intestinal disinfectant since, in contrast to the other rifamycins in use,
it is not absorbed along the gastro-intestinal tract.

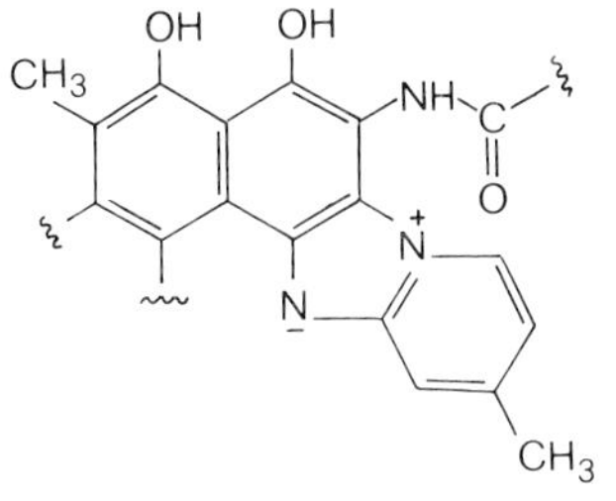

FIG.6. Structural formula of rifamycin L 105.

All the structures mentioned above belong to naphthalenic ansamycins. Among the benzenoid ansamycins only a few X-ray crystal structures have been determined. They are the structures of the antiprotozoal geldanamycin (Wang and Paul 1976), of the antitumor drug maytansine (Bryan et al. 1973), and of the herbicidal herbimycin A (Furusaki et al. 1980).

In conclusion, with regard to the prospects for X-ray crystallography being applied to the study of drug action, we suggest that the scheme successfully adopted for penicillins, cephalosporins and expecially for rifamycins should be used for other drugs. Furthermore, we feel there is a need for the channeling of efforts in such a way as to obtain crystals of drug-receptor complexes, or at least of models of the same, which may enable us to obtain a deeper insight into the biological problem.

REFERENCES

β-Lactam Antibiotics

ABRAHAMSON, S, HODGKIN, D.C., and MASLEN, E.N. (1963). Biochem.J. 86, 514.

DEXTER, D.D. and VAN DER VEEN, J.M. (1978). J. Chem. Soc., Perkin 1, 185.

DOBSON, C.M., FORD, L.O., SUMMERS, S.F., and WILLIAMS, R.J.P. (1975). J.-Chem. Soc., Faraday Trans.2, 1145.

GHUYSEN, J.(1981). In Topics in Antibiotic Chemistry (ed. P.G. Sammes) Vol. 5, pp.15-117. Ellis Horwood Limited, Chichester.

GHUYSEN, J.M., FRERE J.M., LEYH-BOUILLE, M., PERKINS, H.R., and NIETO, M. (1980).Phil. Trans. Roy. Soc. Lond. B. 289, 285.

GREEN, G.F.H., PAGE, J.E., and STANIFORTH, S.E. (1965). J.Chem.Soc., 1595.

PITT, G.J. (1952) Acta Crystallogr. 5, 770.

TIPPER, D.J. and STROMINGER, J.L. (1965). Proc.Nat.Acad. Sci. 54, 1133.

SWEET, R.M. (1972). In Cephalosporins and Penicillins: Chemistry and Biology (ed. E.H. Flynn), p. 280. Academic PressNew York.

-- and DAHL, L.F. (1970).J.Amer. Chem. Soc. 92, 5489.

Ansamycins

ARORA, S.K. (1981). Acta Crystallogr. Sect. B. 37, 152.

BELLOMO, P., MARCHI, E., MASCELLANI, G., and BRUFANI, M.(1981). J.Med.

Chem. 24, 1310.

--, BRUFANI, M., MARCHI, E., MASCELLANI, G., MELLONI, W., MONTECCHI, L., and STANZANI, L. (1977).J. Med. Chem. 20, 1287.

BRUFANI, M. (1977). In Topics in Antibiotic Chemistry (ed. P.G. Sammes) Vol.1, pp.91-217. Ellis Horwood Limited, Chichester.

--, CERRINI, S., FEDELI, W., and VACIAGO, A. (1974) J. Mol. Biol. 87, 409.

--, FEDELI, W., GIACOMELLO, G., and VACIAGO, A. (1964). Experientia 20, 339.

--,--,--,--. (1967). Experientia 23, 508.

--, CELLAI, L., CERRINI, S., FEDELI, W., and VACIAGO, A., (1978). Mol. Pharmacol. 14, 693.

--, CELLAI, L., CERRINI, S., FEDELI, W., SEGRE, A., and VACIAGO, A., (1982). Mol.Pharmacol. 21, 394.

BRYAN, R.F., GILMORE, C.J., and HALTIWANGER, R.C. (1973). J. Chem. Soc. Perkin Trans. 2, 897.

CELLAI, L., CERRINI, S., SEGRE, A., BRUFANI, M., FEDELI, W., and VACIAGO, A., (1982a). J. Chem. Soc. Perkin Trans.2, 1633.

--, CERRINI, S., SEGRE, A., BRUFANI, M., FEDELI, W., and VACIAGO, A. (1982b). J.Org. Chem. 47, 2652.

DAMPIER, M.F. and WHITLOK, H.W., Jr. (1975). J. Amer. Chem. Soc. 97, 6254.

FURUSAKI, A., MATSUMOTO, T., NAKAGAWA, A., and OMURA, S., (1980). J. Antibiot. Tokio 32, 781.

GADRET, M., GOURSOLLE, M., LEGER, J., and COLLETER, J. (1975). Acta Crystallogr. Sect.B 31, 1454.

GHISALBA, O., TRAXLER, P., FUHRER, H., and RICHTER, W.J. (1980). J. Antibiot. Tokio 33, 847.

KAMIYA, K., SUGINO, T., WADA, Y., NISHIKAWA, M., and KISHI, T. (1969). Experentia 25, 901.

LANCINI, S. and ZANICHELLI, W. (1977). In Structure-Activity Relationship among the Semisynthetic Antibiotics (ed. D. Perlmam) pp.531-600. Academic Press, New Jork.

MAGGI, N., PASQUALUCCI, C.R., BALLOTTA, R., and SENSI, P. (1966). Chemotherapy 11, 285.

MARCHI, E., MASCELLANI, G., MONTECCHI, L., BRUFANI, M., and CELLAI, L. (1982). Chemoterapia supplement to n.4; Vol.1 p. 106.

OPPOLZER, W., PRELOG, V., and SENSI, P. (1964). Experientia 20, 336.

SENSI, P., MARGALITH, P., and TIMBAL, M.T. (1959). Il Farmaco, Ed. Sci. 14, 146.

--, BALLOTTA, R., GRECO, A.M., and GALLO, G.G. (1961). Il Farmaco, Ed. Sci. 16, 165.

WANG, A.M.J. and PAUL, I.C. (1976). J.Amer. Chem. Soc. 98, 4612.

--, PAUL, I.C., RINEHART, K.L., Jr., and ANTOSZ, F.J. (1971). J.Amer. Chem. Soc. 93, 6275.

WEHRLI, W. and STAEHELIN, M. (1969). Biochim. Biophys. Acta 182, 24.

22 Steroid conformation, receptor binding, and hormone action

W.L. Duax, J.F. Griffin, and D.C. Rohrer

Minor changes in the composition of steroid hormones result in significant variation in their chemical and biological activities. In addition
to having a direct influence on solubility, membrane transport, and protein binding, structural changes affect activity by altering the overall shape of the molecule, disturbing the balance between conformational
isomers (Wellman and Djerassi 1965), and inducing conformational transmission effects (Barton 1955). The exceptionally large body of accurate data on steroid conformations provided by over 500 X-ray crystal
structure determinations (Duax and Norton 1975; Duax, Griffin and Weeks
1983a) is useful for the identification of substituent effects, the geometric details of conformational transmission, and the structural basis
for hormone action.

1. MOLECULAR MECHANICS CALCULATIONS

Because the active form of a hormone may not be its minimum energy
form, it is important to have reliable information concerning relative
energies of metastable states and populations of conformational isomers
in various environments. If the crystallographically observed distribution of molecular conformers associated with a particular structural
feature is not systematically altered by packing forces, it should be a
useful guide for the prediction of relative energies, molecular flexibility, and solution conformer populations. For these reasons we have
begun a systematic comparison of the geometry of the crystallographically observed structures with that obtained from empirical force field
calculations (Duax, Griffin and Rohrer 1981a).

1.1 Comparison with neutron diffraction data

The high degree of accuracy in the geometric parameters for hydrogen and nonhydrogen atoms obtained from the low temperature neutron
diffraction determination of 20-methyl-5-pregnene-3β,20-diol (Fig. 1.1)
provides an excellent model structure for testing the accuracy of
various force field programs currently in use. The bond lengths, bond

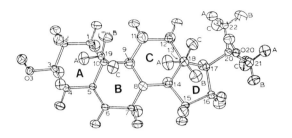

FIG. 1.1.   The molecular geometry of 20-methyl-5-pregnene-3β,20-diol
at 123°K showing the steroidal numbering scheme.   Thermal ellipsoids
enclose 50% probability.

by the PROPHET/Wipke (Wipke 1980), QCFF/MCA (Huler, Sharon and Warshel
1977), MMP1 and MM2p* (Allinger 1976; Allinger and Yuh 1981) programs in
Fig. 1.2, 1.3, and 1.4.   For purposes of comparison the observed neutron
structure values are subtracted from the calculated values and the geometric parameters involving nonhydrogen atoms only are shaded.   The
neutron diffraction results include correction for the effects of thermal
motion.   The ranges of estimates of standard deviations in the neutron
diffraction determination were 0.001 to 0.002Å for distances, 0.07 to
0.09° for angles, and 0.2 to 0.3° for torsion angles involving nonhydrogen atoms and 0.002 to 0.005Å for distances, 0.1 to 0.4° for angles
and 0.3 to 0.7° for torsion angles involving hydrogen atoms.

    Figures 1.3 and 1.4 clearly show that the PROPHET/Wipke program
fails to reproduce the crystallographically observed bond lengths and
angles with the accuracy achieved by the other three programs. The structure calculated using the MM2p program has the smallest deviation from
the neutron results and the differences are normally distributed.   The

---

*The MM2p program is an amended version of the MM2 program which
contains the VESCF (Variable Electronegativity Self-Consistent Field)
portion of the MMP1 program for properly treating the electronics of a
conjugated system.   The corresponding MMP2 program is not available
from the Quantum Chemistry Program Exchange, thus making development
of this version necessary.

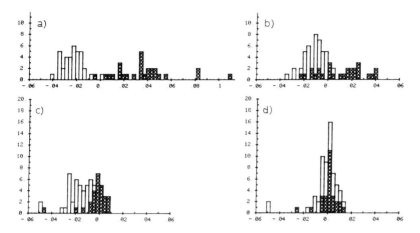

FIG. 1.2. Graphic representation of the differences between bond
lengths observed in the neutron diffraction determination of 20-MPD
and those calculated by (a) PROPHET/Wipke, (b) QCFF/MCA, (c) MMP1 and
(d) MM2p molecular mechanics programs. Numbers of observation are plotted versus magnitude of difference. The values involving all nonhydrogen atoms are shaded and in all cases the values are of the theoretical value minus the neutron observation.

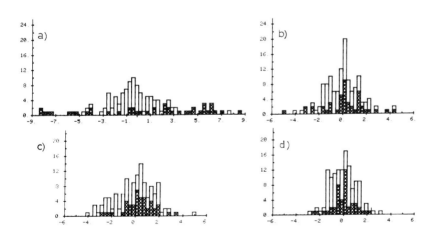

FIG. 1.3. Comparison of the observed and calculated bond angles as
in Fig. 1.2.

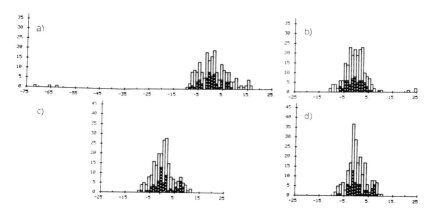

FIG. 1.4.   Comparison of the observed and calculated torsion angles as
in Fig. 1.2.

PROPHET/Wipke, QCFF/MCA and MMP1 programs systematically underestimate
the bond lengths involving hydrogen atoms and the PROPHET/Wipke program
systematically overestimates the bond lengths involving the nonhydrogen
atoms. The maximum differences in bond length in the MM2p comparison
involve the hydroxyl hydrogens and oxygens, and these differences may be
related to the hydrogen bonding in the crystal. Because the differences between observed and calculated bond angles are well distributed
about zero even in the inaccurate PROPHET/Wipke calculation, systematic
errors such as those seen in the bond length comparison are not apparent unless a systematic analysis is carried out. The fact that the
MM2p calculations are   much closer to the experimentally determined
values than those obtained with the PROPHET/Wipke and QCFF/MCA programs leads us to believe that an even greater correspondence to
the neutron diffraction determinations is possible.

The torsion angles provide the most sensitive measure of molecular
conformation. The force field programs predict minimum energy conformations that differ significantly from the neutron diffraction determination. Calculations using the PROPHET/Wipke program exhibit a bias
toward overestimating torsion angles involving hydrogens. Torsion
angle differences from calculations using the other three programs are
smaller and require careful analysis.

Of the six nonhydrogen atom torsion angles that show a 6° to 10°
discrepancy between the QCFF/MCA calculations and the neutron determination, five are in the B ring and may be due to the inadequate
treatment of the unsaturated ring. When the MM2p calculation is compared with the neutron determination, the A, B, and C ring torsion
angles are in good agreement and the principal discrepancy is in the
C(17) side chain orientation and the D-ring. The orientation of the
C(17) side chain in the MM2p structure is rotated plus 7° about the
C(17)-C(20) bond and the D-ring shifts toward a 13β,14α-half chair
from a 13β-envelope conformation. In contrast, the orientation of the
17β-side chain and D-ring conformation in the QCFF/MCA minimization
closely approximates the neutron diffraction determination.

The torsion angles calculated by the PROPHET/Wipke and QCFF/MCA
programs are compared to those calculated by the MM2p program in Fig.
1.5. Since the MM2p program generated the most reliable bond lengths
and angles, it is used as the standard in this comparison. The calculated structures are found to differ from one another as much as
they differ from the neutron diffraction determination. For this
reason it is important to determine which, if any, of the programs is
properly handling the parameters that govern torsion angle geometry and
overall conformation. The accumulated effect of errors of even 4° in
a series of torsion angles could result in totally incorrect predictions
concerning the overall shape and relative conformational energy of
flexible molecules.

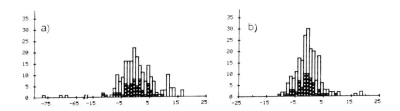

FIG. 1.5. Graphic presentation of differences between torsion angles
for the minimum energy conformation of 20-MPD calculated by MM2p and
those calculated by (a) the PROPHET/Wipke and (b) the QCFF/MCA programs.
The values are the PROPHET/Wipke or QCFF/MCA minus the MM2p calculation.
Values involving all nonhydrogen atoms are shaded.

In the next sections we attempt to determine whether the observed
variations in torsion angles are random or systematic in nature.
While random variations would be inconclusive, systematic differences
would offer the means of uncovering the cause of variation. In order
to do this we compared the calculated and observed torsion angles for a
great many structures having a common 17β-side chain. Here we are re-
lying on the consistency of the data provided by a large sample rather
than the precision of a single carefully conducted experiment in order
to identify exploitable patterns.

1.2 Progesterone side chain conformation

Wellman and Djerassi (1965) concluded that for steroids having the
D ring and side chain of progesterone, the two most stable side-chain
conformations are those shown in Fig. 1.6. Modeling studies and CD
spectra indicated that the conformer with a C(16)-C(17)-C(20)-O(20)

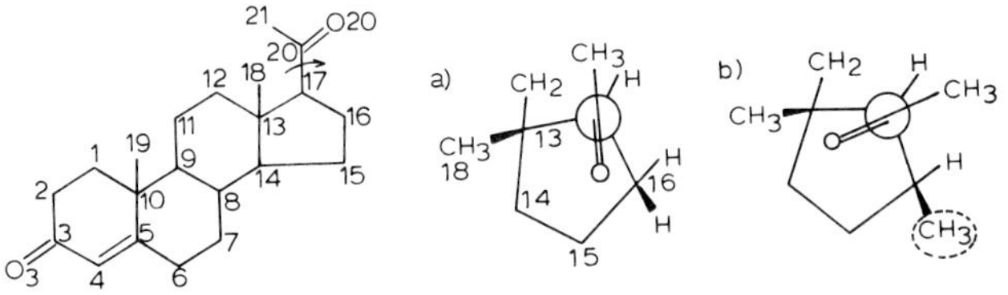

FIG. 1.6. Chemical formula for progesterone and Newman projections,
C(20)-C(17), illustrating two conformations of the progesterone side
chain C(16)-C(17)-C(20)-O(20) = -30° (a) and -90° (b).

torsion angle, τ, of approximately -30° (Fig. 1.6a) was 1.1 Kcal/mole
lower in energy than the conformer where τ is -90° (Fig. 1.6b). In the
case of the 16β-methyl-substituted structures, double maxima in the CD
spectra suggested an equilibrium between the two conformers, with a
slight preference for conformer b (Fig. 1.6).

Examining the crystal structure data for 77 pregn-20-one analogs
without a 16β-H substituent has revealed a conformational preference for
a single orientation of the 17β side group (Duax et al., 1981a). The
C(16)-C(17)-C(20)-O(20) torsion angle, τ, in each of these structures

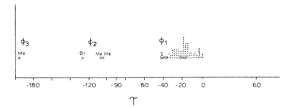

FIG. 1.7.   The observed values of the C(16)-C(17)-C(20)-O(20) torsion
angles in the 85 crystal structures determinations of pregnanes having
a 20-one substituent (x =16β-substitutent other than hydrogen).

is in the range 0° to -46°, See Fig. 1.7. Furthermore, the conformations appear to be normally distributed about the average of -21° which
is within 10° of the position predicted by Wellman and Djerassi (1965).
These observations suggest that this orientation of the side chain represents a well-defined energy minimum conformation which is significantly better than any other minimum and that the barrier to rotation
about the C(17)-C(20) bond is high enough to hold the side group within this minimum.

     The distribution of observed conformations of the eight 16β-sub-
stituted analogs is somewhat different (Duax *et al.*, 1981a). Four of
these structures have 17β-side-group orientations at the low extreme in
the range of the unsubstituted analogs (mean τ = 43 ± 3°, conformer ϕ_1),
see Fig. 1.7, but in three other structures the mean value of τ is
-115 ±11° (conformer ϕ_2), and in one case τ = +162° (conformer ϕ_3).
These side-chain orientations are illustrated in Fig. 1.8. While the

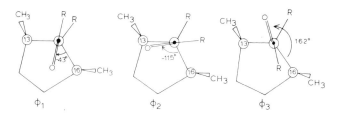

FIG 1.8.   Three conformational isomers of the progesterone side chain
of the 16β-methyl-substituted steroids viewed in Newman projection
(C(20)→C(17)).

number of observations is limited, the addition of the 16β substituent
seems to shift the location of the best minimum energy orientation and
reduce the difference between minima so that additional orientations
have significant populations. This is also consistent with the earlier
proposal by Wellman and Djerassi from CD spectra that several conformations of the 17β side group were present. Although the actual conformations predicted may not have been correct, conformers ϕ_1 and ϕ_2
are most probably the forms responsible for the double maxima noted.

   The structure of 18-deoxyaldosterone (Fig. 1.9) provides the first
example of a 16β-hydrogen pregnane having a τ value in the ϕ_2 range,
-132° (Duax, Griffin, Strong, Ulick and Funder 1982). This analog
crystallizes with two molecules in the asymmetric unit, with the second
molecule having τ in the normal range, -28°. Epoxide formation between
C(11) and C(18) draws the C(18) hydrogens away from the C(17) position.
This removes steric hindrance to population of the second conformer and
makes the τ = -132° and τ - -28° conformers of comparable energy leading
to their co-crystallization.

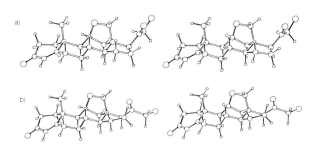

FIG. 1.9. Stereo ORTEP views of the two molecules of 18-deoxy-
aldosterone illustrating differences in the 17β-side-chain orientations, (a) τ = -28, (b) τ = -132.

   The structure of 21-fluoro-17α-acetoxy-4,9(11)-pregnadiene-3,20-
dione (Fig. 1.10) was found to have an even more unusual and totally
unexpected side-chain conformation in the ϕ_3 region, τ = 173.2° (Duax
1983). Comparison of this structure with the closely related structure
of 17α-acetoxy-progesterone indicates that the remarkable conformational

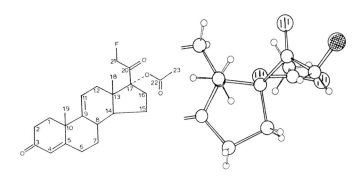

FIG. 1.10. Chemical structure and 17-side-chain orientation of 21-
fluoro-17α-acetoxy-4,9(11)-pregnadiene-3,20-dione.

difference was brought about by a combination of the double bond at
C(9)-C(11) and the fluorine substituent. In this structure, the
C(21)-F(21) bond is coplanar with the carbonyl. In addition, the
C(9)-C(11) double bond shifts the location of the 12β hydrogen relative to where it would be in the saturated ring. These structural
features combine to make the τ = -21 orientation energetically less
favorable due to the close non-bonded contact, 2.2Å, between the
hydrogens on C(21) and C(12). The strain associated with this close
contact was avoided by rotation to the ϕ_3 conformation.

Using MM2, Profeta, Kollman and Wolff (1982), have calculated the
energy as a function of τ for progesterone and 16β-methyl-progesterone.
They find that for progesterone the energy of the local minimum at τ =
-120° is just 0.3 Kcal/mole above the global minimum at τ = 0°. They
conclude that the failure to observe the higher energy conformation
in 77 crystallographic structures is due to the fact that the τ = 0°
conformation points the carbonyl away from the molecule and allows
"favorable" intermolecular contacts with this group, while the τ =
-120° conformation makes the carbonyl less accessible to intermoleclar
interactions. Careful analysis of the intermolecular contacts including those used by Profeta raises serious questions concerning the
validity of this argument.

(1) Analysis of 80 crystal structures failed to show an apparent
correlation between intermolecular hydrogen bonding to the carbonyl, a

favorable interaction, and the 17β-side-chain orientation (Duax *et al.*, 1981a). In the case of the less specific types of intermolecular inter- actions, no general trend in the arrangement of the molecules in the crystal, e.g. head to tail ribbons or stacked columns of molecules, could be found which could explain the preference for a single confor- mation of the side chain.

(2) The intermolecular contact between the 21-methyl and O(3) of an adjacent molecule in the progesterone crystal lattice (Campsteyn, Dupont and Dideberg 1972) when τ = -7 was offered as a demonstration of the importance of an exposed carbonyl. However, in this case no favor- able contact involving the carbonyl is found but rather a contact in- volving the methyl.

(3) It is true that rotation of the side chain to τ = -120° would result in a very close contact, 3.02Å, with O(3) on an adjacent molecule in the crystal lattice. However, there is no reason to presume that a molecule with the side chain rotated to the τ = -120° position would crystallize in a manner isomorphous with the conformer having τ = -7°.

(4) The 18-deoxyaldosterone crystal structure illustrates that the equally favorable intermolecular contacts involving the carbonyl and adjacent molecules in the crystal lattice can be made with the molecule in either the τ = -132° or τ = -28° orientations.

(5) If there is a preference for the carbonyl to be more exposed forming favorable contacts in the crystal lattice, then the distribu- tion of conformations should be skewed toward τ = 0° in the crystal structures. The MM2 calculations seem to show that there are no intra- molecular reasons why the distribution should not be closer to zero. In fact, the crystal structure results show a systematic difference in the negative direction relative to the MM2 calculations, see Table 1.1, which seems to indicate that the MM2 potentials are overemphasizing the energy reduction associated with the τ = 0° conformation.

Since we find no evidence that packing forces restrict the con- formation to a single range as observed in the crystal structures of 77 pregn-20-ones, perhaps the energy differences between the conformations account for the preference. If this is the case, the 0.30 Kcal/mole energy difference calculated between the τ = 0° and τ = -120° confor- mations must result in better than a 77:1 ratio of conformers. How- ever, the Boltzman distribution for the progesterone energy diagram indicates that the distribution would be a 4:3 ratio. Furthermore,

TABLE 1.1

Comparison of observed and calculated values for τ,
C(16)-C(17)-C(20)-O(20)

Structure*	X-Ray	MM2p	X-MM2p**
Progesterone	−9.6†	−4.2	−5.4
11α-Methylprogesterone	−20.4	−4.3	−16.1
21-Hydroxyprogesterone	−11.1	−7.4	−3.7
11β,21-Dihydroxyprogesterone	−13.6	−5.2	−8.4
17α-Methyl-19-nor-4,9-pregnadiene-3,20-dione	−6.2	−7.0	0.8
17α-Hydroxyprogesterone	−22.7	2.9	−25.6
17α-Hydroxy-6α-methylprogesterone	−14.0	2.7	−16.7
11β,17,21-Trihydroxyprogesterone	−31.1	1.0	−32.1
17α,21-Dihydroxy-1,4-pregnadiene-3,11,20-trione	−33.2	−2.4	−30.8
17α-Acetoxyprogesterone	−19.1	−8.0	−11.1
6α-Methyl-17α-acetoxyprogesterone	−19.0	−8.5	−10.5
21-Acetoxy-17α-hydroxy-4-pregnene-3,11,20-trione	−31.9	−18.7	−13.2
17α,21-Dihydroxyprogesterone 21-acetate	−29.6	−18.8	−10.8
11β,16α,17,21-Tetrahydroxy-1,4-pregnadiene-3,20-dione	−34.7	−6.4	−28.3
16α-Ethyl-21-hydroxy-19-norprogesterone	−25.7	−12.8	−12.9
3β-Acetoxy-16β-methyl-5-pregnen-20-one	−45.8	−34.7	−11.1
18-Deoxy aldosterone	−28.4	−6.1	−22.3
3β-Acetoxy-16β-methyl-5-pregnen-20-one	−111.4	−109.9	−1.5
18-Deoxy aldosterone	−132.0	−133.2	1.2
21-Fluoro-17α-acetoxy-4,9(11)-pregnadiene-3,20-dione	173.2	−179.6	−7.2

*Full references for X-ray determinations appear in (Duax *et al.*, 1981a).
** X-MM2 is the signed magnitude of the X-ray value for the torsion angle minus the value calculated using the MM2p program.
†Average of τ's observed in three crystallographic modifications of progesterone.

these results are not compatible with the structural data and energy
calculations for 16β-methyl-progesterone. In this case, the observed
3:4 distribution of conformers and co-crystallization of a 1:1 ratio
of conformers in the crystal structure of 3β-acetoxy-16β-methyl-5-
pregnen-20-one would suggest that the energy difference between conformers is nearly zero. The calculations indicate that the energy
differences are similar to those of progesterone, but with the τ = -120°
conformer preferred. Thus this explanation is neither internally consistent nor consistent with the expected distribution of conformers
using the energy diagrams calculated by Profeta *et al.*, (1982) and does
not provide a suitable explanation for these observations.

The MM2p program has been used to minimize the energy of 20 progesterone related structures starting with the crystallographic observations (Duax, Fronckowiak, Griffin and Rohrer, 1982b) see Table 1.1. With
the exception of conformers in the ϕ_2 and ϕ_3 ranges (τ = -120° and 180°
the 17-side chains in the minimized structures are rotated an average of
15° toward the point where the carbonyl eclipses the C(17)-C(16) bond.
The published calculations (Profeta, *et al.*, 1982) on a smaller subset of
steroid data heavily weighted by cortisol derivatives, but including
eight structures not included in Table 1.1, reveal similar systematic
differences of smaller magnitude. It was suggested that the systematic
7° difference between observed and calculated values could be dismissed
because it was "the same as the mean deviation of τ for the 21 unsubstituted progesterone crystal structures". This is incorrect. The 21
crystal structures referred to are not unsubstituted progesterone crystal
structures. They are structures that have the same D-ring and 17-side-
chain composition as progesterone but differ in the composition of their
A, B and C rings. Ample evidence has been presented (Duax *et al.*, 1981a)
to support long range conformational transmission effects from the A, B
and C rings causing variations in τ of this magnitude. We, therefore,
conclude that just as MM2 has improved upon previous programs with regard to correctly optimizing bond lengths and angles, even more reliable
calculations of the minimum energy positions of torsion angles and
overall conformation can be achieved.

2. RECEPTOR BINDING

The characteristic responses of steroidal hormones require their
binding to specific receptor proteins in target tissue (Jensen and

Jacobson, 1962). While response clearly depends upon the interaction
of the receptor-steroid complex and nuclear chromatin, the precise details of this interaction and the role played by the steroid in this
process remain undetermined (King and Mainwaring, 1974; O'Malley and
Birnbaumer, 1978; Milgrom, 1981). Structural details undoubtedly have
a direct bearing upon receptor affinity and will directly or indirectly
influence receptor activation, transport, and nuclear interaction. The
existence of antagonists that compete for the steroid binding site of
the receptor with high affinity demonstrates that the phenomena of
binding and activity are at least partially independent. If agonists
and antagonists compete for the same site on a receptor a comparison of
their structures should make it possible to identify which structural
features are responsible for binding and which control activity.

2.1. Progestin binding

Examination of the chemical structures of steroids whose affinity
for the progesterone receptor in rabbit, mouse, sheep or human uterus
is equal to or higher than that of progesterone itself (Smith, H. E.,
Smith, R. G., Toft, Neergaard, Burrows and O'Malley 1974; Raynaud,
Philibert and Azadian-Boulanger 1973; Terenius 1974; Kontula, Janne,
Vijko, de Jager, de Visser and Zeelen 1975) indicates that extensive
structural variation is compatible with high affinity binding (Duax,
Cody, Griffin, Rohrer and Weeks 1978). The only structural feature
common to all compounds with high affinity for the progesterone receptor is the steroid ring system and 4-en-3-one composition. We have
proposed that binding requires a tight complementary fit between the
structures of the receptor and the steroids 4-en-3-one A ring. However, many steroids that have the 4-en-3-one composition have little or
no affinity for the uterine progesterone receptor (i.e., testosterone).
Since a 4-en-3-one A ring appears to be required but not sufficient for
high affinity binding to the progesterone receptor, we examined the conformations of the 4-en-3-one A rings of the highest affinity compounds in
search of some unusual electronic, geometric, or sterochemical features
that might explain their enhanced binding.

Crystallographic studies of a number of synthetic steroids having
exceptionally high affinity for the progesterone receptor revealed the
presence of an unusual "inverted" A-ring conformation (Fig. 2.1b) in
which the 2β hydrogen atom is flipped into an equatorial position rather

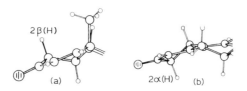

FIG. 2.1. A comparison of the normal A-ring conformation (a) and
the unusual 1β,2α inverted half-chair conformation (b).

than the normal axial position (Fig. 2.1a). The high progesterone receptor affinities of steroids having the inverted A-ring conformation
suggest that this conformation is optimal for binding. The most potent
progestins appear to be constrained to the conformation that permits
strongest association with the receptor, and consequently, they are
observed in this conformation in the solid state. We propose a highly
stereospecific association between the receptor and the A-ring region
that is best satisfied by a 4-en-3-one ring in the inverted conformation. Association between the receptor and the D-ring end of the
steroids either does not occur, or is far less stereospecific.

2.2. Estrogen binding

Compounds that bind to the estrogen receptor exhibit remarkable
variability in composition and stereochemistry (Fig. 2.2). They include nonsteroidal compounds, semi-synthetic steroids having unnatural
chirality, clinically useful anticancer agents, suspected carcinogens,
and simple one or two ring compounds. The only structural element that
all of these compounds have in common is a phenolic ring.

When the phenol rings of the nonsteroidal estrogen diethylstilbestrol, the mycotoxin *trans*-zearalenone and the semi-synthetic estrogen 17-keto-9β-estrone, all of which bind to the estrogen receptor, are
superimposed upon the A ring of estradiol, significant differences in
the D-ring region of the molecules are observed (Fig. 2.3). Hence, as
with the progestins, tight contact between the estrogens and receptor
would appear to be limited to the A and possibly the B rings. Since
the 3-hydroxyl of estradiol can serve as a donor or acceptor in

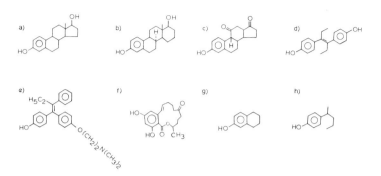

FIG. 2.2.   Compounds that bind to the estrogen receptor with varying
degrees of affinity include (a) estradiol, (b) 8α-D-homoestradiol, (c)
11-keto-9β-estrone, (d) diethylstilbestrol (DES), (e) monohydroxy-
trans-zearalenone, (g) tetrahydronaphthol and (h) *p*-sec-amyl phenol.

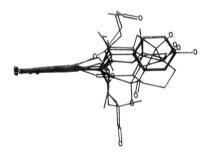

FIG. 2.3.   Superposition of the phenol rings of six estrogens
demonstrates the variability in D-ring orientation that is compatible
with receptor binding and some degree of activity.

hydrogen bonds (Duax and Weeks 1980), it can account for 6 kcal/mole of
binding energy (Duax, Smith, Swenson, Strong, Weeks, Ananchenko and
Egorova 1981b).

3.   HORMONE ACTION

3.1   Agonists and antagonists

In those cases where antagonists act by direct competition

for the hormone binding site on the receptor, it has been possible
to identify which structural features are responsible for binding to
that site and which control agonist versus antagonist response.

The structures of agonists of three of the principal classes of
steroid hormones (estrogens, mineralocorticoids, and glucocorticoids)
are compared with structures of antagonists of these hormones (Borgna
and Rochefort 1981; Karmin and Brown 1972; Pons and Simons 1981) in
Fig. 3.1. The high degree of similarity at the A-ring ends of the
agonists and antagonists would appear to be a major factor accounting
for their competition for binding to a common receptor site. The
differences at the D-ring end of the steroids almost certainly account
for the agonist *versus* antagonist response that follows binding (Duax,
Griffin, Rohrer and Weeks 1982c).

3.2.   D-Ring control of activity

The structural features that are required for binding to the
progestin and estrogen receptors, and the structural features that
differentiate agonism from antagonism, suggest that the A ring plays
the primary role in binding, whereas the D ring plays the primary role
in controlling activity. The possible means by which the steroid D
ring might control activity include one or more of the following (*i*)
inducing or stabilizing an essential conformational state in the receptor (allostery), (*ii*) influencing the aggregation state of the receptor, or (*iii*) participating in a direct interaction with DNA or
chromatin (Duax, Griffin, Rohrer, Weeks and Ebright 1983b) (see Fig.
3.2).

In model (*i*), contact between the bound steroid and a distant part
of the receptor (Fig. 3.2a) induces and/or stabilizes a change in the
receptor conformation. If the appropriate D-ring substituent is missing,
the steroid will bind, but the conformational change in the receptor
will not occur and the hormone will be more readily released. Such a
mechanism would be compatible with the rapid off-rate of estrone that
accounts for its weaker (than estradiol) binding to the estrogen receptor (Weichman and Notides 1980). Alternatively (*ii*), the functional groups on the D rings might stabilize or destabilize the formation of different multimeric forms of the receptors. The nuclear
processing steps essential for the expression of estrogenic activity,
which are partially or completely impaired in estrogen antagonists

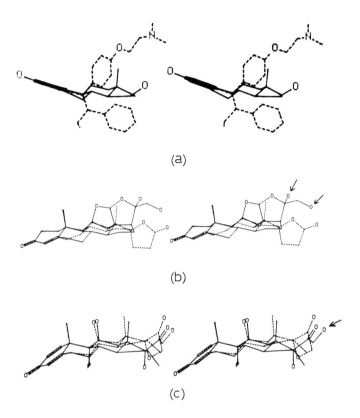

(a)

(b)

(c)

FIG. 3.1. Comparison of the structures of agonists (solid) and antagonists (dashed lines) for three classes of steroid hormones. (a) The endogenous estrogen estradiol (Busetta, Courseille, Geoffre and Hospital 1972) and its antagonist *trans*-tamoxifen (Precigoux, Courseille, Geoffre and Hospital 1979). (b) The endogenous mineralo-corticoid aldosterone (Duax and Hauptman 1972) and its antagonist canrenone (Weeks, Hazel and Duax 1976). (c) The potent glucocorticoid dexamethasone (Dupont, Dideberg and Campsteyn 1974) and its antagonist dexamethasone oxetanone (Duax, unpublished data).

(Horwitz and McGuire 1978), could be the stabilization either of a con-formational state (*i*) or of an aggregate state of the receptor (*ii*) by the D ring. Tseng and Gurpide (1976) have found that only phenolic steroids possessing a 17β-hydroxy group compete with estradiol for nuclear binding.

When the receptor-steroid complex interacts with DNA, the steroid
D ring may be sufficiently exposed to contact the DNA directly (*iii*).
A possible model for such an interaction is provided by the crystal
complex of deoxycorticosterone and adenine (Weeks, Rohrer and Duax
1975), in which the carbonyl and hydroxyl substituents on the corti-
coid D ring form hydrogen bonds to the two nitrogens of adenine that
would normally be involved in Watson-Crick base-pairing. Such contacts
might be critically involved, either in DNA sequence recognition, or in
the activation of transcription by the steroid-receptor complex.

We presently favor the notion that the D ring has at least two
roles: (1) stabilizing a specific conformational state in the receptor
(allostery), and (2) participating in a direct steroid-DNA contact. We
postulate a model whereby the D ring mediates both effects simultaneously (Fig. 3.2).

4.   SUMMARY

Steroid conformations observed in neutron and X-ray crystal structure determinations have been compared with structures calculated using
three molecular mechanics programs, PROPHET/Wipke, MM2p and QCFF/MCA.
Analysis of the data for a very accurate low temperature (123°K) neutron diffraction determination of 20-methyl-5-pregnene-3β,20-diol indicates that the maximum differences between observed and calculated
nonhydrogen atom bond lengths and angles are 0.12Å and 8° for PROPHET/
Wipke, 0.04Å and 5° for QCFF/MCA calculations and 0.02Å and 2.5° for
MM2p calculations. While MM2p calculations adequately reproduce bond
lengths and angles, they do not reproduce the observed torsion angles
and overall conformations of the steroids examined. The narrow range
of side-chain conformations seen in 85 pregnane structures having a
20-one substituent, strongly suggests that crystallographically observed
conformers seldom deviate from lowest minimum energy positions regardless of hypothetical broad energy minima, small differences in energy
between metastable states, and small barriers to rotation. A comparison of observed and calculated values for the orientation of the 17β-
side chain of progesterone and related 20-one steroids indicates a
significant bias in the calculations, a systematic rotation of the
side chain about the C(17)-C(20) bond of approximately 15°. This dis-
parity is probably a result of inappropriately weighted non-bonding

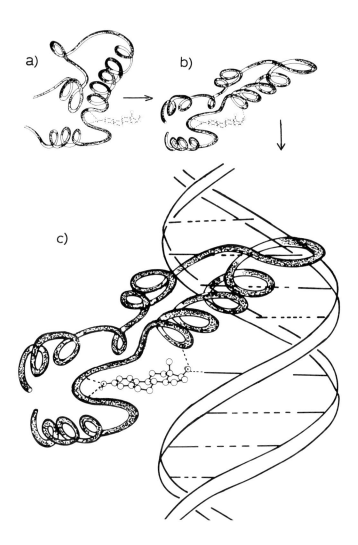

FIG. 3.2. Hypothesis: (a) Receptor binding involves tight contacts only to the steroid A ring. The receptor is in the inactive conformation. (b) Substituents on the steroid D ring induce or stabilize an essential conformational state in the receptor (allostery). (c) The steroid D ring contacts a DNA base. This event is essential, either for DNA-sequence recognition and/or the activation of transcription.

DUPONT, L., DIDEBERG, D., and CAMPSTEYN, H. (1974). Acta Cryst. B30, 514.

FRONCKOWIAK, M. D., Doctoral Thesis, State University of New York at Buffalo (1982).

HORWITZ, K. B. and MC GUIRE, W. C. (1978). J. Biol. Chem. 253, 8185-8191.

HULER, E., SHARON, R. and WARSHEL, A. (1977). Q.C.P.E. 11, 325.

JENSEN, E. V. and JACOBSON, H. I. (1962). In Recent Progress in Hormone Research (ed. R. O. Greep). pp. 387-414. Academic Press, New York.

KARMIN, A. and BROWN, E. A. (1972). Steroids 20, 41-62.

KING, R. J. B. and MAINWARING, W. I. P. (1974). Steroid-Cell Interaction. University Park Press, Baltimore.

KONTULA, K., JANNE, O., VIJKO, R., de JAGER, E., de VISSER, J., and ZEELEN, F. (1975). Acta Endocrinol. 78, 574-592.

MILGROM, E. (1981). In Biochemical Actions of Hormones (ed. G. Litwack). Vol. 8, pp. 465-492. Academic Press, New York.

O'MALLEY, B. W. and BIRNBAUMER, L. (1978). Receptors and Hormone Action. Vol. 1. Academic Press, New York.

PONS, M. and SIMONS, S. (1981). J. Org. Chem, 46, 3262-3264.

PRECIGOUX, G., COURSEILLE, C., GEOFFRE, S., and HOSPITAL, M. (1979). Acta Cryst. B35, 3070.

PROFETA, S. JR., KOLLMAN, P. A., and WOLFF, M. E. (1982). J. Am. Chem. Soc. 104, 3745.

RAYNAUD, J. P., PHILIBERT, D., and AZADIAN-BOULANGER, G. (1973). In Physiology and Genetics of Reproduction (eds. E. Coultinka and F. Fuchs). Part A, pp. 143-160. Plenum Press, New York.

SMITH, H. E., SMITH, R. G., TOFT, D. O., NEERGAARD, J. R., BURROWS, E. P., and O'MALLEY, B. W. (1974). J. Biol. Chem. 249, 5924-5932.

TERENIUS, L. (1974). Steroids 23, 909-918.

TSENG, L. and GURPIDE, F. (1976). J. Steroid Biochem. 7, 817-822.

WEEKS, C. M., ROHRER, D. C., and DUAX, W. L. (1975). Science 190, 1096-1097.

WEEKS, C. M., HAZEL, J. P., and DUAX, W. L. (1976). Cryst. Struct. Comm. 5, 271.

WEICHMAN, B. M. and NOTIDES, A. C. (1980). Endocrinology 106, 434-439.

WELLMAN, K. M. and DJERASSI, C. (1965). J. Am. Chem. Soc. 87, 60.

WIPKE, T. W. (1980). Described in PROPHET Molecules (eds. W. P. Rindone and T. Kush). pp. 2-21 to 2-29. Bolt Beranek and Newman, Inc. Cambridge, MA.

23 Protein crystallography, interactive computer graphics, and drug design

I.J. Tickle, B.L. Sibanda, L.H. Pearl, A.M. Hemmings, and T.L. Blundell

There is now reliable information concerning the three-dimensional structures of more than one hundred proteins, some of which are of great interest to the pharmaceutical industry. They include human hormones and growth factors such as insulin which has recently been made in bacteria using recombinant DNA techniques. They also include enzymes such as renin and dihydrofolate reductase which are the targets of inhibitors which could be, and in some cases already are, used as clinically useful drugs.

Some of these proteins, such as dihydrofolate reductase and insulin, have been studied directly by amino acid sequencing and X-ray crystallography. For most, however, the three-dimensional structures of closely homologous proteins have been defined and interactive computer graphics have been used to model the relevant molecule by modifying the sequence, assuming a very similar tertiary structure. These techniques have been successfully used for renin (Blundell and Sibanda, 1983) and the insulin-like growth factors (Blundell et al, 1978). In some cases – and increasingly often – the protein primary structure will be defined by sequencing the cDNA.

A detailed knowledge of structure function relationships and enzyme mechanism in conjunction with X-ray studies of protein inhibitor complexes can help to define the interaction of the protein receptor and its true substrate. Interactive computer graphics are then used to model the complex, generate complementary surfaces and examine docking of the two molecules together. This information provides a good basis for a rational approach to drug design.

The new approach to rational drug design has been brought about not only by an increased knowledge of the proteins of pharmaceutical

interest, but also by the introduction of improved interactive computer
graphics systems.   New calligraphic or vector drawing systems can simultaneously display many thousands of vectors and transform them in real
time.   Thus complete proteins with ∿3000 atoms can be displayed as line
drawings or as "dot" or "net" surfaces (Tickle, 1982; Langridge et al,
1981; Pearl and Honegger, 1983; Honegger and Blundell, 1983).   The
protein molecules can be manipulated in real time by movement of fragments
or rotations around torsion angles, and substrates or inhibitors can be
docked into active sites (Jones, 1978; Tickle, 1982; Busetta et al,
1983).

       We have previously described the use of such computer graphics
techniques to the study of insulin-like growth factors and their receptor
interactions (Honegger and Blundell, 1983).   We now discuss the applications of the methods to the design of inhibitors of renin, which may be
important in the control of hypertension.

COMPUTER GRAPHICS MODELLING OF RENIN

       Renin is an aspartyl proteinase which catalyses the first and rate
limiting step in the conversion of angiotensinogen to the hormone angiotensin II.   The catalysis is highly specific, and plays an important
physiological role in the regulation of blood pressure.   The complete
cDNA (Pathier et al, 1982) and amino acid sequences (Misono et al, 1982)
of the mouse submaxillary gland renin, which is very similar to the human
enzyme (Hirose et al, 1977) show that these enzymes are homologous to
pepsin, chymosin and other aspartyl proteinases.   Nevertheless, Misono
et al (1982) note that the extremely high specificity and the neutral
optimal pH of renin are not immediately evident from its primary structure.

       The three-dimensional structure of endothia pepsin, refined at
2.1Å resolution (Pearl et al, 1983) was used as a starting point for
modelling of the structure of mouse submaxillary renin.   This was jus-
tified by the close homology of renin to the other enzymes (Misono et al,
1982) and the observation that the three-dimensional structures of all
aspartyl proteinases defined by X-ray analysis to this date are remarkably
similar (Subramanian et al, 1977; Tang et al, 1978).   We have used the
interactive computer graphics program FRODO (Jones, 1978) modified for
an Evans and Sutherland Picture System II by T. A. Jones and I. J. Tickle.
In this analysis we have used the residue numbering of pepsin (Tang et
al, 1973) to facilitate comparison between the enzymes.   Most of the
main-chain and homologous side-chains were first identically positioned,

and then the other side-chains of the core were replaced so that they
occupied closely similar orientations.   The few insertions in the main-
chain required, for example at the disulphide loops involving cystine
residues 45,50 and 206,210 (pepsin numbering) were easily accommodated
with "allowed" main-chain torsion angles.   An extension of the β-turn
in the region 279-280 accommodated the insertion of the four residues
including those cleaved out in the conversion of the zymogen to the two
chain active enzyme (Pathier et al, 1982; Misono et al, 1982).   This
showed that the core of the tertiary structure is retained as hydrophobic
with the same volume occupation, consistent with the main-chain conformation being largely conserved in renin as in the other aspartyl proteinases.   Finally the remaining and often variable side-chains were placed
on the surface in positions which optimised tertiary interactions.

The model shown in Figure 1 was then displayed using the program
MIDAS (Tickle, 1982).   Thus renin is a bilobal enzyme with a deep and
extended cleft.   The two lobes are topologically similar and each provides an aspartate (Asp 32 and Asp 215) important to the catalytic activity in the centre of the cleft.   Figure 2 shows the arrangement of
these aspartates in endothia pepsin (Pearl et al, 1983).   The groups
are related by the pseudo-dyad which also relates the two lobes of the
enzyme (Tang et al, 1978; Pearl et al, 1983).   The aspartates are almost exactly coplanar, hydrogen bonded together and to a water molecule
which lies in the plane of the aspartates.   The aspartates are further
hydrogen-bonded to the main-chain NH groups of residues 34 and 217, and
to the side-chain hydroxyls of Ser 35 and Thr 218.   These residues are
identical or conservatively varied in all aspartyl proteinases including renin, as shown in Table 1.   This symmetrical environment is also

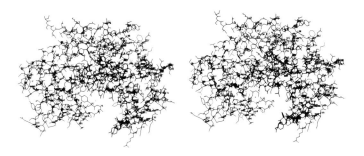

FIGURE 1.   A stereo view of the proposed three-dimensional structure
proposed for mouse submaxillary renin.

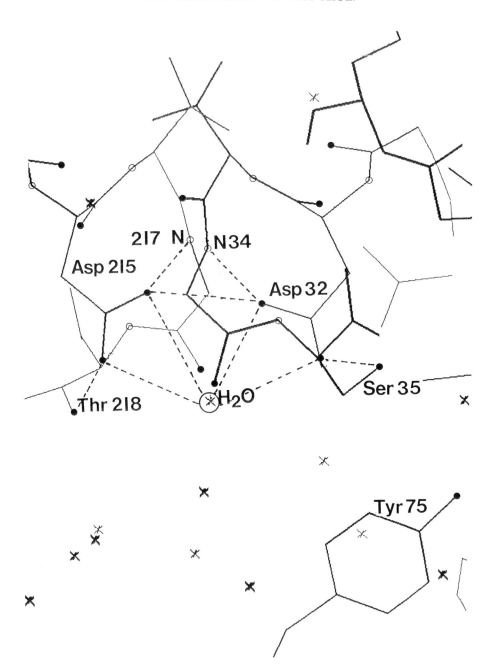

FIGURE 2.   The active site of endothia pepsin

TABLE 1. (N‑terminal lobe)

	11	12	13	30	32	33	34	35	36	37	39	73	75	76	77	111	112	115	116	117	118	120
Endothia pepsin	D	A	A	D	A	D	T	G	S	S	W	I	Y	G	D	F	T	S	T	I	D	L
Chymosin	D	S	Q	L	D	T	G	S	S	D	W	I	Y	G	T	F	T	A	E	F	D	I
Penicillo pepsin	D	E	E	N	D	R	G	S	A	D	W	I	Y	G	D	F	Q	T	N	N	D	L
Porcine pepsin	D	T	E	I	D	T	G	S	S	N	W	I	Y	G	T	F	L	A	P	F	D	I
Renin	N	S	Q	Q	I	D	T	G	A	N	W	I	Y	G	S	F	–	A	Q	F	D	V
Residue Number (Pepsin scheme)	11	12	13	30	32	33	34	35	36	37	39	73	75	76	77	111	112	115	116	117	118	120
(Pepsin scheme) subsite	P₃	P₃	P₁	P₁				P₂'	P₃'		P₁		P₁		P₂	P₁						P₁

Active site Aspartates (↑ at residue 30 / 32)

TABLE 1. (C‑terminal lobe)

	187	188	189	213	215	216	217	218	219	220	222	296	297	298	299	301	304
(Pepsin scheme) subsite	P₃'	P₃'	P₁'					P₂	P₃		P₁'						P₁'
(Pepsin scheme) Residue Number	187	188	189	213	215	216	217	218	219	220	222	296	297	298	299	301	304
Renin	T	D	S	V	D	T	G	S	S	F	S	T	G	P	V	V	A
Porcine pepsin	D	G	Y	I	D	T	G	T	S	L	T	S	G	E	L	I	D
Penicillo pepsin	Q	G	F	I	D	T	G	T	T	L	L	G	I	G	F	I	D
Chymosin	Q	Q	Y	I	D	T	G	S	K	V		S	–	Q	K	I	D
Endothia pepsin	Q	G	F	I	D	T	G	T	T	L	Y	G	I	G	I	I	D

TABLE 1. The residues in the active site clefts of endothia pepsin (V. Pedersen and L. H. Pearl, unpublished results); bovine chymosin (Foltmann and Pedersen, 1977); porcine pepsin (Tang et al, 1973); penicillopepsin (Hsu et al, 1977) and mouse submaxillary renin (Pathier et al, 1981; Misono et al, 1981). The numbering is based on the pepsin sequence. The residues are arranged so that topologically equivalent residues on the two lobes related by a pseudo‑dyad are in the same column. Subsites are indicated according to the scheme of Schechter and Berger (1967).

TABLE   2

The sequences of horse (Skeggs et al, 1957), human
(Tewksbury et al, 1981) and rat (Bouhnik et al, 1981)
angiotensinogens which bind to the active site cleft
of renin.   The residues (S_n) and the complementary
enzyme subsites (P_n) are labelled according to the
scheme of Schechter and Berger (1967)

Enzyme	P_4	P_3	P_2	P_1	$P_1{}'$	$P_2{}'$	$P_3{}'$
Substrate	S_4	S_3	S_2	S_1	$S_1{}'$	$S_2{}'$	$S_3{}'$
Horse	Pro	Phe	His	Leu	Leu	Val	Tyr
Human	Pro	Phe	His	Leu	Val	Ile	His
Rat	Pro	Phe	His	Leu	Leu	Tyr	Tyr

tions bind well.

The residues in subsite P_3 are also similar; however, residue 189 in subsite $P_3{}'$ is remarkably different.   In all other aspartic proteinases sequenced this residue is large and aromatic, either phenylalanine or tyrosine, whereas in renin it is a serine, which is small and hydrophilic.   This may account for the fact that renin prefers large groups at $S_3{}'$ either tyrosine or histidine, which may hydrogen bond to the serine.

At the edges of the active site cleft there are some remarkable differences between renin and the homologous enzymes.   These are illustrated in Figure 4.   They include a surface loop with proline at 293, 295 and 298.   The external surface of the "flap", residues 71-83 which lies across the cleft, becomes highly basic with Lys 81, Arg 79 and His 74.   There is a further surface region involving Lys 239, Lys 241, Arg 242 and His 244; none of these residues is basic in any other aspartyl proteinase.   In addition in the zymogen there is a β-loop easily available on the surface containing three basic groups from which two arginines are cleaved in the conversion to an active enzyme.

This unusual surface region may have three important functional roles.   First it may facilitate recognition of the renin zymogen by the important conversion enzymes.   The resulting residues and the other surface loops may then play a role in binding angiotensinogen which in rat is a protein of molecular weight ∿57000 (Bouhnik et al, 1981); it would be surprising if renin were specific only for residues in the

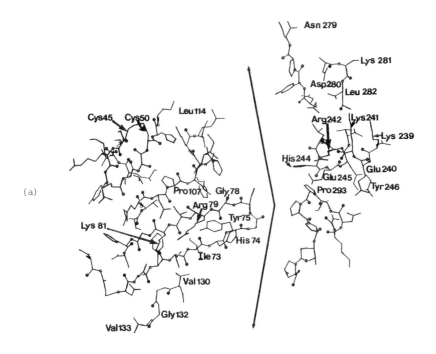

(a)

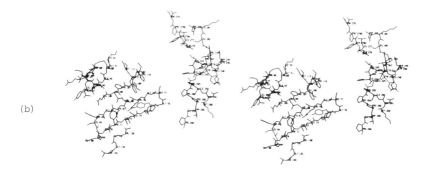

(b)

FIGURE 4.   The arrangement of the residues on the edge of the active
site cleft of renin, viewed in the same direction as Figure 2.

        (a)   diagramatic arrangment

        (b)   stereo view

locality of the scissile bond.   Finally, all of these residues may move
down on to the substrate once it is in the cleft by an induced fit
mechanism; this would allow an increased specificity giving interactions
with the substrate particularly at S_4, S_2, S_2' and S_4'.   These processes would contribute towards an increased activity of the enzyme compared to the zymogen.

MODELLING THE RENIN-SUBSTRATE COMPLEX

We have investigated the use of two programs, DOCKER (Busetta et
al, 1983) and FRODO (Jones, 1982, modified by I. J. Tickle and T. A.
Jones) in docking substrates into the active site cleft.   Figure 5 shows
a preliminary but plausible arrangement of the substrate, rat angiotensinogen, into the active site cleft of mouse submaxillary renin.   The
side-chains are docked into the specificity pockets described above.
This arrangement allows a number of interesting hydrogen bonds to be
formed between the extended main-chain of the substrate and the enzyme.
Of particular interest is the antiparallel β-sheet arrangement of the
NH_2-terminal half of the substrate with the main-chain of residues 217
to 219 affording two hydrogen bonds $\left[NH(S_1) \text{---} O=C \ (217) \text{ and } C=O(S_3) \text{---} \right.$
$\left. H-N \ (219) \right]$.   A similar but parallel β-sheet arrangement of the COOH-
terminal half of the substrate giving a hydrogen bond $\left[N-H(S_2') \text{---} O=C \right.$
$\left. (34) \right]$ with the dyad related and therefore topologically equivalent
stretch of main-chain is possible.

With the substrate side-chains in specificity pockets and the
hydrogen bonds formed with the enzyme, the peptide of the scissile bond
is brought very close to the water molecule bound to the two active site
aspartates.   In fact there are then two possibilities.   First the carbonyl oxygen of the scissile peptide may displace the water molecule.
Although this is stereochemically feasible, the carbonyl oxygen would
be close to a net negative charge on the aspartates in a position which
was previously occupied by a water molecule with two hydrogens capable
of forming hydrogen bonds.   A more feasible arrangement would leave the
water molecule in the position found in the native enzyme.   This water
molecule is hydrogen bonded, it would be partly negatively charged and
more nucleophilic than a free water.   This could therefore be the
nucleophile which attacks the scissile peptide bond; in fact the model
shown in Figure 4 indicates that the water oxygen to carbonyl carbon is
correct for a covalent bond and that the peptide can be placed in an
orientation that leads directly to a tetrahedral transition state.   We

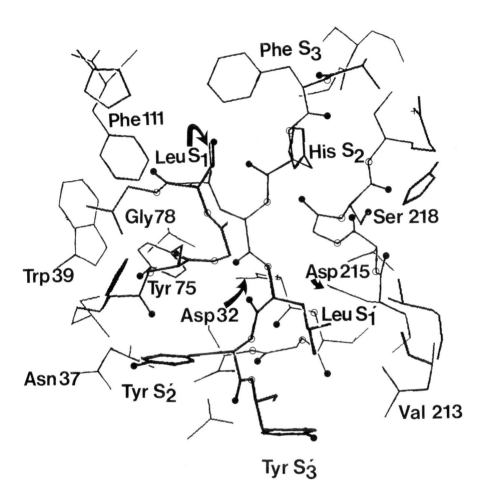

FIGURE 5.   A model of the active site of renin with a fragment of
angiotensinogen bound as substrate.

assume that one of the water protons is transferred to the NH of the
peptide (Pearl et al, 1983).

The model study shows that the complex is more easily formed if
the planarity of peptide bond is lost, a development which would favour
the formation of the tetrahedral state.   We note that this model is
similar to the arrangement of the pepstatin statine hydroxyl in the
complexes with rhizopus pepsin (Bott and Davies, 1983).   It provides
support for the synthesis of peptide analogues with a reduced peptide

bond (Szelke et al, 1982).

THE DESIGN OF NEW INHIBITORS

The model for the renin angiotensinogen complex provides a basis for the rational approach to the design of renin inhibitors. We can first energy minimise the renin substrate complex to model the changes in enzyme conformation induced by the presence of the substrate. We can then generate surfaces to define the enzyme active site using either BILBO (Blundell and Honegger, 1983) or MIDAS (Tickle, 1982) and endow this with indications of the hydrophobicity, charge distribution, hydrogen bonding capacity, etc. We can also describe our molecular model of the substrate in the active site in the same way. We then proceed to examine ways by which a substrate analogue could form a stronger complex with the enzyme. This might involve three steps :

(i) replacement of the scissile bond by transition state analogues resistant to hydrolysis;

(ii) replacement of other parts of the substrate with a chemistry which has a more rigid stereochemistry characteristic of the substrate bound state to minimise the unfavourable loss of entropy on formation of the complex;

(iii) replacement of the peptide groups with others which make a tighter complex with the enzyme. This would be investigated by docking of the putative substrate analogue into the active site cleft to examine complementarity of fit. The most obvious developments would be to investigate increasing the size of the side-groups at S_1 and S_1', and extension of the substrate chain to provide further interactions at the ends of the active site cleft.

These experiments are now under way and will of course be modified to take into account the sequence of human renin when this is available.

ACKNOWLEDGEMENTS

We thank Professor T. Inagami for sending us a preprint of his paper on the amino acid sequence of renin. We are grateful to the UK Science and Engineering Research Council for the support of the Evans and Sutherland Picture System II.

REFERENCES

BLUNDELL, T. L., JENKINS, J. A., KHAN, G., CHOWDHURY, P. R., SEWELL, T., TICKLE, I. J. and WOOD, E. A. (1978) FEBS Letts. 52, 81-94

BLUNDELL, T. L., PEARL, L. H., JONES, H. B., TAYLOR, G. L., JENKINS, J. A., SEWELL, B. T. (1980) in Enzyme Regulation and Mechanism of Action (eds. Mildner, P. and Ries, B.) Pergamon Press, Oxford, pp. 281-288

BLUNDELL, T. L., BEDARKAR, S., RINDERKNECHT, E. and HUMBEL, R. E. (1978) Proc. Natl. Acad. Sci. USA, 75, 180-184

BLUNDELL, T. L. and SIBANDA, B. L. (1983) Nature, in Press

BOTT, R. and DAVIES, D. R. (1983) J. Mol. Biol., in Press

BOUHNIK, J., CLAUSER, E., STROSBERG, D., FRENOY, J-P., MENARD, J. and CARVOL, P. (1981) Biochem. 20, 7010-7015

BURTON, J., CODY, R. J., HERD, J. A. and HABER, E. (1980) Proc. Natl. Acad. Sci. USA, 77, 5476-5479

BUSETTA, B., TICKLE, I. J. and BLUNDELL, T. L. (1983) J. Appl. Cryst., in Press

FOLTMANN, B. and PEDERSEN, V. B. (1977) Adv. Exp. Med. Biol. 95, 3-22

HIROSE, S., WORKMAN, R. J. and INAGAMI, T. (1977) Circ. Res. 45, 275-279

HONEGGER, A. and BLUNDELL, T. L. (1983) in Insulin-like Growth Factors (ed. Spencer, M.), in Press

HSU, I. N., DELBAERE, L. T. J., JAMES, M. N. G. and HOFMANN, T. (1977) Nature, 266, 140-145

INAGAMI, T., MURAKAMI, K., MISONO, K., WORKMAN, R. J., COHEN, S. and SUKETA, Y. (1977) in Acid Proteases (ed. Tang, J.) Plenum, New York, pp. 225-247

JAMES, M. N. G., DELBAERE, L. T. J. and HSU, I. N. (1977) Nature, 267, 808-813

JAMES, M. N. G., HSU, I. N., HOFMANN, T. and SIELECKI, A. (1981) in Structural Studies on Molecules of Biological Interest (eds. Dodson, G., Glusker, J. P. and Sayre, D.) Clarendon, Oxford, pp. 350-389

JAMES, M. N. G., SIELECKI, A., SALITURO, F., RICH, D. H. and HOFMANN, T. (1982) Proc. Natl. Acad. Sci. USA, 79, 6137-6141

JAMES, M. N. G. (1982) Unpublished results

JONES, T. A. (1978) J. Appl. Cryst., 11, 268-272

LANGRIDGE, R., FERRIN, T. E., KUNTZ, I. D. and CONNOLLY, M. L. (1981) Science, 211, 661-666

MISONO, K. S. and INAGAMI, T. (1980) Biochem. 19, 2616-2622

MISONO, K. S., CHANG, J-J. and INAGAMI, T. (1982) Proc. Natl. Acad. Sci. USA, 79, 4858-4862

PATHIER, J-J., FOOTE, S., CHAMBRAUD, B., STROSBERG, P. V. and ROUGEON, F. (1982) Nature, 298, 90-92

PEARL, L. H., SEWELL, B. T., JENKINS, J. A. and BLUNDELL, T. L. (1983) Unpublished results

PEARL, L. H. and HONEGGER, A. (1983) J. Molec. Graphics, 1, in Press

POWERS, J. C., HARLEY, A. D. and MYERS, D. V. (1977) in Acid Proteases, Structure, Function and Biology (ed. Tang, J.) Plenum, New York, pp. 141-157

SCHECHTER, I. and BERGER, A. (1967) Biochem. Biophys. Res. Commun. 27, 157-162

SKEGGS, L. T., KHAN, J. R., LENTZ, K. and SHUMWAY, N. P. (1957) J. Exp. Med. 106, 439-453

SUBRAMANIAN, E., SWAN, I. D. A., LIU, M., DAVIES, D. R., JENKINS, J. A., TICKLE, I. J. and BLUNDELL, T. L. (1977) Proc. Natl. Acad. Sci. USA, 74, 559-566

SZELKE, M., LECKIE, B., HALLETT, A., JONES, D. M., SEIRAS, J., ATRASH, B. and LEVER, L. A. F. (1982) Nature, 299, 555-557

TANG, J., SEPULVEDA, P., MARCINISZYN, J., CHEN, K. C. S., HUANG, W. Y., TOO, N., LIU, D. and LANIER, J. P. (1973) Proc. Natl. Acad. Sci. USA, 70, 3437-3439

TANG, J., JAMES, M. N. G., HSU, I. N., JENKINS, J. A. and BLUNDELL, T. L. (1978) Nature, 271, 618-621

TEWKSBURY, D. A., DART, R. A. and TRAVIS, J. (1981) Biochem. Biophys. Res. Commun. 99, 1311-1315

TICKLE, I. J. (1982) Molecular Interactive Display and Selection, Birkbeck College, London

24 Drug design: principles and techniques
V. Austel

1. PROBLEMS OF PRACTICAL DRUG DESIGN

Almost all of the presently used types of drugs have been found fortuitously. Even lead optimization which is aimed at improving the therapeutic properties of compounds known to be active, still relies heavily on chance. Why are there hardly any reports about successful targeted search for new drugs? Clearly, the reason for this phenomenon is the complexity of the biological system. This complexity has prevented the derivation of structure-biological property-relationships which are valid for a larger number of unrelated chemical structures. The advent of QSAR-methods which allowed biological properties to be quantitatively correlated with physicochemical or structural properties of chemical compounds had initially awakened hopes that therapeutically useful new structures might become predictable. However, the excitement was soon attenuated as the practical application has met with rather limited success.

Yet if drug research were to rely completely on chance it would soon run into unsurmountable problems. Thus at present a newly synthesized compound has an estimated chance of about 10^{-4} to be therapeutically valuable. Since the standards which new drugs have to fulfill are rising this chance inevitably decreases. The concomitant drop of the innovation rate can in principle be overcome by increasing the experimental capacity. But there are serious financial limitations to such an approach. On the other hand innovation in drug research is still desparately needed.

Thus for the majority of the presently known about 30.000 diseases no drug treatment is available to date. In addition, many medicaments used for treating widespread deseases are far from ideal, consider for example the serious side effects which still have to be tolerated in cancer chemotherapy, or the very small therapeutic index of cardiac glycosides which are widely used for treating heart failure.

How can medicinal chemists contribute to upholding a reasonable innovation rate in drug therapy? The main point of concern in this respect is indeed the role of chance. No one can deny that we are far from being able to conceive new drugs on a merely theoretical basis. Rather one still has to rely on empirical procedures. This does, however, not imply that chance is the only determinant of success. Instead chance can be controlled to a certain extent by adopting a systematic approach. Such an approach is characterized by two features, i.e.

1) Experiments, including the structures of test compounds, are designed so that a maximum of structure-activity-information is obtained from a given experimental expense (optimization of the information-expense-ratio).

2) The information content of a series of experiments must be extracted efficiently and completely.

This brings us back to the role of structure-activity-relationships in the search for new drugs. Structure-activity-relationships express the information contained in the biological and chemical data of the test compounds. This information is required in order to design new test compounds properly.

Even though systematic procedures cannot preclude chance from playing a role in drug design they can at least ensure that experimental capacity is employed economically.

Using (empirically derived) structure-activity-relationships in the design of test compounds can reduce the significance of chance but may also introduce bias. Bias is the most serious obstacle on the way to new drugs. Unfortunately structure-activity-relationships are frequently used in a way which optimizes personal bias more than the biological properties of the lead structure. The danger of running into this trap is especially great if structure-activity-relationships are applied in order to predict the "best" compound. Such an application is only safe under the following conditions:

- the predicted structure lies within the structural area for which the respective structure-activity-relationship is valid.

-every possible alternative structure-activity-relationship has been disproven

Judging by the usual data sets reported in the literature the second condition seems to be normally ignored. Such misuse especially of quantitative-structure-activity relationships has contributed a great deal to their failure in practical drug design.

How can such traps be avoided? This is possible first by carefully designing test series so that their composition does not already reflect a certain bias. Secondly, compounds which deviate from a structure-activity-relationship more than can be expected from experimental error ought to be given particular attention rather than being ignored as outliers. Thirdly, by examining the test series for physicochemical or structural properties which might not have been accounted for by the structure-activity-analysis.

A systematic procedure which takes all these considerations into account will be dicussed in the next paragraph. Since this procedure relies heavily on the quality of test series, this aspect will be given special emphasis.

2.STRATEGY AND TECHNIQUES OF PRACTICAL DRUG DESIGN

2.1.Search for a lead structure

The search for new drugs is normally guided by a medicinal objective, which specifies the biological and therapeutical properties expected of a new drug.

As an example from our own work on new cardiotonic agents, acceptable compounds were required to

1) have a higher intrinsic activity and

2) a considerably larger therapeutic index than cardiac glycosides,

3) have no pronounced effect on heart rate in therapeutic doses,

4) show a long lasting effect,

5) be orally effective.

How can one go about searching for a compound which meets the demands of the medicinal objective? A three step procedure can be envisaged.

1) determination of a structural field which presumably contains suitable compounds. Subsequently a preliminary investigation is carried out with the aim of finding a lead compound.

2) determination of a structural area around the lead compound. By investigating this area one hopes to find compounds which comply in every respect with the medicinal objective

3) completion of the investigation. The aim is to make sure that the probability of finding any (more) interesting compounds in the area is less than the corresponding a priori probability in an arbitrary alternative area.

In many cases a lead compound is already available in which case the first step becomes obsolete.

How are the initial structural fields chosen? Quite frequently they are derived from chemical concepts. These may be based on a particular reaction, on a certain structural system which the medicinal chemist has made accessible or on a starting material which has become available in larger quantities.

Finding new drugs via this approach is completely a matter of chance and the probability of succeeding is correspondingly low. However, lead compounds obtained in this way are frequently more promising for optimization purposes than known active drugs.

Thus the benzodiazepine tranquillizers were discovered by applying a purely chemical concept. In searching for new tranquillizers Sternbach (1978) selected a structural field which had hardly been investigated before but which he assumed to be readily accessible to him, i.e. the benzoxadiazepines 1. However, these compounds turned out to be quinazoline-N-oxides 2. The synthetic work included the introduction of a dialkylamino group ($X = NR_2$) via a nucleophilic displacement of chlorine ($X = Cl$). None of these compounds showed the desired pharmacological properties. However, on attempting to exchange chlorine for monoalkyl amino groups an unexpected rearrangement took place leading to benzodiazepines 3, whose biological properties were significantly superior to the reference compounds.

      1   2   3

The role of chance can be reduced considerably by choosing structural fields on the basis of biomolecular processes which are involved in the respective diseased states. In this context information about the structure of receptors and enzymes and of their respective complexes with substrates or effectors would be particularly valuable. Unfortunately, such knowledge is still scarce. Recent advances in the techniques of isolating receptors and enzymes and in the elucidation of their structures justify the prediction that biomolecular aspects will become more and more important in conceiving new drugs. With enzyme inhibitors in particular, known structures of respective enzyme-substrate complexes have already been used in order to determine structural fields.

Thus captopril 5, a new antihypertensive drug has been developed from such a structural field (Cushman, Ondetti, Cheung, Sabo, Antonaccio and Rubin 1980). Captopril blocks angiotensin converting enzyme (ACE) which removes a dipeptide from the carboxyl end of angiotensin I. The reaction product is angiotensin II a potent natural pressor substance. It was hypothesized that the catalytic unit of ACE resembles that of carboxypeptidase A, the structure of which in the presence of a ligand had previously been elucidated by X-ray crystallography. From these data a structural field characterized by general structure 4 was derived. It comprises three essential features, i.e. a negatively charged oxygen (thought to interact with a positively charged arginine in the enzyme), a hydrogen bond acceptor (V) and a group Z capable of interacting with the Zn^{++}-ion of the catalytic site. The distance between the positively charged arginine and the Zn^{++}-ion was assumed to correspond to the length of a dipeptide unit (in contrast to carboxypeptidase A in

which this distance only accounts for the one amino acid which is
cleaved off the substrate)

After having chosen a structural field one can start searching for a
suitable lead compound. At this stage it is advisable to select the test
compounds from various structurally different areas of the field. As an
example, in our search for new cardiotonic drugs one of the structural
fields which we had decided to invesitgate was characterized by moiety
$\underline{6}$. The initial test sets which were selected from this field covered

among others the areas shown in Table 1.

The investigation is continued until a reasonably active compound has
been obtained. This compound becomes the new lead. In most cases the
lead is unsuitable as a new drug, usually because of intolerable side
effects or unwanted pharmacokinetic properties.

In our example a new lead compound has been found in the subarea of
the benzimidazoles. This compound ($\underline{7}$) meets the requirements with respect to intrinsic activity, therapeutic index and effect on heart rate.
However, relative potency was too low. Moreover the compound was ineffective orally and on i.v. administration effects were very short-lived.

TABLE 1

Examples for areas of the structural field 6

	Area	Example
Open chain enamines		
Pyridines		
Thiazines		
Quinazolines		
Benzimidazoles		

Shortcomings of lead compounds can frequently be abolished through suitable structural variations. The search for such variations is generally referred to as lead optimization.

2.2. Lead optimization

Targeted lead optimization is possible but mostly confined to some specific improvements of pharmacokinetic properties. A well known method in this category is the prodrug approach. This approach utilizes the organism's ability to convert inactive compounds (prodrugs) into active drugs. Sometimes active molecules contain moieties which are responsible for undesirable pharmacokinetic properties such as insufficient oral absorption or too rapid excretion. Yet these moieties are essential for the molecule's pharmacodynamic properties and can therefore not be replaced by other groups except if the organism is capable of reconverting them into the essential moiety. Thus alcoholic or phenolic groups can be changed into ethers or esters in order to improve e.g. gastro-intestinal absorption. The prodrug concept has been described in detail in recent reviews (Sinkula and Yalkowsky 1975, Pitman 1981) and shall not be discussed here any further.

Generally, however, a lead compound requires primarily pharmacodynamic improvement. In addition, pharmacodynamically acceptable analogues do not always contain groups which lend themselves to a prodrug approach. In such cases the lead structure needs to be optimized empirically. A pertinent systematic procedure is outlined in the following paragraph.

General procedure. Systematic lead optimization starts with defining a suitable structural area around the lead compound. This area is characterized by a basic skeleton, which is derived from the lead compound. The basic skeleton incorporates or has attached to it certain points of variation. These variations can, but need not, be specified in terms of real structures. The compounds which lie within the structural area form a set of primarily infinite size. Reduction to a finite size is achieved by specifying the envisaged types of variations.

As an example, a structural area around lead compound 7 could be defined as shown in formula 8. The basic skeleton is the same as in compound 7 except that it includes one position of variation (X).

$$\underline{8}$$

R^1 and R^2 are arbitrary substituents and the phenyl ring may carry additional substituents.

How does one search for the new drug(s) which are supposedly located in the structural area around the lead compound? It is, of course, impossible to synthesize and test every single element of this area (basic set). Therefore representative sets of samples (test compounds) must be selected from the set. The test sets are subjected to structure-activity-analyses. The resulting structure-activity-relationships form the basis for choosing the next test set. Again, a structure-activity-analysis is carried out and the result is compared with the previously derived relationships. If there is a deviation a corresponding correction is made and a next test set is selected etc. until consistency is achieved.

Additionally, one has to make sure that the final structure-activity-relationships are unique, i.e. that there are no alternatives which could explain the biological data equally well. Only under these circumstances is the area completely investigated. If a suitable compound has been found, one has achieved one's aim.

If no such compound has been encountered and if the structure-activity-relationships suggest a very low probability of finding one within the structural area, the search within this area should be discontinued. This iterative optimization procedure is visualized in Fig. 1.

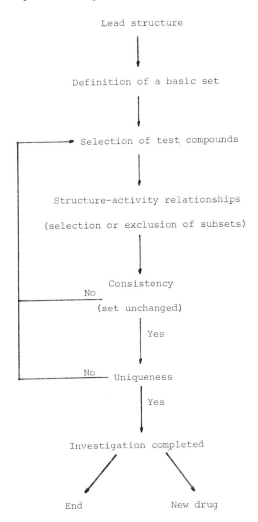

FIG. 1 Systematic lead optimization

What types of structure-activity-relationships does one need for lead optimization? One can distinguish three types of structure-activity-relationships. They differ with respect to precision and to the way in which the chemical structures are described:

1) structure-activity-classifications by qualitative comparison of structural moieties (qualitative structure-activity-relationships)

2) structure-activity-classifications by physicochemical and structural descriptors.

   In both cases the original set is divided into a few subsets which differ with respect to the biological properties of their members.

3) quantitative structure-activity-relationships, which express a biological property in terms of a mathematical function of physicochemical and structural descriptors.

In the early stages of an optimization procedure the use of qualitative structure-activity-relationships is sufficient, whereas in the advanced stages classifications of the second type become important. Quantitative structure-activity-relationships can become helpful in the final stages in order to determine consistency and uniqueness.

Structure-activity-relationships are an indispensable constituent of systematic drug design. With their aid the factors (physicochemical or structural properties or parameters) can be detected which govern the biological properties of a set of compounds. This information is needed in order to design new test series and to get an estimate as to how thoroughly a structural area has been investigated. Structure-activity-relationships also define promising and less promising subsets of the basic set. By excluding the latter ones from further examination the basic set becomes gradually reduced. If no further reduction is possible, the corresponding area can be considered sufficiently investigated.

Design of test sets. The reliability of structure-activity-relationships and hence of all conclusions drawn from them is heavily dependent on the composition of the parent test sets. If these test sets are not representitive for the respective basic or reduced set, erroneous

structure-activity-relationships may result. Consequently valuable experimental capacity is wasted. But how does one know whether or not a
test set is representative for a certain basic set and how can one design appropiate test sets? An answer can be found by characterizing sets
a little more precisely than just by some basic skeleton. One needs to
specify the physicochemical or structural features which presumably
influence the biological properties under consideration. In addition,
these features must be represented by suitable parameters. These include
indicator variables, physicochemical parameters such as linear free
energy parameters, quantum mechanical parameters, connectivity idices,
distances and angles, etc. One can now proceed in two ways, i.e.

1. by specifying the ranges which the parameters ought to cover within
   the basic set
2. by specifying all the elements of the basic set. Subsequently the
   corresponding ranges of the parameters are determined.

The parameters determine a parameter space and the respective ranges
define a certain area within this space. This area is associated with
the basic set. Every element (compound) of the basic set corresponds to
a point in the associated area. A test set is considered representative
if the points corresponding to its members are evenly distributed over
the whole associated area. With two-dimensional parameter spaces the
evenness of distribution can be checked by visual inspection. In multi-
dimensional cases numerical controls are needed. In our laboratory we
use orthogonality and shortest relative distance between two test compounds for this purpose. We assume a test set to be sufficiently orthogonal as long as none of the eigenvalues of the correlation matrix is
below about 0.3. The shortest distance is measured in % of the maximum
distance within the area in parameter space (scaled so that every parameter assumes values from -1 to 1). Normally this distance should not be
much lower than 15 %.

   There are two ways of forming test sets, i.e. either by designing
appropriate structures (if the area is defined according to procedure 1)
or by selecting suitably located elements from the basic set (in case of
procedure 2). The latter alternative is more commonly applied even
though it may cause problems. Thus, the elements of a basic set do sometimes not cover the whole associated area. Consider for example the
basic set of substituents shown in Fig. 2.

In the two-dimensional parameter space with the axes π and MR (lipophilicity and polarizability) the associated area ranges from -1.58 to
0.71 in π-direction and from 0.92 to 13.82 in MR-direction. From Fig. 2
it becomes evident that this set is incomplete because it leaves the
lower left part of the associated area completely blank.

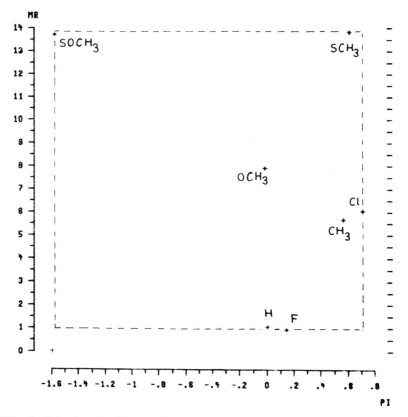

FIG. 2 Set of substituents in a two-dimensional parameter-space. Broken
lines enclose the associated area (values from Hansch and Leo 1979)

Obviously it is impossible to select evenly distributed test sets from
such incomplete basic sets.

    A number of methods for designing or selecting test sets have been
reported in the literature. Some can be applied manually, e.g. the
Topliss (1972, 1977) methods and the Craig (1971) graphical procedure,
whereas others require computer facilities. Well known examples are a

method based on cluster analysis, as reported by Hansch, Unger and Forsythe (1973), and one which uses nonlinear mapping (Wootton, Cranfield, Sheppey and Goodford 1975). The Craig procedure yields well spread test sets and the two other methods also frequently give satisfactory results provided they are applied to complete basic sets.

If, however, the basic set is incomplete. This may cause problems since the latter two methods do not allow one to detect such a situation.

We have therefore developed a method for designing test sets which can handle incomplete basic sets as well. This method does not only reveal incompleteness but it also allows compounds to be designed which properly amend the basic set. The method is applicable to both ways of composing test sets, i. e. by design and by selection, and can be used manually as well as with computer assistance. There are no limitations neither with respect to structural types and variations nor with regard to the number and types of parameters. The method is based on 2^n-factorial experimental design (detailed descriptions e.g. by Cochran and Cox 1968) using Yates-schemes (Yates 1937). For series design purposes the columns of these schemes refer to physicochemical or structural parameters and the rows to subareas of the associated area in parameter space. The plus and minus signs denote the levels which the parameters adopt in the respective subarea. Schemes for two- and three-dimensional cases are shown in Table 2a and b respectively.

TABLE 2

2^n-Factorial schemes with a central area for 2 (a) and 3 (b) factors (parameters).

a) Subarea

No.	A	B
1	−	−
2	+	−
3	−	+
4	+	+
5	0	0

b) Subarea

No.	A	B	C
1	−	−	−
2	+	−	−
3	−	+	−
4	+	I	−
5	−	−	+
6	+	−	+
7	−	+	+
8	+	+	+
9	0	0	0

Every scheme has been augmented by a central subarea in which all the parameters adopt medium (0) values. In the manual version the levels are specified by certain ranges of parameter values. These ranges can in principle be chosen freely. We usually identify the upper, middle and lower third of the total range which a parameter covers within the basic set with the plus, medium and minus levels respectively. Using this specification and the 2^2-factorial scheme in Table 2a the area associated with the basic set in Fig. 2 can be subdivided as shown in Table 3.

TABLE 3

Subareas determined by applying the 2^2-factorial scheme of Tab. 2a to the basic set of Fig. 2. The last column contains substituents which represent the respective subarea. Subareas no. 1 and 5 are not covered by the basic set.

Subarea No.	π (A)	MR (B)	Substituent
1	< -0,82	< 5.22	-
2	> -0.05	< 5.22	F
3	< -0.82	> 9.52	SOMe
4	> -0.05	> 9.52	SMe
5	-0.82 < π > -0.05	5.22 < MR < 9.52	-

In trying to represent any one of these subareas by one element of the basic set we note that this is not possible for subareas 1 and 5. Hence, the basic set is incomplete. The location in parameter space of the blank parts is defined by the factorial scheme (Tab. 3) Therefore appropriate substituents for completion of the set can be determined directly. In the present case NO_2 and NH_2 are suitable. These substituents complete the set as can be confirmed graphically (Fig. 3). Fig. 3 illustrates the geometric significance of the subareas defined in Tab. 3.

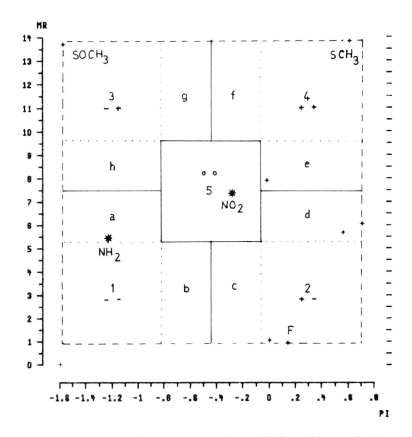

FIG. 3 Basic set of Fig. 2 completed by addition of NH_2 and NO_2.

The parts labelled a-h do not fit into this scheme but can be added to the previously defined subareas as shown in Fig. 3. Test sets are assembled by selecting one element out of each subarea. Compounds which occupy the border line regions a-h should only be taken if no other alt- ernative is available or feasible. In our example the following test set can be selected (subarea in brackets): NH_2 (1), F (2), SOMe (3), SMe (4), NO_2 (5). This test set is rather well spread as one can see by eye inspection and as can be confirmed by the eigenvalues of the cor- relation matrix (1.12, 0.88). The shortest distance between two selected substituents is 31 % of the longest possibe distance within the assoc- iated area. The latter distance is the one between two opposing corners of the area in scaled units.

Lead optimization with the aid of 2^n-factorial schemes. 2^n-factorial schemes cannot only serve as a means of series design, they are also applicable to data analysis. This renders them a valuable tool for lead optimization. For practical drug design purposes one can adopt the following pragmatical procedure:

1) determination of subareas
2) selection or design of a test set as outlined above
3) exclusion of all subareas (subsets) whose representatives are not sufficiently active (this step corresponds to a structure-activity-classification followed by a reduction of the basic set).
4) selection of a test set from the remaining subarea(s) as under 1) etc.

A simplified example for such a procedure is provided by the optimization of the previously mentioned cardiotonic lead compound 7 for which for example the structural field characterized by 8 can be chosen. The following structural features are considered potentially relevant for the cardiotonic activity of the compounds:

1) nature of X, represented by an indicator variable I_X which is 1 for X=N and 0 for X= CH
2) nature of R^1, represented by an indicator variable (I_R) which is 1 for R^1=H and 0 for R^1=H
3) possibility of a hydrogen bond between one N of the imidazole and R^2, represented by an indicator variable I_H which is 1 if such a bond is possible and 0 otherwise.

Assigning values of 1 to the plus and of 0 to the minus levels and using a 2^3-factorial scheme a set of test compounds can be designed as shown in Table 4. Since the structures are described by binary indicator variables only a central area cannot be defined.

The compounds were tested in vitro (guinea pig atria) and in vivo (anesthetized cat i.v.). In both models compound no. 6 was significantly better than the lead (5) and the other compounds. Therefore the structural area around compound no. 6 was given priority for further

TABLE 4

Test set designed with the aid of a 2^3-factorial scheme for the basic set characterized by $\underline{8}$

No.	I_X (A)	I_R (B)	I_H (C)	X	R^1	R^2	others
1	−	−	−	CH	H	H	3,4-di-OMe
2	+	−	−	N	H	H	−
3	−	+	−	CH	Me	OMe	4-OMe
4	+	+	−	N	a)	H	−
5	−	−	+	CH	H	OH	4-OH
6	+	−	+	N	H	OMe	4-OMe
7	−	+	+	CH	Me	OH	4-OMe
8	+	+	+	N	b)	OH	4-OMe

a) morpholino-propyl   b) 4-methoxy-phenyl

investigation. The latter structural area (basic set) can be characterized by general structure $\underline{9}$.

$$\underline{9}$$

In the previous step no effort was made to limit the size of the basic
set and the test set was formed by design. For the following step we
shall confine ourselves to a finite set by identifying R^1 with Cl, F,
H, OH, OMe, SMe and R^2 with Cl, F, H, Me, OMe, SOMe, SMe. The test set
is selected from the basic set formed by the 6x7 = 42 combinations of
these substituents. Three properties (parameters) are considered relevant, i.e. overall lipophilicity ($\Sigma\pi$ for R^1 and R^2), steric effect of
R^1 [represented by S_b (Austel, Kutter, Kalbfleisch 1979)] and bulk

of R^2 (approximated by MR). The associated area in this three-dimensional parameter space extends, from -2.25 to 1.42 in $\Sigma\pi$-direction, from
0 to 2.2 in S_b-direction and from 0.92 to 13.82 in MR-direction. Using
a 2^3-factorial scheme with the central area (Tab. 2b) and taking again
the upper, middle, and lower third of the range of each parameter as the
plus-, zero- and minus-level, respectively, we obtain the nine subareas
shown in Tab. 5.

TABLE 5.

Subdivision of the area associated with the basic set characterized by
structure 9

Sub-area No.	$\Sigma\pi$ (A)	S_b (B)	MR (C)
1	< -1.03 (-)	< 0.7 (-)	< 5.22 (-)
2	> 0.20 (+)	< 0.7 (-)	< 5.22 (-)
3	< -1.03 (-)	> 1.5 (+)	< 5.22 (-)
4	> 0.20 (+)	> 1.5 (+)	< 5.22 (-)
5	< -1.03 (-)	< 0.7 (-)	> 9.52 (+)
6	> 0.20 (+)	< 0.7 (-)	> 9.52 (+)
7	< -1.03 (-)	> 1.5 (+)	> 9.52 (+)
8	> 0.20 (+)	> 1.5 (+)	> 9.52 (+)
9	$-1.03 < \Sigma\pi$ < 0.20 (0)	$0.7 < S_b$ < 1.5 (0)	$5.22 < MR$ < 9.52

Three of these (1-3) do not contain any element of the basic set. One
could represent these subareas by substituents taken from border line
areas (analogous to areas a-h in Fig. 3): R^1=H, R^2=OH (1); R^1=H,
R^2=Me (2); R^1=OMe, R^2=OH (3). However, only the second compromise
compound lies close enough to the corresponding subarea to be acceptable. In order to achieve a better representation of the remaining two
subareas (1 and 3) the basic set needs to be amended, e.g. by R^1=H,
R^2=NH$_2$ (1) and R^1=OMe, R^2=NH$_2$ (3). Both compounds occupy border line areas but are very close to the respective subareas. The test
set selected after completion (Tab. 6) of the basic set represents the

TABLE 6

Test compounds selected from the basic set characterized by 9. a) compounds which had to be added in order to complete the original set b)
this compound lies in one of the border line areas added to subarea 2

Sub-area No.	R^1	R^2	$\Sigma\pi$	S_b	MR
1	(H	$NH_2)$ [a)]	-1.23	0.0	5.42
2	H	Me	0.56	0.0	5.65 [b)]
3	(OMe	$NH_2)$ [a)]	-1.25	2.0	5.42
4	SMe	H	0.61	2.2	1.03
5	H	SOMe	-1.58	0.0	13.70
6	H	SMe	0.61	0.0	13.82
7	OMe	SOMe	-1.60	2.0	13.70
8	OMe	SMe	0.59	2.0	13.82
9	OH	OMe	-0.69	1.0	7.87

associated area very well [eigenvalues of the correlation matrix; 1.24,
0,98, 0.79; smallest distance between two test compounds (1,2): 28 % of
the maximum distance]. Three of these compounds (3,7,8) exert very good
in vitro and in vivo activity. Compound no. 7 (AR-L 115, Vardax[R], 10)
also fulfills all the other requirements of the medicinal objective,
i.e. it has a high therapeutic index, does not significantly accelerate
heart rate in therapeutic doses, is orally active and gives reasonably
long lasting effects. The compound is now under advanced clinical investigation.

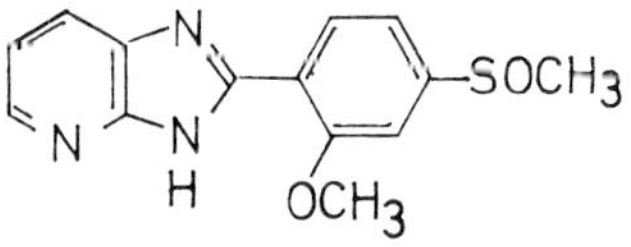

10

REFERENCES

AUSTEL, V., KUTTER, E. and KALBFLEISCH, W. (1979). Arzneim. Forsch.
　　29 (I), 585

COCHRAN, W.G. and COX G.M. (1968). Experimental Design (2nd edn)
　　pp 148-182, J. Wiley, New York

CRAIG, P.N. (1971). J. Med. Chem. 14, 680

CUSHMAN, D.W., ONDETTI, M.A., CHEUNG, H.S., SABO, E.F., ANTONACCIO, M.J.
　　and RUBIN, B. (1980). In Enzyme Inhibitors as Drugs (ed. M. Sandler)
　　pp. 231-247. Macmillan, London

HANSCH, C. and LEO, A. (1979). Substituent Constants for Correlation
　　Analysis in Chemistry and Biology. J. Wiley, New York

HANSCH, C., UNGER, S.H., and FORSYTHE, A.B. (1973). J. Med. Chem. 16,
　　1217

PITMAN, I.H. (1981), Med. Res. Rev. 1, 189

SINKULA, A.A. and YALKOWSKY, S.H. (1975). J. Pharm. Sci. 64, 181

STERNBACH, L.H. (1978). Progress in Drug Research 22, 229

TOPLISS, J.G. (1972). J. Med. Chem. 15, 1006

TOPLISS, J.G. (1977). J. Med. Chem. 20, 463

WOOTTON, R., CRANFIELD, R., SHEPPEY, G.C., and GOODFORD, P.J. (1975).
　　J. Med. Chem. 18, 607

YATES, F. (1937). Design and Analysis of Factorial Experiments, Imperial
　　Bureau of Soil Science, London

25 Molecular modelling by computer techniques and pharmacophore identification

J.P. Tollenaere, H. Moereels, and L.A. Raymaekers

1. INTRODUCTION

It is now widely accepted that conformation is an extremely important aspect of a drug molecule since it strongly influences molecular interactions with the receptor and therefore should provide valuable information about the three-dimensional anatomy of molecules and topographical details of processes taking place at the receptor. A detailed study of the topographical characteristics of drug molecules belonging to a given pharmacological class or putatively interacting with the same receptor may furnish a set of structural requirements necessary to elicit a given pharmacological response. The structural requirements common to a set of compounds acting at the same receptor may then be used to define a pharmacophoric pattern of atoms or groups of atoms mutually oriented in space. In its turn, the pharmacophore may subsequently be used to either propose new molecules featuring the pharmacophore embedded in their structure or to rationalize the SAR of a set of compounds belonging to a pharmacological class.

As conformation is not the sole determinant of biological activity, it should be borne in mind that although molecules may possess the proper pharmacophore they may be endowed with additional properties which may render the new molecules unsuitable for further consideration. For example, isosteric replacement of a methylene group by an ether oxygen will considerably reduce the lipophilicity of the resulting molecule. Replacement or addition of atoms in the neighbourhood of a basic nitrogen atom may result in an appreciable change of the basicity of that nitrogen and a concomitant change in the ratio neutral: protonated species at physiological pH. What at first sight may look like a harmless structural modification may lead to a pronounced departure from the op-

timal charge distribution in a molecule which may have grave consequences for the Coulombic contribution to the drug-receptor interaction
energy. In other words, structural modification almost always leads inevitably to changes in lipophilicity (log P), basicity (pKa, proton affinity) and charge distribution (μ). Stated otherwise: molecular properties are not orthogonal. Nevertheless, despite this non-orthogonality
and with due respect for the complications arising from to the non-
orthogonality, conformational analysis and subsequent pharmacophore
identification is a rich source of knowledge and a powerful tool in
gaining a better understanding of the SAR and in the rational design of
new, more potent and/or more specific drugs.

The problem facing the medicinal chemist involved in rational drug design is two-fold. The first bears on the conformational analysis itself.
The second deals with the methods and techniques to be used for the handling and the interpretation of the data obtained from the conformational
analysis for the proposal of a possible pharmacophore. In other words,
conformational analysis is not a goal in itself but an essential and
necessary prerequisite for the construction of a hypothetical pharmacophoric pattern.

This contribution addresses the techniques of conformational analysis
and secondly will deal with the results of conformational analysis and
how they can be exploited towards a better understanding of the SAR of
4-anilinopiperidine derivatives.

2. CONFORMATIONAL ANALYSIS

In practice, three different methods are available to determine the
conformation of relatively large molecules such as drugs. These are:
X-ray diffraction analysis for the solid state, NMR spectroscopy for the
liquid or dissolved state and quantum chemical or empirical molecular
mechanics calculations for the isolated state (Tollenaere, 1981).

2.1. X-ray diffraction analysis

Single-crystal X-ray diffraction analysis is the method of choice
for determining the three-dimensional structure of a molecule. Although
a very precise description of the molecular geometry is obtained with
today's modern equipment, X-ray diffraction usually reveals only one
conformation a molecule may adopt and tells us nothing about other, pos-

sibly equienergetic or other low energy conformations and their relative stabilities. Inspection of the molecular packing arrangement, however, may yield some valuable information about intermolecular contacts, sites of intermolecular hydrogen bonds and the relative orientation of the molecule and solvent molecules if present in the unit cell. An additional strong point in favour of X-ray diffraction analysis is that the atomic coordinates thus obtained constitute in many cases the primary input data for quantum chemical and molecular mechanics calculations.

2.2. Nuclear Magnetic Resonance spectroscopy

In principle, NMR analysis yields the conformation of a molecule in solution. In general, the structural complexity of a drug molecule renders the interpretation of an NMR spectrum rather difficult and laborious. More often than not, NMR analysis yields partial answers with respect to the conformational aspects of molecules in solution. Evidence has accumulated over the years that in many cases one particular conformation found in solution corresponds to the conformation found in the solid state. In those cases where a thorough NMR analysis has been made it is often possible to determine the rotamer ratio at a given temperature. The latter may yield valuable information regarding the conformational flexibility of the molecule.

2.3. Quantum chemical and molecular mechanics calculations

Provided sufficient computer resources are available, quantum chemical calculations yield the whole conformational domain of a molecule in the isolated state, the relative stabilities of the various conformers, the energy levels, the charge distributions and other quantities that can be derived from the wave functions. It is outside the scope of this contribution to discuss the merits and shortcomings of the various quantum chemical formalisms currently in use. Over the last 10 years or so, however, the semi-empirical all-valence electron method PCILO (Perturbative Configuration Interaction using Localized Orbitals) of Diner et al. (1969) has been used extensively in many laboratories including our own (Tollenaere et al., 1980) because it offers a good comprise between speed of computation and reliability of prediction of conformational preference.

Notwithstanding the increased speed of calculation over many other quantum chemical calculations, PCILO may still be too slow for the exploration of high-dimensioned conformational hyperspaces. In those cases molecular mechanics or empirical force-field calculations become a

necessary choice and in most instances a valid and reliable alternative.
Molecular mechanics calculations are based on the summation of empirically fitted potential energy functions of the Lennard-Jones 6-12 type
for non-bonded interactions, of harmonic functions for atomic orbital
interactions, of Hooke functions for bond stretching and angle bending,
of Coulombic charge-charge interactions and special hydrogen-bond functions. Molecular mechanics methods pioneered and developed by Allinger
(1976), Momany et al. (1975), Andose and Mislow (1974), Hopfinger
(1973), Stuper et al. (1979) and Weiner and Kollman (1981) are fast,
easily programmed and yield reliable solutions to complex conformational
problems.

   Theoretical conformational analysis is a combinational problem. Taking into account all potential functions mentioned above in the exploration of a complicated conformational hyperspace may make molecular
mechanics calculations as prohibitively time consuming as the more ri-
gourous computational schemes. Therefore, depending on the answers one
is looking for and the complexity of the molecule one is dealing with
one may be forced to resort to simplifying assumptions. One such assumption is to perform calculations by keeping bond lengths and bond angles
fixed at standard or experimentally observed values. Under this assumption the shape of a molecule is solely determined by rotations around
single bonds and consequently the combinational problem is reduced to
$N = (360/I)^n$ where I is the rotational increment and n the number of
rotatable bonds. A typical example where $I = 10°$ and $n = 3$ yields $N =$
46 656 conformations to be explored. However, when $I = 10°$ and $n = 4$ the
number of conformations jumps to 1,679,616. It is clear that even molecular mechanics calculations might present non-trivial computational
problems!

2.4. Comments on conformational analysis

   Modern conformational analysis places at our disposal three techniques permitting the determination of the conformation(s) of a molecule
in three different environments or aggregation states. As long as we are
ignorant about the details of the exact environment prevailing at the
receptor site there is no a priori reason to assume that any one of
those three experimentally accessible aggregation states is a good
approximation of the environment at the receptor, at the so-called medicinal chemist's "fourth aggregation state". For instance, it may be
questioned whether the isolated state with the implicit assumption of an

isotropic surrounding environment reflected by $\epsilon = 1$ is a not too gross a simplification of the medium in which drug interactions are assumed to take place. Numerous reports in the literature suggest that ϵ values somewhere between 2 and 10 for the intervening medium are more appropriate (Go et al., 1971; Kier and Aldrich, 1974; McGuire et al., 1972; Weintraub and Hopfinger, 1973). NMR results based on aqueous or chloroform solutions are equally suspect if water and chloroform have to serve as model compounds mimicking the receptor environment.

Possible discrepancies between the results of the theoretical approach and the two experimental methods are not necessarily a reflection on the reliability of the former. For instance, both ab initio and PCILO calculations predict that the intrinsic conformational preferences for both the mono- and dications of histamine are those as found in the solid state. The NMR results, however, indicate different conformations for histamine in solution. By using the supermolecule approach (Port and Pullman, 1973), the theoretical results are in close agreement with the NMR results (Pullman and Port, 1974). A similar discrepancy between experiment and theory for serotonin has been described and successfully explained by specifically taking into account in the calculations the environmental factors prevailing in solution (Pullman et al., 1974). Similarly, the disagreement between the solid-state conformation for adrenaline and that predicted by PCILO and ab initio calculations can be traced back to the crystalline environment or the absence of it in the theoretical calculations (Caillet et al., 1977).

3. COMPUTER GRAPHICS

As conformational analysis becomes more and more a matter of routine and our knowledge about the 3-D aspects of molecules grows hand in hand with developing computer technology, a strong need is felt to visualize the results of conformational analysis and to be able to manipulate structures in a more sophisticated way than can be done using the popular hand-held molecular models. Though physical models of the stick or space filling variety are still valuable, they are cumbersome, easily transformed by curious hands into rather doubtful conformations and do not lend themselves to superposition one upon the other. With the advent of computer aided design systems and more specifically computer graphics many of these and other problems have been elegantly overcome. Unfortunately, computer graphics has some of its own problems, of which overcoming the difficulty of displaying in a convincing way a three-dimen-

sional shape onto a two-dimensional screen is not the most trivial. To a
large extent how comfortable a medicinal chemist is in front of the
graphics screen is determined by the commitment that has been made in
the new technology (Weintraub, 1979; Gund et al., 1980; Dyott et al.,
1980; Langridge et al., 1981). A minimum commitment beyond which the
extra expense is not compensated by the benefits is necessary for there
to be a significant improvement over the old methods of computational
batch procedures and hand-held models.

The computer graphics system presently being built in our laboratory
is the offspring of our ideas of how we could offer a contribution to
the ultimate goal of more rational drug design. Some years ago, our calculation procedures in conformational analysis and pharmacophore identification were batch procedures using a wide range of routines to generate sets of coordinates and associated energies calculated by PCILO.
Separate routines such as BMFIT (Nyburg, 1974) enabled us to superimpose
two structures. Structural representation came as line printer drawings
(Tollenaere et al., 1977) and later as ORTEP (Johnson, 1965) drawings.
Gradually the need was felt to link some of these batch programmes and
to transfer some of the results to an interactive typewriter terminal.
Nevertheless, as time went on the package evolved into an unmanageable
conglomerate of routines which made it difficult to work with and almost
impossible to modify and expand. This untenable situation led to the
decision to build a completely new system.

AIDA (Aid in Interactive Drug Analysis) is an interactive system
using a Tektronix 4054 graphics microcomputer with 56 Kbytes of RAM memory and an additional 32 Kbytes of dynamic memory dedicated to the cre-
ation and display of refresh objects independent of the 4054's microprocessor. By current standards the Tektronix 4054 graphics system is at
the lower end of the scale both in price and sophistication of its hardware. The architecture of AIDA (Fig. 1) and its software at present
being developed, guarantees easy additions and enhancements. Furthermore,
due to its highly modular structure AIDA is designed for maximum hardware independence.

AIDA allows the manipulation, modelling and display of relatively
small molecules containing a maximum of about 280 atoms. The molecular
file contains the atomic cartesian coordinates and connection tables of
about 1000 biologically and pharmacologically important molecules
(neuroleptics, analgesics, anticholinergics, antihistaminics, anti-
inflammatories, sympathomimetics, α- and β -blocking agents, antimycotics,

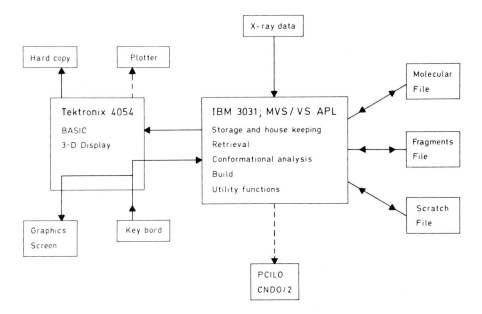

FIG. 1. Logical architecture of the AIDA system.

antiprotozoals, etc.). The primary input for the molecular file comes from X-ray data transformed by AIDA into cartesian coordinates. The H atom coordinates are calculated according to a standard geometry procedure (Fig. 2) by the utility function SETH.

In addition to the molecular file, a fragments file is available for retrieval of chemical fragments or moieties frequently used in the BUILD routine. The scratch file temporarily contains all the output data generated by the utility functions during an AIDA session. At any one time during an AIDA session two molecules are available for display due to the dual molecule workspace architecture of the system. This facilitates independent manipulation and modelling of two molecules and allows flexible and easy geometric matching and inter-molecular interaction simulations (not yet implemented) between two molecules.

The Tektronix 4054 microcomputer with its built-in matrix ROM packs handles all 3-D display aspects and enables the user to perform global rotation, translation and scaling (see footnote a of Table 1) of the molecular display by means of the 20 function keys. Fig. 3 shows a typical example of a molecular display. Simple commands generate displays of a molecule with or without hydrogens, atomic labels, numbering, double bonds etc.

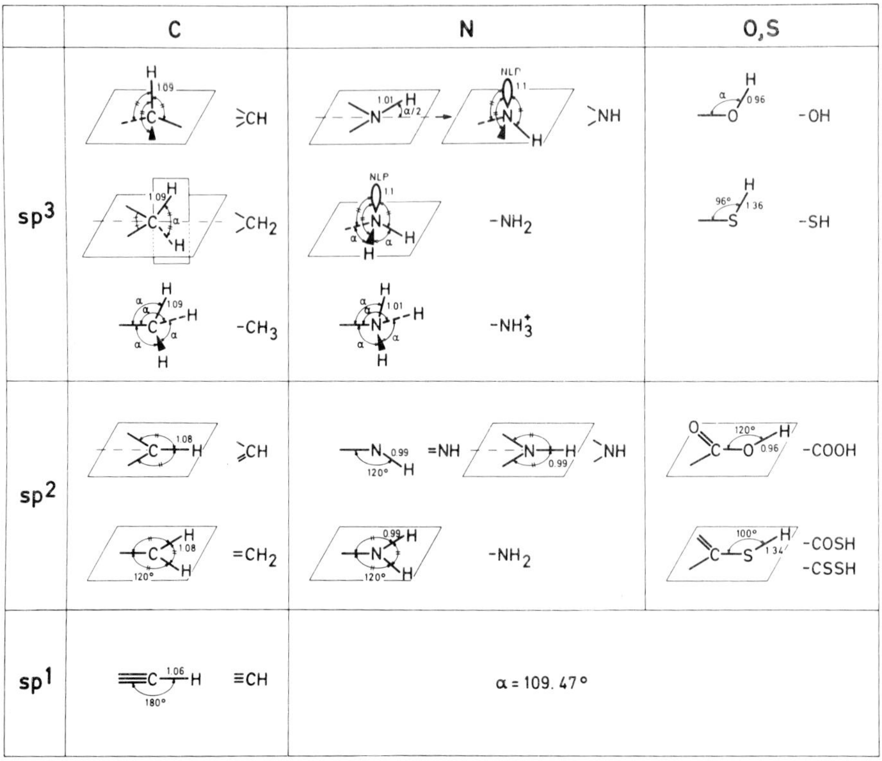

FIG. 2. Geometry used for the calculation of the H-atom coordinates employed by the utility function SETH.

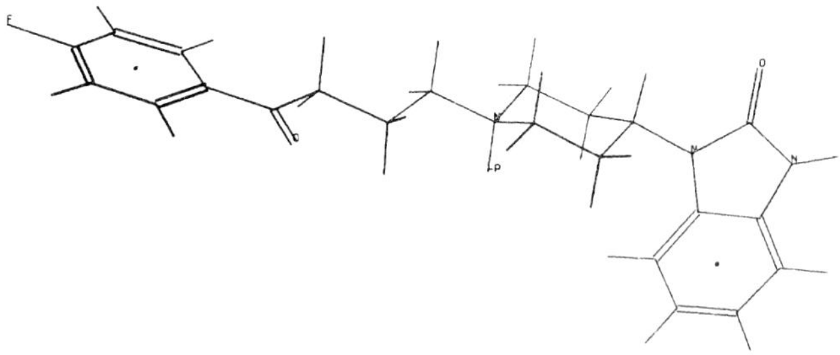

FIG. 3. Example of a molecule drawn on the Tektronix 4054 screen by AIDA.

The general utility routines driven by the IBM host computer are activated by key bord commands the most important of which are summarized in Table 1.

TABLE 1

Main Utility Functions of AIDA

---

1. Operations on Molecules Commands[a] (OMC)

RETRIEVE   retrieves molecule according to pharmacological class number, clinical file number, internal sequential number, empirical formula or generic name

INV   creates the mirror image of the molecule

SAVE   saves the molecule in the workspace or fragments file

ROTRAN   rotates and translates the molecule according to user specifications

DELMOL   deletes the molecule from the workspace

BMFIT   superimposes two molecules

2. Internal Coordinates Commands (ICC)

DIST[b]   calculates (bond) distances

ANGLE[b]   calculates (bond) angles

TORS[b]   calculates torsion angles

CRTORS   changes torsion angles of groups on ring systems

DISTMAT   calculates (reduced) distance matrix

DROT   calculates distances as a function of a rotational angle

BADCON   reports all non-bonded 1, $\geqslant$ 4 distances up to a user specified value

MNBDIST   reports all minimum non-bonded 1, $\geqslant$ 3 distances up to a user specified value

XYZCOR   generates new structures from bond lengths, bond angles and torsion angles supplied by the user

SETH[c]   calculates the coordinates of all H-atoms

3. Picture Definition Commands[d] (PDC)

STEREO   displays a stereo representation of the molecule

SPACE   displays a space-filling representation of the molecule

ORTEP   displays an ORTEP-like representation of the molecule

4. Conformational Analysis Commands (CAC)

SCAN   calculates all conformational energies according to a

TABLE 1 continued

grid specified by the user

RAND   calculates the energies of N randomly generated con
                formations

MINIM   searches for the minimum energy conformation(s) with
                a minimization algorithm

POT   calculates the energy of a single conformation

5. Data Report Commands (DRC)

LOGP, PK, DM,   reports log P, pKa, dipole moment, empirical

EMFOR, CAS   formula, CAS registry number of the molecule

REF, RVAL, COM   reports the literature reference of the X-ray data,
                 the R value of the determination and possible
                 comments

DATALL   reports all the data

---

a
Global rotation, translation and scaling is under function key control as well as swap (interchanging molecules in the dual workspace)
and split screen (display of both molecules of the dual workspace)

b
CDIST, CANGLE, CTORS changes bond lengths, bond angles and torsion
angles to the value specified by the user; BONDALL, ANGLALL and
TORSALL reports all bond lengths, bond angles and torsion angles,
respectively

c
From the number of ligands and the hybridization code (sp^1 = 1,
sp^2 = 2, sp^3 = 3, aromatic = 4, quaternary = 5) of each non-hydrogen atom, SETH calculates the H-coordinates according to standard
geometry (see Fig. 2)

d
Depending on the argument 1 (on) or 0 (off) the system will display
or delete H (hydrogens), LP (lone pairs), PC (ring centres), LA
(atomic labels), DT (double/triple bonds), NO (atom numbers), HB
(dashed lines of H-bonds).

Although a more complete molecular mechanics or force field calculation
package is currently under development it was considered part of our
research to investigate how the most simple conformational energy calculation based on eq (1) would perform in a truly interactive system such
as AIDA specifically designed for the medicinal chemist. In eq (1) used
for the rigid rotation the conformational energy, E_{conf} is expressed
as the sum of non-bonded interactions and the standard cosine

relationship used for the torsional energy contribution

$$E_{conf} = \Sigma \left(\frac{b_{ij}}{r_{ij}^{12}} - \frac{a_{ij}}{r_{ij}^{6}} \right) + \frac{U_o}{2} \left[1 + \cos(n\tau - \delta) \right] \tag{1}$$

where b_{ij} is the repulsive, a_{ij} the attractive constant and r_{ij} the distance between atoms i and j. The constant U_o represents the rotational barrier, τ the torsional angle, n the symmetry number of the appropriate bond and δ a phase angle. The constants a_{ij} and b_{ij} have been taken from Stuper et al. (1979).

Although eq (1) may be criticized because of its oversimplification it will be shown in the following sections of this contribution that it performs well enough in providing reliable answers. In fact, our experience obtained at minimal computer time expense, is that the main conclusions derived from the conformational analysis based on eq (1) are not fundamentally different from their PCILO counterparts.

4. MOLECULAR MODELLING AND PHARMACOPHORE IDENTIFICATION

The comparison of the three-dimensional characteristics of structurally related drug molecules is relatively easy if one only has to list a set of common geometrical parameters viz. distances between key atoms, torsion angles defined by atoms i, j, k, l or the height of let us say a nitrogen atom above the plane of an aromatic ring (Reboul and Cristau, 1977a, 1977b; Reboul and Cristau, 1978a, 1978b). This rather static approach badly fails when one considers a series of compounds containing several structural subsets i.e. when the set of compounds is structurally non-homogeneous but assumed to interact at the same receptor. The matter becomes even more complex when the molecules under investigation are conformationally flexible. Molecular modelling using theoretical and experimental conformation analysis, employing the AIDA system, may enable us to deduce a more detailed pharmacophoric pattern of atoms or groups of atoms.

Comparison of conformational profiles of various molecules may reveal those overlapping or intersecting conformational spaces which are energetically acceptable i.e. within a reasonable range of let us say 3 to 5 Kcal/mole above the minimum conformational energy. Let it be assumed that compound A of Fig. 4 exhibits high biological activity in a given test system and that molecule B is less active than A in the same test. Let it further be assumed that the secondary minimum A2 of compound A is energetically unacceptable i.e. that A2 lies about 20 or more

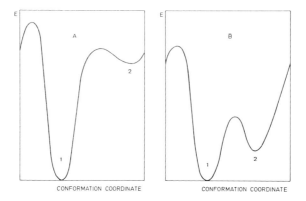

FIG. 4. Conformational profile of structurally related molecules A and
B. The energy minima A1 and B1 are assumed to pertain to the biologically relevant conformation.

Kcal/mole above A1 which is hypothesized to be the biologically relevant
conformation, that is the conformation in which the atoms constituting
the pharmacophore are in the correct mutual orientation. From Fig. 4 it
is seen that A1 and B1 are the overlapping conformational spaces from
which it follows that B2 is a biologically irrelevant conformation. If,
however, B2 is assumed to be for example 0.5 Kcal/mole less stable than
B1 an appreciable amount of compound B will be in the B2 conformation.
Under this assumption and using the Boltzmann factor (ignoring entropic
effects) it can be calculated that at T = 310°, some 31 % of B will be
present in the B2 conformation. In other words, B being less active than
A may be accounted for by the fact that only 69 % of B is available in
the correct conformation required for recognition of and interaction
with the receptor*. The intersection of all conformational maps of molecules A, B, C... then yields the possible pharmacophoric pattern common
to the whole series of compounds under examination.

In the following section examples will be presented showing how conformational analysis and conformational profiles comparison may contribute to pharmacophore identification in the field of neuroleptics.

* Other factors such as unfavourable lipophilicity and pK differences
between A and B may also contribute or could even be the sole reason
for the diminished activity of B. As a matter of fact, in SAR studies
wisdom dictates that one should take into account as many factors as
possible!

4.1. Benperidol, spiperone and related compounds

Benperidol and spiperone, both derivatives of p-F-butyrophenone are among the most potent neuroleptics known (Janssen and Van Bever, 1975, Leysen, 1981). Furthermore, as can be seen from Table 2, they possess similar bulk properties as far as log P and pK is concerned.

TABLE 2

Log P, pK and dipole moments[a] of benperidol, spiperone and related compounds

Compound	log P	pK	μ (D)
Benperidol	3.64	7.90	3.11
Cl-derivative of benperidol	~4.1	~7.90	—
Timiperone	~4.2	~7.90	—
Droperidol	3.75	7.64	3.07
Spiperone	3.65	8.31	3.34
F-derivative of spiperone	~3.8	~8.3	—

[a] Measured in this laboratory; estimated log P and pK values (~) are based on additivity principles.

Hence, using these compounds as template molecules should increase the chances of success of attempts to deduce a reliable neuroleptic pharmacophoric pattern. The p-F-fluorobutyrophenone moiety being common to both molecules compelled us to investigate the conformational characteristics of the 4-anilinopiperidine fragments.

Benperidol. The potential energy curves of benperidol, represented by the model compound 1 (Fig. 5) using the geometry based on the crystal structure of benperidol (Declercq et al., 1973) except for the H atoms which were calculated by AIDA (see Fig. 2) are shown in Fig. 6.

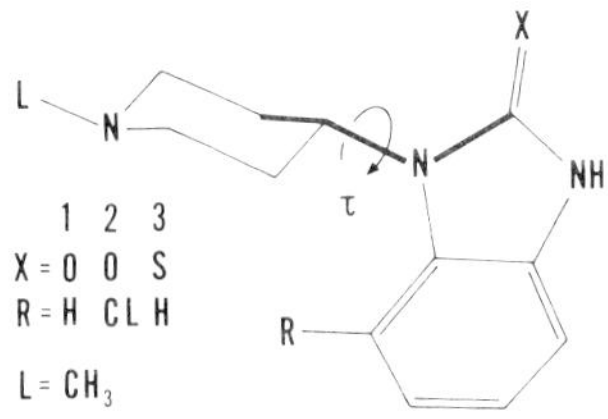

FIG. 5. Model compounds used for the calculation of the potential energy based on the X-ray data of benperidol (L = p-F-butyrophenone; X = O and R = H). The structurally related neuroleptic timiperone is represented by L = p-F-butyrophenone; X = S and R = H.

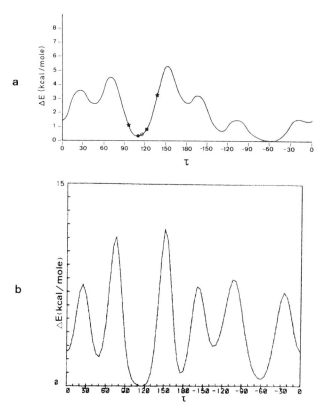

FIG. 6. Potential energy curve of model compound 1 calculated by PCILO (a) and AIDA (b). Experimental conformation of benperidol, ●; pimozide. HCl,   R 24 763 ■ ; clopimozide, o; halopemide molecule 1 and 2 in the unit cell, ✦ .

The two energy minima are separated by a fairly complex rotation barrier of about 5 to 7 Kcal/mole according to PCILO. Although AIDA predicts much higher barriers, the overall conclusion* of both calculations, how-ever, is identical viz. the benzimidazolinone appears to be clearly con-fined to two regions i.e. one with the aromatic ring in a cis configur-ation, the other in a trans configuration relative to the nitrogen lone pair of the piperidine ring. Experimentally, all known benzimidazolinone piperidine moieties adopt the cis configuration (Tollenaere et al., 1979).

---

* AIDA takes 10 sec and PCILO 305 sec CPU time for the calculation of
  the 36 conformations.

As both computational procedures essentially predict an equally popula-
ted mixture of rotamers, the question arises as to which one of these
two is biologically relevant. Although merely a guess based on all
available crystallographic evidence, it was hypothesized that the cis
configuration was the biologically relevant one.

Chloro derivative of benperidol. One way of testing this hypothesis is
to substitute the aromatic ring in that position which would force the
benzimidazolinone moiety out of the cis configuration thereby leading to
a considerable loss of activity. In fact, as can be seen from the AIDA
curve of Fig. 7, substitution as in compound 2 (Fig.5) completely destroys the conformational domain corresponding to the cis configuration.
The single minimum energy conformation of compound 2 is solely restricted
to the trans configuration shown in the geometrical fit presented in
Fig. 8. The fact that this chloro derivative of benperidol proved to be
virtually inactive as a neuroleptic gives a strong indication of the
correctness of the proposed pharmacophore.

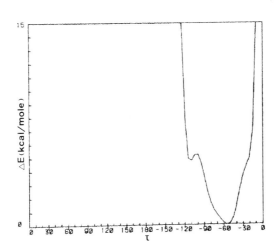

FIG. 7. Potential energy curve of model compound 2 calculated by AIDA.

Timiperone. Timiperone, the thiocarbonyl analog of benperidol is about 6
to 14 times more potent than benperidol in the apomorphine-induced
stereotyped behaviour test and is reported to be a very potent neuroleptic (Sakurai et al., 1977). Timiperone therefore, is ideally suited to
test our neuroleptic pharmacophore and incidentally serves to study the
effects of the isosteric replacement of the carbonyl oxygen by sulfur.

The potential energy curve of compound 3 based on PCILO calculations using the crystal data of benperidol except for the C = S bond taken at the standard bond length of 1.71 Å is depicted in Fig. 9.

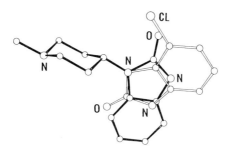

FIG. 8. Geometric fit between model compound 1 in its experimentally observed conformation of benperidol and the minimum energy conformation of compound 2 (see FIG. 7).

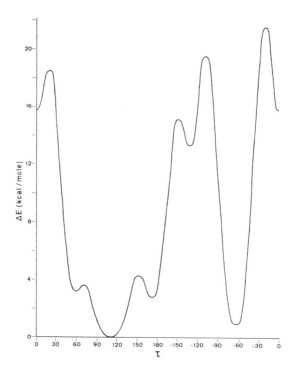

FIG. 9. Potential energy curve of model compound 3 calculated by PCILO.

The results clearly show that the cis configuration is now definitely favoured over the trans configuration in contrast to the benperidol results shown in Fig. 6 where a virtually 1:1 cis/trans rotamer mixture is calculated. It is interesting to note that the replacement of the carbonyl oxygen by the larger sulfur atom entails much higher rotation barriers. The overall results thus suggest that the preferred conformational domain of timiperone is much more confined to the putative biologically relevant conformation i.e. the cis configuration.

Droperidol. Droperidol represented by the model compound shown in Fig. 10, is the dehydropiperidine derivative of benperidol and is reported

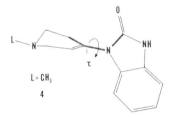

FIG. 10. Model compound used for the calculation of the conformational energy based on the X-ray data of droperidol (L = p-F-butyrophenone).

to have about half the neuroleptic activity of benperidol (Janssen and Van Bever, 1975; Leysen et al., 1978). This and the anticipated structural effect of the presence of the double bond in the piperidine ring renders droperidol a probe molecule for our pharmacophore hypothesis. From the PCILO results (Fig. 11) based on the model compound (Fig. 10) using the X-ray data of droperidol (Blaton et al., 1980) it is seen that its conformational profile departs quite

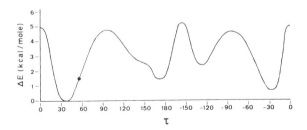

FIG. 11. Potential energy curve of compound 4 calculated with PCILO. Experimental conformation of droperidol, ● .

considerably from that of benperidol and timiperone. The extent of conformational departure is depicted in Fig. 12 showing the superimposition
of benperidol in its crystal structure conformation and for example droperidol in its PCILO conformational minimum ($\tau = 40°$). The energy penalty involved for droperidol to adopt a conformation ($\tau = 110°$) maximally resembling (Fig. 13) that of benperidol is about 4 Kcal/mole

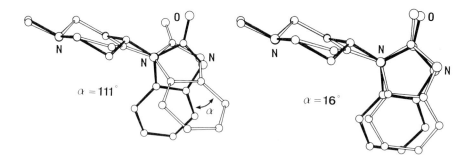

FIG. 12. Geometric fit between model compound 1 in the conformation
observed in benperidol and the
PCILO calculated minimum energy
conformation of compound 4.

Fig. 13. Geometric fit between compound 1 in the conformation observed
in benperidol and compound 4 in a
conformation which maximizes the fit
with compound 1.

according to PCILO. Although the latter value is far too high to explain
the observed activity difference between benperidol and droperidol, it
should be recalled that rigid rotation in general predicts too high energy differences and secondly that mutual structural adaptation between
drug and receptor may alleviate to a certain extent the steric strain in
the resulting drug-receptor complex. Taking these factors into account
the present comparison between the conformational profiles of the two
molecules rationalizes the fact that droperidol is less active than benperidol.

Spiperone. Having established that the cis configuration of the aromatic
ring with the nitrogen lone pair of the piperidine is most probably an
essential feature of the neuroleptic pharmacophore, the question remains
as to the orientation of the phenyl plane relative to the piperidine
mean plane. This question can be answered by the calculation of the
rotational behaviour of the phenyl ring of spiperone represented by the

model compound shown in Fig. 14. The potential energy curves (Fig. 15) calculated with PCILO and AIDA using the crystal structure of spiperone (Koch, 1973) reveal that the minimum energy conformation is identical with that observed in the solid state conformation of spiperone (Koch, 1973), spirilene (Koch and Evrard, 1973), R 5174 (Koch and Dideberg, 1973), R 48 455 and R 49 399 (Durant, 1980). The narrow energy minima and steep rotational barriers suggest that the phenyl ring is firmly held in the given orientation resembling that of benperidol. Consequently, it is expected that the observed conformation is also the biologically relevant conformation.

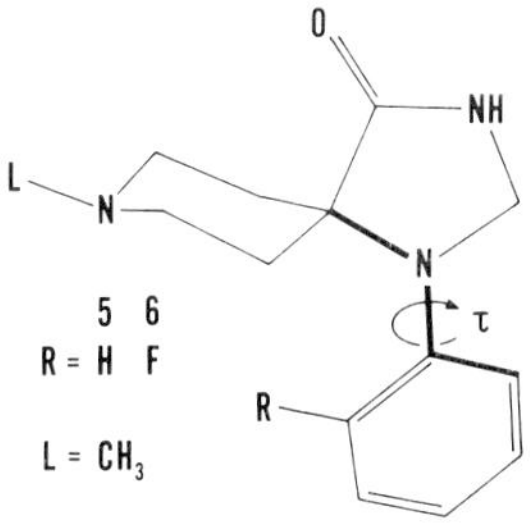

FIG. 14. Model compounds used for the calculation of the potential energy based on the X-ray data of spiperone (L = p-F-butyrophenone, R = H).

2-Fluorospiperone. The introduction of a substituent at the ortho position of the aromatic ring of spiperone may be expected to alter the conformational behaviour of that phenyl ring. In fact, ortho fluoro substitution effectively destroys the conformational domain of interest as can be seen from Fig. 16. The almost 1000-fold observed activity decrease of 2-fluorospiperone again strongly supports the validity of the proposed pharmacophoric pattern.

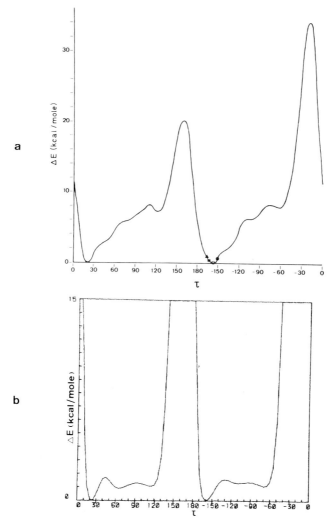

FIG. 15. Potential energy curve of model compound 5 calculated by PCILO
(a) and AIDA (b). Experimental conformation of spiperone, spirilene, ● ;
R 5174, o; R 48 455, ▲ ; R 49 399, ■ .

4.2. Conclusion

The overall experimental and theoretical evidence presented above
thus quite strongly suggests that an aromatic ring whose centre is about
5.8 Å away from the tertiary nitrogen atom of the piperidine ring should
be in a cis conformation with the nitrogen lone pair and should bisect
the mean plane of the piperidine ring.

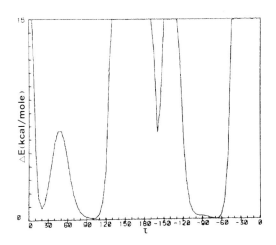

FIG. 16. Potential energy curve of model compound 6 calculated by AIDA.

5. CONCLUDING REMARKS

Conformational analysis using experimental and theoretical methods is a powerful tool towards the identification of pharmacophores essential for a given pharmacological activity. The sheer wealth of information regarding the 3-D aspects of drug molecules accumulated over the last ten years or so, calls for appropriate techniques and procedures to handle and exploit this huge amount of data. It is a firmly held belief in many laboratoires including our own, that molecular graphics packages will contribute substantially to the exploitation of the ever growing 3-D information on drug molecules and related compounds towards a rational design of new and more specific drugs.

The AIDA system has been shown to be capable of handling the available 3-D data on neuroleptics for example, and to interactively provide answers regarding the putative pharmacophore of neuroleptics. The speed and user-friendliness of AIDA holds great promise for the future. It is noteworthy that AIDA's highly modular architecture permits easy-to-install additions to the system which is continually evolving into a research tool enabling medicinal chemists to test some of their ideas before the actual experiments in the laboratory.

ACKNOWLEDGEMENTS

The authors sincerely thank Mr. L. De Bie for his relentless efforts in
the design and programming of the AIDA package into a medicinal chemist-
friendly research tool. Thanks are due to Mr. P. Van de Schans who is
currently programming the algorithms used in the conformational analysis
module.

REFERENCES

ALLINGER, N.L. (1976). Adv. Phys. Org. Chem. 13, 1-82.

ANDOSE, J.D. and MISLOW, K. (1974). J. Am. Chem. Soc. 96, 2168-2176.

BLATON, N.M., PEETERS, O.M. and DE RANTER, C.J. (1980). Acta Crystallogr.
B36, 2828-2830.

CAILLET, J., CLAVERIE, P. and PULLMAN, B. (1977). Acta Crystallogr.,
A33, 885-889.

DECLERCQ, J.P., GERMAIN, G. and KOCH, M.H.J. (1973). Acta Crystallogr.
B29, 2311-2313.

DINER, S., MALRIEU, J.P., JORDAN, F. and GILBERT, M. (1969). Theor.
Chim. Acta 15, 100-110, and references cited therein.

DURANT, F. (1980). Private communication.

DYOTT, T.M., STUPER, A.J. and ZANDER, G.S. (1980). J. Chem. Inf. Comput.
Sci. 20, 28-35.

GÓ, M., GÓ, N. and SCHERAGA, H.A. (1971). J. Chem. Phys. 54, 4489-4503.

GUND, P., ANDOSE, J.D., RHODES, J.B. and SMITH, G.M. (1980). Science
208, 1425-1431.

HOPFINGER, A.J. (1973). Conformation Properties of Macromolecules,
Academic Press, New York.

JANSSEN, P.A.J. and VAN BEVER, W.F.M. (1975). In "Current Developments
in Psychopharmacology" (eds. W.B. Essman and L. Valzelli) Vol. 2,
pp. 167-184.

JOHNSON, C.K. (1965). ORTEP (Oak Ridge Thermal Ellipsoid Program),
Report ORNL-3794 revised, Oak Ridge National Laboratory, Oak Ridge,
Tennessee.

KIER, L.B. and ALDRICH, H.S. (1974). J. Theor.Biol. 46, 529-541.

KOCH, M.H.J. (1973). Acta Crystallogr. B29, 379-382.

KOCH, M.H.J. and DIDEBERG, O. (1973). Acta Crystallogr. B29, 2969-2970.

KOCH, M.H.J. and EVRARD, G. (1973). Acta Crystallogr. B29, 2971-2973.

LANGRIDGE, R., FERRIN, T.E., KUNTZ, I.D. and CONNOLLY, M.L. (1981).
Science 211, 661-666.

LEYSEN, J.E. (1981). In "Clinical Pharmacology in Psychiatry". (Eds. E. Usdin, S.G. Dahl, L.F. Gram and O. Lingjoerde) pp. 35-62. MacMillan Publishers Ltd.

LEYSEN, J.E., NIEMEGEERS, C.J.E., TOLLENAERE, J.P. and LADURON, P.M. (1978). Nature, 168-171.

MCGUIRE, R.F., MOMANY, F.A. and SCHERAGA, H.A. (1972). J. Phys. Chem. 76, 375-393.

MOMANY, F.A., MCGUIRE, R.F., BURGESS, A.W. and SCHERAGA, H.A. (1975). J. Phys. Chem., 79, 2361-2381.

NYBURG, S.C. (1974). Acta Crystallogr. B30, 251-253.

PORT, G.N.J. and PULLMAN, A. (1973). Theor. Chim. Acta 31, 231-237.

PULLMAN, B. and PORT, G.N.J. (1974). Mol. Pharmacol. 10, 360-372.

PULLMAN, B., COURRIERE, P. and BERTHOD, H. (1974). J. Med. Chem. 17, 439-447.

REBOUL, J.P. and CRISTAU, B. (1977a). Eur. J. Med. Chem. 12, 71-75.

REBOUL, J.P. and CRISTAU, B. (1977b). Eur. J. Med. Chem. 12, 76-79.

REBOUL, J.P. and CRISTAU, B. (1978a). J. Chim. Phys. 75, 1109-1118.

REBOUL, J.P. and CRISTAU, B. (1978b). Ann. Pharm. Fr. 179-189.

SAKURAI, T., KOJIMA, H. and KASAHARA, A. (1977). Jap. J. Pharmacol. (Suppl), 124 P.

STUPER, A.J., DYOTT, T.M. and ZANDER, G.S. (1979). In "Computer-Assisted Drug Design" (eds. E.C. Olson and R.E. Christoffersen) ACS Symposium Series 112, pp. 383-414.

TOLLENAERE, J.P. (1981). Trends in Pharmacol. Sci. 2, 273-275.

TOLLENAERE, J.P., MOEREELS, H. and KOCH, M.H.J. (1977). Eur. J. Med. Chem. 12, 199-211.

TOLLENAERE, J.P., MOEREELS, H. and RAYMAEKERS, L.A. (1979). Atlas of the Three-Dimensional Structure of Drugs. Elsevier/North-Holland Biomed. Press. Amsterdam.

TOLLENAERE, J.P., MOEREELS, H. and RAYMAEKERS, L.A. (1980). In "Drug Design" (ed. E.J. Ariëns) Vol. 10, pp. 71-118.

WEINER, P.K. and KOLLMAN, P.A. (1981). J. Compt. Chem. 2, 287-303.

WEINTRAUB, H.J.R. (1979). In "Computer-Assisted Drug Design" (eds. E.C. Olson and R.E. Christoffersen) ACS Symposium Series 112, pp. 353-370.

WEINTRAUB, H.J.R. and HOPFINGER, A.J. (1973). J. Theor. Biol. 41, 53-75.

26 Receptor-bound conformation of drugs

G R Marshall

Information derived from x-ray crystal structure determinations can
contribute significantly to molecular pharmacology when used judicious-
ly. An obvious example would be the crystal structure of an enzyme such
as angiotensin converting enzyme, where the size lends assurance that
the conformation in the crystal is relevant to that in vivo. A simple
view is that the ratio between volume and surface determines the degree
of internal stabilization compared with external perturbations. The
situation is obviously much different when one considers drugs which
would inhibit this enzyme. Here surface interactions would predominate,
and crystal packing forces including desolvation may well perturb the
solution conformer, or ensemble of conformers.

Since the primary concern in a drug-receptor interaction is the
bound conformation of the drug and those interactions with the receptor responsible for activity, one must ignore information which is
unlikely to be relevant. This includes most of the studies focusing
on the minimum energy conformation, whether one is speaking of theoretical in vacuo calculations, solution conformations from NMR, or crystal structures of small molecules. Because of the binding process, a
significant amount of interaction between the receptor and the drug
must occur. The receptor is asymmetric and presents a potentially
dominant perturbation to the conformational equilibrium in the adjacent solution. Affinities of receptors for natural transmitters are
under evolutionary control as exemplified in the large differences in
affinity of pre- and post-synaptic receptors for neurotransmitters such
as dopamine. One mechanism to modulate the affinity is to bind a
higher energy form of the transmitter.

Other areas of invaluable contributions from X-ray crystallography
include assignment of absolute stereochemistry and structural determination. This can be critical as most receptor systems show extreme
stereoselectivity. Of increasing value is the use of crystal structures (Burgi and Dunitz 1983) as a basis for the determination of
molecular mechanics force field parameters directly. This is, of
course, in addition to crystal structures as the criteria for valida-
tion of theoretical calculations. While crystal structures are not the

final answer in the drug-receptor puzzle, they provide an essential
information base and an excellent starting point for structure-activity
analysis.

COMPUTATIONAL APPROACH

Computational chemistry is an area governed by compromise between
the desire to simulate physical reality in detail and the limitations
on computational power available for such a task. Inevitably, the
practical considerations of the time prevail and physical chemistry
suffers. A case in point involves the extensive use of minimization
methodology when faced with a problem with multiple variables, i.e. the
minimum energy conformer of a molecule of interest. In any minimization method, one locates the nearest minimum, and in spite of extensive
research, most procedures fail to detect the global minimum in a multi-
dimensional problem. This basically implies that the answer one obtains
is very much dependent on the form of the question, i.e. the starting
conformation.

The source of difficulty in examining the conformational energy
surface available to a molecule arises from its complexity. Because
of the divergent nature of molecular forces (Burkett and Allinger 1982;
Niketic and Rasmussen 1977; Boyd and Lipkowitz 1982) and the number of
degrees of freedom, or variables, for any real molecule, a highly con-
voluted, multidimensional energy surface is the rule rather than the
exception. This can be compared to the contour map of a mountainous
region such as the Alps or the Rockies where one is trying to determine
the lowest valley by hiking. The rule of most minimizers is to only
walk downhill, which would result in only finding a local minimum,
unless one happened to start the hike in the valley containing the
global minimum. Many times, efforts to find the global minimum will be
made by coupling random starting points with minimizers. Consider the
number of hikers following the minimization rule necessary to find the
lowest point in Switzerland if given random starting points. It is
obvious that a large number would be necessary to be assured that the
lowest point were found. It is not clear that a random procedure would
be more efficient than a grid search if one were truly interested in
finding the lowest point, and not just statistically determining the
average height above sea level.

Various procedures such as molecular dynamics (McCammon and Karplus 1980) and Monte Carlo simulation have been developed to address
this problem, at least statistically, in predicting ensemble properties. Systematic search is an alternative general approach which has
its own limitations, but whose accuracy is known. For example, if one
systematically examines every conformation differing by 1° increments
and locates the minimum energy conformer, one is assured of being
within 1° of the minimum.

The difficulty with a systematic examination is simply cost,
whether one considers systematic search of an energy surface, or the
production of a contour map of Switzerland. One compromise that is
often used to reduce the number of variables in systematic search is
the assumption of fixed bond angles and bond lengths, leaving only
torsional variables to specify a conformation. The importance of
reduction of variables is due to the combinatorial nature of the problem. For example, a molecule with six degrees of torsional freedom
such as chlorpromazine has 72^{6} or 139,314,069,504 possible conformations at a 5° increment. In other words,

$$\text{theoretical number of conformers} = (360/\text{increment})^{R}$$

where R = number of torsional degrees of freedom

If one considers that for each conformation, the computer must generate the appropriate coordinates for each atom and then evaluate energy
terms which at a minimum are a function of each pair of atoms (number
of energy evaluations = N(N-1)/2, where N = number of atoms), one begins
to appreciate the computational complexity of the problem.

To illustrate the obvious futility of a direct, brute-force attack,
let us examine some typical problems assuming that an energy evaluation can be accomplished in one microsecond. For chlorpromazine with
42 atoms at a 5° increment, then 861 energy evaluations for each of
the theoretical conformations would have to be performed. Assuming
that the coordinates for each atom could also be calculated in one
microsecond, then

$$72^{6} \cdot N \cdot N(N-1)/2 \qquad \text{where } N = 42$$

microseconds would be required. This equals 1,399,409.7 hours or
58,308.7 days or 159.7 years. Clearly, one does not approach systematic search with strictly brute force.

There are some very obvious simplifications that one can make.
If one examines chlorpromazine, several sets of atoms, for example,
the phenothiazine ring, do not change their relationship by varying
the designated torsional variables, and, therefore, do not have to be
evaluated more than once for any possible conformation, i.e. their
relationship is fixed throughout the problem. At most, this reduces
the factor N in N(N-1)/2 by a factor of 2 or the problem by a factor
of 4. The reduction does not extend to the coordinate update as each
atom must still be calculated in order to examine its changed relationship to some other atom in the molecule. Furthermore, one does not
need to continue the evaluation of a particular conformation if any
two atoms are clearly in such an unfavorable relationship that the conformation in question would be energetically disallowed. This arises
from strong Van der Waals repulsion because two atoms are basically
occupying the same space. This allows one to prune non-productive
branches of the combinatorial tree as soon as they are detected. Even
with these and other algorithmic improvements, practical implementa-
tions of systematic search are usually limited to ten or less rotatable
bonds. While this may seem insignificant when considering proteins or
nucleic acids, the majority of compounds of pharmacological interest
fit within this restriction, especially considering that other simplifications such as treating a methyl group as a united atom can be used.
Because of the combinatorial nature of systematic search, the strategy
in analyzing a problem becomes an important factor by introducing constraints as early as possible in the analysis.

DRUG-RECEPTOR INTERACTIONS

One basic problem facing the medicinal chemist is the absence of
information regarding his target, the receptor. Since he must design
new compounds, he often modifies known active structures as the most
likely strategy to retain the desired activity. Only by insight into
the nature of the drug-receptor interaction can he transcend the constraints of known structures and design truly novel compounds with
sufficient probability of activity. The basis for this insight is the
concept of pharmacophore, i.e., those features common to a set of drugs

acting at the same receptor which are responsible for recognition and transduction of the appropriate response.

With this simplifying assumption, a basis (Marshall, Barry, Bosshard, Dammkoehler and Dunn 1979) for computer aided drug design is established. First is the problem of determining possible three-dimensional arrangements of features common to the set of active drugs. Each of these common arrangements becomes a candidate pharmacophore which can be rejected because of its incompatibility with other observations (inconsistency) or by its lack of predictability of activity of new compounds. What are these features which should be considered? These can be the traditional concepts of medicinal chemistry, heteroatoms, lone pairs, pi centers, aromatic planar groups, hydrogen-bonding functions, etc.; or more sophisticated physical chemical concepts such as fragment dipole moments, electrostatic potential properties, lipophilicity, etc. If one makes the further assumption that the receptor-bound conformation is the one of lowest energy, then the problem becomes straightforward. For each analog, one determines the appropriate conformer and simply looks for common three-dimensional features present in the set of active analogs. Programs for detecting such common features have been described by Gund (1977). Unfortunately, efforts to correlate activity with conformers seen in crystal structures, in solution by NMR, or by theoretical calculation have been disappointing. The physical chemical basis for this lack of correlation should be clear, these procedures ignore the perturbation caused by interaction between the drug and the asymmetric force field represented by the receptor. In other words, there are no a priori reasons to assume that a receptor binds the low energy conformer of a drug, and considerable evidence exists to suggest just the opposite in several well-documented cases such as ribonuclease S (Bierzynsk, Kim and Ballwin 1982) and NAD (Rossman, Liljas, Branden and Banaszak 1975).

It is, therefore, imperative to consider the ensemble of conformers accessible by perturbation of the solution ensemble upon binding to the receptor. This requires an estimate of the energy difference between the minimum energy conformer in solution and that bound to the receptor. Since some receptors are capable of breaking carbon-carbon bonds, the difference can be quite great in the case of enzymes; and a continuum probably exists. The presence of cis-peptide bonds in the crystal structure of proteins suggests a 10-12 Kcal/mol perturbation to a minimum structure is a feasible upper limit for non-enzymatic systems. It would, therefore, seem advisable to examine all conformers available to each active analog within such a limit for the presence of common pharmacophore candidates. Systematic search is the obvious candidate for generating the possible set of conformers.

By considering the possible pharmacophores in orientation space (Marshall 1979), a relative distance space, one eliminates the need for a common frame of reference between molecules, and simplifies the dimensions of the problem. For example, a candidate pharmacophore can often be represented by three or four distances, where the conformation of the molecule might require specification of eight or more torsional angles plus the parent molecule.

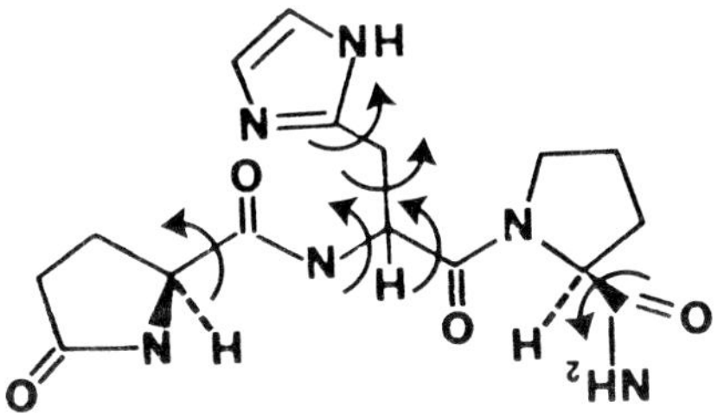

FIG. 1. Structure of thyroliberin (TRH) showing rotatable bonds considered in conformational search.

As an example of the use of this approach, the study (J. Font, A. G. Hortmann and G. R. Marshall, unpublished results) on thyroliberin (TRH) is appropriate. Examination of the structure-activities of this molecule led to the hypothesis that the recognition features are the carbonyl of the pyrrolidone ring, the C-terminal carboxamide group, and a planar hydrophobic group in the sidechain of residue two, preferably aromatic. Six analogs of TRH with conformational differences were chosen for intensive study. The six rotatable bonds indicated on Figure 1 were searched at $10°$ increments for each analog, and the distances necessary to position the three candidate pharmacophoric groups recorded. Intersection of the resulting orientation maps led to three candidate pharmacophores. Novel non-peptide structures which

hold these three groups in the correct position for pharmacophore 1
have been designed and are currently being synthesized. One such candidate structure and a comparison of its minimized structure with TRH
in the conformation associated with pharmacophore 1 is shown in Figure
2.

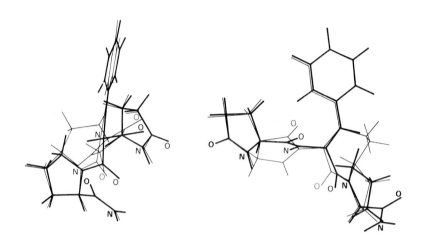

FIG. 2. Orthogonal views of minimized structure of candidate (light
lines) shown overlapped with $[\triangle Phe^2]$ -TRH (dark lines) in conformation associated with pharmacophore one.

Several points need to be emphasized. First, it is important to
extract three-dimensional information from structure-activity data.
The six analogs used in this study were designed to have conformational
restrictions, and not to perturb other parameters such as lipophilicity
or charge. Second, it is important to know that the features proposed
for recognition are consistent with the structure-activity data in
three-dimensional terms. Third, it is crucial to know how many three-
dimensional arrangements of the features are consistent with the observations. In the TRH case, three arrangements make it feasible to
distinguish between them synthetically. If twenty or more arrangements
had resulted, then further analogs would have been necessary prior to
the attempt to cross congeneric series with totally novel analogs to
ascertain the receptor-bound conformation.

RECEPTOR MAPPING

There are   other criteria besides the presence of the correct
pattern of electronic distribution, or pharmacophore, which must be
satisfied before a successful interaction between drug and receptor
can take place. Sufficient space at the receptor must be available to
allow the correct presentation of the electronic pattern. Steric repulsion can preclude appropriate orientation and interaction, and the
presence of the pharmacophoric pattern is, therefore, a necessary but
not sufficient condition for activity. This provides both a criteria
for consistency as well as an opportunity to derive relevant information from inactive compounds.

For example, if one were to use the pharmacophore as a common
frame of reference for active molecules, one can assume that the combined volume for the active molecules must be available at the receptor since they are active. If an inactive molecule can fit within
that union of active volumes while presenting the pharmacophore, then
a logical inconsistency has been detected. One might consider an alternative assignment of features, multiple modes of action of the drugs
under investigation, or errors in the pharmacological data. In most
of the examples studied, inactive compounds capable of presenting the
pharmacophore require novel volume, some of which can be assumed to be
occupied by the receptor, thus precluding a productive interaction with
the receptor. By examining a variety of such compounds, a cast of the
receptor can be obtained. The first example mapped in this way was the
methionine binding site of the enzyme S-adenosyl ATP transferase
(Sufrin, Dunn and Marshall 1981). Recent evidence for similar sites of
action for the convulsants picrotoxinin and gamma-butyrolactone analogs
has resulted from independent analysis of the data of the two series
and comparison of the resulting site maps by Klunk et al (1983). The
stereospecificity of semi-rigid dopamine agonists and antagonists such
as apomorphine, octoclothepin, and butaclamol is also consistent with
such an analysis (Marshall, Barry and Humber 1978). Traditional caveats
regarding the use of inactive compounds due to multiple possible reasons for inactivity must always temper ones enthusiasm for such studies.
In addition, inactivity may be due to conflict between the drug and the
receptor of only a small portion of the volumes discovered by such
studies. Nevertheless, a methodology for predicting inactivity due to
steric factors is evolving.

CAVEATS

Several implicit assumptions underlie the static pharmacophore
approach outlined above which can be seriously challenged. First, the
assumption of a common binding mode is clearly oversimplified. Receptor sites are multifaceted and capable of generating comparable binding
affinities by interaction with combinations of subsites. One is forced,
however, by Occam's Razor (Russell 1945) to assume the simplest site
until the accumulated data forces a more complicated interpretation.
In the case of the opioid mu-receptor, the accumulated structure-
activity data supports multiple binding modes to a more complicated
receptor model (Humblet and Marshall 1981) than that originally proposed by Beckett and Casy (1954). Second, receptors appear to be proteins which are clearly more flexible than this method would appear to
accommodate. The accumulated evidence from protein NMR relaxation
studies as well as molecular dynamics simulations clearly points to
a flexible structure of the receptor. One can argue, however, that
the static state exemplified by protein crystallography may indeed
represent the low energy state which is statistically significant. In
addition, agonists require both binding and transduction. One would
assume that only a limited set of protein conformers are capable of
undergoing the transduction transformation after binding the drug.
This is analogous to the transition state of a chemical reaction, and
agonists are to be distinguished from antagonists which can function
equally well by binding to multiple conformations of the receptor
as long as the transduction transformation is not induced.

One advantage which arises from this approach is the lack of
emphasis of accurate energy calculations. Difficulties arise when one
is attempting to determine the details of molecular energetics due to
limitations in force field parameters, inappropriate treatment of the
dielectric constant (Greenberg, Barry and Marshall 1978), and lack of
consideration of solvation and entropic effects.

SUMMARY

A molecular understanding of drug-receptor interactions is clearly
desirable as a basis for drug design. Lack of correlation between
experimental and theoretical studies and observed activities can be
seen to be due to over-simplified assumptions, technical limitations

such as the local minima problem, and lack of detailed information about the receiver itself. The use of systematic search and the assumption of a pharmacophore provides a rational basis on which to examine simple hypotheses regarding features responsible for a given activity. Consistency and predictability are the primary criteria by which one can judge such simplified ideas. While the scientific method requires the use of such a simple-minded approach, one must keep in mind that evolution has had multiple opportunities to develop complicated mechanisms. In other words, Mother Nature never shaved with Occam's razor; and we must view our simplified hypotheses as a means to gain insight into the more complicated mechanisms which underlie both molecular recognition and pharmacological activity.

REFERENCES

Beckett, A. H. and Casy, A. F. (1954). $\underline{J}$. $\underline{Pharm}$. $\underline{Pharmac}$. 6, 986-1001.

Bierzynsk, A., Kim, P. S. and Ballwin, R. L. (1982). $\underline{Proc}$. $\underline{Natl}$. $\underline{Acad}$. $\underline{Sci}$. $\underline{USA}$ 79, 2470-2474.

Boyd, D. B. and Lipkowitz, K. B. (1982). $\underline{J}$. $\underline{Chem}$. $\underline{Educ}$. 59, 269-274.

Burgi, H. B. and Dunitz, J. D. (1983). $\underline{Acc}$. $\underline{Chem}$. $\underline{Res}$. 16, 153-161.

Burkett, U. and Allinger, N. L. (1982). $\underline{Molecular}$ $\underline{Mechanics}$ ACS Monograph 177, Washington, D.C.

Greenberg, D. A., Barry, C. D. and Marshall, G. R. (1978). $\underline{J}$. $\underline{Amer}$. $\underline{Chem}$. $\underline{Soc}$. 100, 4020.

Gund, P. (1977). $\underline{Prog}$. $\underline{Mol}$. $\underline{Subcell}$. $\underline{Biol}$. 5, 117-143.

Humblet, C. and Marshall, G. R. (1981). $\underline{Drug}$ $\underline{Development}$ $\underline{Res}$. 1, 409-434.

Klunk, W. E., Kalman, B. L., Ferrendelli, J. R. and Covey, D. F. (1983). $\underline{Mol}$. $\underline{Pharmacol}$. 23, 511.

Marshall, G. R., Barry, C. D., Bosshard, H. E., Dammkoehler, R. A. and Dunn, D. A. (1979). $\underline{Computer-Assisted}$ $\underline{Drug}$ $\underline{Design}$ (ed. E. C. Olson and R. E. Christoffersen) ACS Symposium 112, Washington.

Marshall, G. R., Barry, C. D. and Humber, L. G. (1978). $\underline{Abst}$. $\underline{Metrochem}$. '78 $\underline{Reg}$. $\underline{ACS}$ $\underline{Meeting}$ 7.

McCammon, J. A. and Karplus, M. (1980). $\underline{Ann}$. $\underline{Rev}$. $\underline{Phys}$. $\underline{Chem}$. 31, 29.

Niketic, S. R. and Rasmussen, K. (1977). $\underline{Lecture}$ $\underline{Notes}$ $\underline{Chem}$. 3, Springer-Verlag.

Rossman, M. G., Liljas, A., Branden, C. I. and Banaszak, L. J. (1975). $\underline{The}$ $\underline{Enzymes}$. 61, 102.

Russell, B. (1945). A History of Western Philosophy. Simon and Schuster, p. 472. New York.

Sufrin, J. R., Dunn, D. A. and Marshall, G. R. (1981). Mol. Pharmacol. 19, 307.

27 Present and future computer aids to drug design
P. Gund

1. INTRODUCTION

Computer aided design and computer aided manufacturing (CAD/CAM)
are expected to revolutionize the process of creating and manufacturing complex structures such as airplanes, automobiles and ships.
While it is premature to expect CAD/CAM production of medicines, many
of the same computer capabilities may prove valuable for aiding drug
design.

Thus, for CAD/CAM, inventory systems track parts and materials,
while graphics systems help designers visualize the final product and
assure that components will fit together properly. Modeling software
allows testing of a potential product's performance before it is
built. When the design is complete, CAM systems can control the
machine tools that manufacture the product components. Computerized
equipment may monitor the performance of the manufactured product,
and these results may be used to refine the computer model.

Similarly, computer systems store information on structures and
properties of chemical compounds, and analysis programs can correlate
chemical structure with physicochemical and biological properties.
Modeling software may simulate the interaction of structures with
biological receptors, and graphics systems aid the medicinal chemist
in recognizing stereochemical relationships. Other systems aid the
chemist in devising a synthesis of a target compound, although the
actual laboratory synthesis and workup remain a more or less manual
process. The synthetic product is often subjected to computer-
controlled systems for determining physical properties (IR, UV, MS,
NMR spectra, crystal structure) and biological properties. Such data
are used to refine the computer models relating structure to chemical
and biological properties.

The remainder of this contribution will discuss some of these
aids to drug design which are presently available, and others which
are being developed. Finally, we consider the question of how these
aids may best be utilized.

2. PRESENT COMPUTER AIDS

We may identify several major types of computer aids to drug

design. Storage and retrieval of chemical information and data have
a relatively long history; chemical companies were sophisticated
users of tabulating and sorting machines in the 1930's, before the
general-purpose computer was developed. Data analysis systems include
statistical techniques which began to be developed in the '40's and
'50's. Information analysis systems fall into the domain of non-
numerical computing and artificial intelligence systems and are just
now becoming well accepted. The modeling of properties goes back at
least to vibrational spectrum simulation, done by desk calculator in
the 1930's, while interactive computer graphics, including molecular
modeling and display, sprang from Project MAC at MIT in the 1960's.
Computer control of instruments and processes is generally thought of
as a separate discipline, perhaps more closely allied to engineering,
and will not be considered further.

2.1. Information and Data Retrieval Systems

    Merck's collection of chemical compounds is managed with the aid
of a structure-based computer system, CSIS (Brown et al., 1976).
This batch system allows retrieval of chemical structures by exact
match or substructure match, with ancillary information about compound
availability, biological testing, sample tracking and batch registration. The system was recently upgraded to allow on-line display of
structures by specification of the L-number (unique Merck identifier),
and review of substructure search results. Other modern systems for
managing in-house chemical information have recently become commer-
cially available (Molecular Design Ltd.; Questel).

    Specialized collections of chemical information are available
from commercial and government sources. The Cambridge Crystallographic
Data Base and Brookhaven Protein Data Bank are familiar to most
members of this conference. Other available collections include six
million substances searchable by substructure (CAS-Online, Questel),
spectral and toxicological data (NIH-EPA Chemical Information System),
and patent and reaction information (Derwent).

2.2. Data Analysis

    Quantitative Structure-Activity Relationships (QSAR) trace their
origins to the Hammett equation of 1937, as further developed by
Hansch, and are familiar to all medicinal chemists (Martin, 1978).
Advances in the field include applications of new statistical techniques,

including pattern recognition; development of new substituent parameters, such as Verloop's steric parameters; and integration of QSAR
with other tools for drug design, especially molecular modeling and
computer graphics (Hansch, 1983).

2.3. Information Analysis and Interpretation

With somewhat greater difficulty, the computer can be programmed
to analyze non-numerical information. Examples of such programs
running at Merck include Simulation and Evaluation of Chemical
Synthesis (SECS: Grabowski et al., 1979; Gund et al., 1980) to aid
synthesis planning, and Program for Analysis of InfraRed Spectra
(PAIRS: Woodruff and Smith, 1980) to aid interpretation of IR spectra.
Some other well-known programs of this type include DENDRAL/CONGEN
and CASE to aid structure elucidation (Lindsay et al., 1980), MYCIN
to aid prescribing of antibiotic regimens, and Corey's LHASA synthetic analysis program.

2.4. Molecular Modeling

The Merck Molecular Modeling System (MMMS) has been described
(Gund et al., 1980), and the field has been reviewed (Humblet and
Marshall, 1981). Briefly, such systems use interactive computer
graphics to aid the chemist in generating accurate three-dimensional
molecular structures, in determining preferred conformations, in comparing the geometry of different molecules having similar biological
activity, and in calculating physical properties. Current hardware
components of MMMS are shown in Fig. 1 and software components are
listed in Fig. 2.

Molecular structures are created by inputting crystal coordinates
(XTAL); by inputting standard bond lengths, angles and dihedral
angles (COORD); by drawing the structure on the screen and performing
an approximate strain minimization (DRAWMOL); by inputting polypeptide
residue codes and torsion angles (ECEPP); or by modifying or merging
previously formed molecules (MOLEDITOR edits molecular files much as
a text editor modifies or merges text files). More accurate geometries
may be derived by a variety of classical mechanical and quantum
mechanical programs. Conformations may be generated systematically
by several procedures, including Prof. Clark Still's RINGMAKER program
(Still and Galynker, 1981). Computer graphics programs are used to
display molecules as stick, ball and stick, and spacefilling figures,

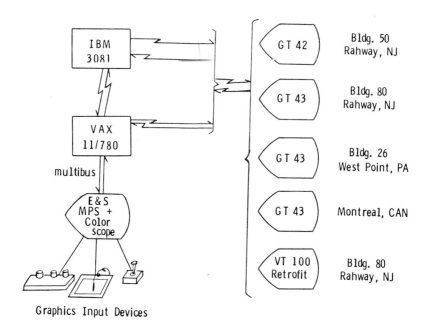

Fig. 1. Merck Molecular Modeling System Network

as contact surface representations, or as superpositions. Excluded
volume maps (Humblet and Marshall, 1981) and orbital and electrostatic
field contours may also be represented. Physicochemical properties,
such as frontier orbitals and charge densities, molecular volume and
surface area, and steric congestion at a reaction center, may be
calculated from the molecular coordinates.

Examples of applications of MMMS to drug design are given in
Gund et al. (1980), and other examples are listed by Marshall (1981).
Molecular modeling aided the development of a potent cyclic hexa-
peptide analog (1) of the metabolically labile tetradecapeptide
somatostatin; 1 and derivatives are being examined for their potential
for improved therapy in insulin-dependent diabetics (Veber et al.,
1981). Modeling at Sandoz (Basel) aided development of a different
octapeptide analog of somatostatin (2), which is also being studied in
the clinic (Bauer et al., 1982). In another example, MMMS studies
aided development of nonpeptidic angiotensin converting enzyme
inhibitors (e.g., 3) as antihypertensives (Thorsett et al., 1983).

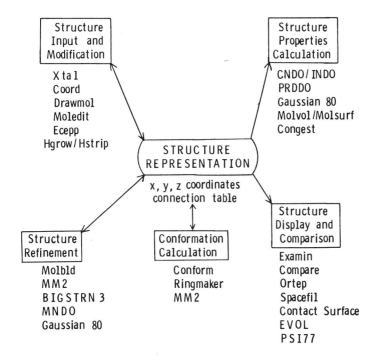

Fig. 2. MERCK MOLECULAR MODELING SYSTEM COMPONENTS

Pro–Phe–D–Trp   H–D–Phe–Cys–Phe–D–Trp
| | | |
Phe–Thr–Lys   Thr(ol)–Cys–Thr–Lys

       1   2

2.5. Crystallographic Analyses

Small-molecule crystallographers at Merck have a fairly standard suite of programs running for diffractometer control, data reduction, and crystal structure solution. In addition, they use the molecular modeling system for examining trial structures, choosing trial hydrogen positions, and studying packing arrangements. Furthermore, solved structures are routinely stored on the system for interested chemists to examine and manipulate interactively. A protein crystallography group at Merck is also making heavy use of our macromolecular modeling graphics facility (described below), especially FRODO.

3. FUTURE COMPUTER AIDS

In a field as fast-moving as this one, prognostication is precarious. However, certain trends are already clear enough to allow some extrapolation.

3.1. Information and Data Retrieval

There is an obvious trend toward the development of "user friendly" systems to be used directly by the chemists; it is safe to predict that the number of scientists using one or more such systems will increase dramatically. New sources of information are appearing; for example, at present there is much effort related to retrieval of chemical reaction information.

3.2 Macromolecular Modeling

There is substantial interest in studying the crystal structures of enzymes and other macromolecules, and modeling their binding of substrates and inhibitors. Based on successful application of our small molecule MMMS, in 1981 we substantially upgraded our modeling capabilities to support macromolecular modeling. At the same time, a protein crystallography effort was begun by the Biophysics department. As diagrammed in Fig. 3, our Macromolecular Modeling Graphics Facility (MMGF) consists of an Evans & Sutherland Multi-Picture System with color monitor, connected by high-speed data link to a VAX 11/780 superminicomputer. A low-speed connection to the corporate IBM computer facilitates off-loading of CPU-intensive jobs.

All MMMS software now resides on the VAX, giving more responsive service and a "friendlier" environment to the chemist-users. The small molecule surface display programs of MMMS run best on the

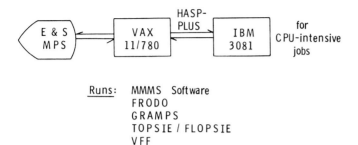

Runs: MMMS Software
         FRODO
         GRAMPS
         TOPSIE / FLOPSIE
         VFF

Fig.3. MERCK MACROMOLECULAR MODELING GRAPHICS FACILITY
(MMGF)

multi-picture system. The protein crystallographers are heavy users
of the MMGF, especially of T. A. Jones' FRODO program for fitting
trial peptide structures to the electron density maps.

     Crystal structures of a number of enzymes are available from the
Brookhaven Protein Data Bank, and the chemists have particularly been
interested in studying available enzyme-inhibitor complexes. Dr. Bruce
Bush at Merck has heavily modified FRODO to facilitate these studies,
including developing a novel method of displaying contact surfaces.
As an alternative strategy, Dr. Graham Smith has used special-purpose
programs to strip off various components of an enzyme crystal structure,
and then used the Olson-O'Donnell GRAMPS package to selectively
display and to animate sequences.

     While graphical examination of enzyme-inhibitor interaction may
provide qualitative indication of binding strength, semi-quantitative
or quantitative estimation requires some sort of energy calculation.
Given the number of atoms involved in an enzyme-inhibitor complex,
some simplifying approximations and assumptions have to be made. One
approach to this problem is being worked out by Prof. Tom Halgren of
City University of New York, who is winding up a sabbatical year at
Merck. An experimental program called TOPSIE (Torsional Optimization
Program for Substrates, Inhibitors, and Enzymes) optimizes the geometry
of substrates in the bound conformation by varying torsion angles.
Preliminary results with TOPSIE are encouraging, but suggest that
fully optimized structures are needed for valid comparisons of energies.
FLOPSIE (Full Optimization Program...) allows such full geometric
optimization of the inhibitor in the active site, and gives better

energies. For selected cases, full geometry optimization of substrate plus active site sidechains, or ultimately of substrate plus entire enzyme, would be desirable: FLOPSIE is presently being extended in these directions. A. Hagler's VFF program offers an alternative approach to these prodigious calculations.

The MMGF has already been used to examine a number of macro-molecular systems. For example we have used it to study the dihydro-folate reductase crystal structures of Matthews and coworkers (Gund, 1982), although far more work along these lines has been performed at Burroughs-Wellcome and at UCSF Graphics Laboratory. More recently, a series of inhibitors bound to thermolysin became available from Prof. Brian Matthews and were studied because of our interest in another zinc protease, angiotensin converting enzyme. The thermo-lysin studies have led to a modified mechanism of substrate cleavage and a detailed model for inhibitor and substrate binding (Hangauer et al., 1983).

3.3. Chemists' Workstations

The plummeting costs and soaring capabilities of microprocessor based systems which will fit on a bench have brought intelligent terminals into the lab. Increasing numbers of chemists are using such systems for experiment control, literature file management, data manipulation and analysis, and report writing. Any system being used by a structure-oriented chemist should preferably have graphics capa-bilities. Although it is not yet a clear trend, some standardization would be useful to preserve system compatibility and enhance the exchange of information and programs. Supply of software for such stations by commercial vendors is likely to become common.

3.4. Networking

There has been increasing proliferation of commercial systems and services, while the chemists are more often "doing their own thing" locally with microcomputers. System incompatibilities have often required different terminals to be obtained and different protocols to be learned for different services. Networking allows, in principle, access to multiple services in a consistent manner through a uniform interface, as envisioned for the U. S. Environmental Protection Agency's Chemical Substances Information System (CSIN: Siegel, 1983). Networking also raises the prospect of allowing the

chemists local capabilities, while retaining access to distributed
stores of information and data according to their needs. Furthermore
the scientist may then conveniently provide access to his or her
private, locally held data to authorized individuals, and ultimately
to a central file for safe storage. Again, the need for standard
networking protocols, and the need to discourage use of terminals and
microprocessors which cannot connect to the network, are apparent.

4.   COMPUTER ASSISTED CHEMISTRY: SYSTEMS FOR THE SPECIALIST?

There is a long-standing controversy in many areas of computer
assisted chemistry as to whether the scientist should use the programs
directly, and perhaps even write his own computer programs. The
specialist has argued that the scientists should not become too
involved because it is not the best use of their expertise; because
scientists usually are not expert enough or frequent enough users of
the systems to obtain best results; because they are likely to misuse
these complex systems and/or misinterpret the results without guidance;
and because there are economies of scale when a specialist handles
all queries. The scientist, on the other hand, argues that specialist
services are often not timely enough; that it is sometimes not clear
that the specialist has properly understood the question being asked;
that the scientist loses the "browsing" advantage by leaving the
searching to an intermediary; that many scientists who have developed
programming skills in their academic training wish to use these
skills for their research; and that they become frustrated when they
cannot generate sufficient organizational support to have a needed
program (which they think they could write themselves) written by
professionals. Professional programmers have a different view; they
consider most scientist- written programs improperly documented and
difficult to maintain or modify.

Such arguments are reminiscent of the classical question of
whether scientists should obtain their own spectra. That question
has largely been resolved in both directions: "human engineered"
spectrometers have enabled scientists to run routine analyses, and
even fairly sophisticated analyses, themselves. On the other hand,
great increases in the power of spectrometers has enabled the specialist
to perform orders-of-magnitude more sophisticated analyses than can
be run on the "hands-on" spectrometers.

Similar progress is being made in other areas. In chemical
information, systems like MACCS (Molecular Design Ltd.) have given
substructure search capabilities to the user, while this and other
systems (e.g., CSIN) enable more sophisticated questions to be posed
by information scientists. In data analysis, systems like RS/1 (Bolt
Beranek & Newman) enable the scientist to organize and analyze his or
her own data, while systems like Statistical Analysis System (SAS
Institute) bring massive statistical capabilities to the specialist
and some end-users. In molecular modeling at Merck, the greatest
impact on drug design has occurred when the chemists performed their
own modeling studies; but in each case additional, more theoretical,
studies by specialists led to revision of some of the initial conclusions. In crystallography, routine analyses have become so
automated that some organic chemists solve structures themselves,
while specialists tackle challenging problems of crystal disorder,
protein structures, and molecular packing.

Perhaps some rivalry is inevitable. As the saying goes, "information is power" -- and no one gives up power voluntarily. But
unlike energy, information is not used up; instead it increases in
value as it is used.

Specialists will continue to design and use systems with greater
capabilities than those currently available. Insofar as such improved
systems are useful, the developers have an obligation to make them
available to other users. Medicinal chemists are increasingly anxious
to make use of such systems to improve their drug design capabilities.
And the proliferation of intelligent graphics terminals, the networking
of systems and databases, and the entry of commercial systems suppliers
into this arena will assure the rapid acceptance of such advances by
the scientist.

Acknowledgment. A great many people have contributed to the systems
being run at Merck. Molecular modeling support has been provided by
Drs. Joseph Andose, Gene Fluder, Gene McIntyre, Bob Nachbar, and
others; applications support by Drs. Graham Smith, Bernie Schlegel
and Jim Snyder; macromolecular modeling capabilities chiefly by
Dr. Bruce Bush; SECS support by Drs. Joseph Andose, Graham Smith and
Dale Hoff; and QSAR analyses by Dr. Hoff. Dr. Horace Brown is chiefly
responsible for creating the environment in which these systems took
shape. And many chemists, including those cited in the references,

demonstrated the utility of these systems for drug design.

REFERENCES.

Bauer, W., Briner, U., Doepfner, W., Haller, R., Huguenin, R.,
    Marbach, P., Petcher, T. J., and Pless, J. (1982). Life
    Sciences 31, 1133.

Brown, H. D., Costlow, M., Cutler, F. A., Jr., DeMott, A. N.,
    Gall, W. B., Jacobus, D. P., and Miller, C. J. (1976).
    J. Chem. Inf. Comput. Sci. 16, 5.

Grabowski, E. J. J., Gund, P., Smith, G. M., Andose, J. D.,
    Rhodes, J. B., and Wipke, W. T. (1979). In Computer Assisted
    Drug Design, ACS Symposium Ser. No. 112 (E. C. Olson and
    R. E. Christoffersen, eds.). American Chemical Society,
    Washington, D.C., p. 527.

Gund, P. (1982). Trends in Pharmacol. Sci. 3, 56.

Gund, P., Andose, J. D., Rhodes, J. B., and Smith, G. M. (1980).
    Science 208, 1425.

Gund, P., Grabowski, E. J. J., Hoff, D. R., Smith, G. M. Andose,
    J. D., Rhodes, J. B., and Wipke, W. T. (1980). J. Chem. Info.
    Comput. Sci. 20, 88.

Hangauer, D. G., Monzingo, A. F., and Matthews, B. W. (1983).
    To be submitted.

Hansch, C. (1983). Presented at DIA Workshop on Computer Assisted
    Chemistry in Drug Design, Philadelphia, PA, Feb. 20-23, 1983.

Humblet, C., and Marshall, G. R. (1981). Drug Devel. Res. 1, 409.

Lindsay, R. K., Buchanan, B. G., Feigenbaum, E. A., and Lederberg, J.
    (1980). Applications of Artificial Intelligence for Organic
    Chemistry: The DENDRAL Project. McGraw-Hill, New York.

Martin, Y. C. (1978). "Quantitative Drug Design." M. Dekker,
    New York.

Siegel, S. (1983). Presented at DIA Workshop on Computer Assisted
    Chemistry in Drug Design, Philadelphia, PA, Feb. 20-23, 1983.

Still, W. C., and Galynker, I. (1981). Tetrahedron 37, 3981.

Thorsett, E. D., Harris, E. E., Aster, S., Peterson, E. R., Taub, D.,
    Patchett, A. A., Ulm, E. H., and Vassil, T. C. (1983). Biochem.
    Biophys. Res. Comm. 111, 166.

Veber, D. F., Freidinger, R. M., Perlow, D. S., Paleveda, W. J., Jr.,
    Holly, F. W., Strachan, R. G., Nutt, R. F., Arison, B. H.,
    Homnick, C., Randall, W. C., Glitzer, M., Saperstein, R., and

Hirschmann, R. (1981). *Nature* 292, 55.

Woodruff, H. B., and Smith, G. M. (1980). *Anal. Chem.* 52, 2321.

Index